# Integrated Ultra-Low Noise Microwave Amplifiers

This book presents the theoretical foundations of low-noise amplifier (LNA) design, with a particular focus on ultra-low-noise amplifiers. It introduces various noise types and sources in semiconductors, examines semiconductor processes (e.g., SiGe, HEMT, MOSFET, SOI, and MMIC) for both passive and active components, and explains noise and impedance-matching techniques, noise measurement methods, possible topologies for ultra-low-noise amplifiers, design and optimization approaches, and relevant design tools. The proposed methodologies are supported by detailed design examples and case studies, including applications for the square kilometer array (SKA).

Features

- Covers the complete design process of low-noise amplifiers
- Updates amplifier design techniques and fabrication processes with real case studies
- Introduces receiver architectures
- Discusses low-noise amplification technology, its evolution, trends, and competing or complementary technologies
- Includes SKA-related design case studies aimed at the radio astronomy community

This book is intended for the scientific research community.

# Integrated Ultra-Low Noise Microwave Amplifiers

Wynand Lambrechts, Saurabh Sinha, and
Abadahigwa Bimana

CRC Press
Taylor & Francis Group
Boca Raton London New York

CRC Press is an imprint of the
Taylor & Francis Group, an **informa** business

Designed cover image: Shutterstock

First edition published 2027
by CRC Press
2385 NW Executive Center Drive, Suite 320, Boca Raton FL 33431

and by CRC Press
4 Park Square, Milton Park, Abingdon, Oxon, OX14 4RN

*CRC Press is an imprint of Taylor & Francis Group, LLC*



ISBN: 9781032527932 (hbk)
ISBN: 9781032604848 (pbk)
ISBN: 9781003459385 (ebk)

DOI: 10.1201/9781003459385

Typeset in Times
by Newgen Publishing UK

# Contents

# Preface

Advances in electronics have shaped the modern world since the independent invention of the triode in 1906 by Lee de Forest, an American engineer, and Robert von Lieben, an Austrian physicist. As the name suggests, the triode is a three-electrode vacuum tube that makes long-distance telephony and amplified radio receivers possible. Almost all aspects of modern life rely on electronics, from its early and simple findings to the most modern and complex developments. Early findings include vacuum tubes and their replacement by transistors and integrated circuits. Modern developments include digital signal processing, software-defined radios, and very high-capacity and very low-latency wireless networks. The most widespread applications consist of computers, communication networks, and multimedia devices, which have experienced a wide, rapid, and continuous expansion in recent decades. Classical computers have become widely accessible while extensive research is done in superconducting quantum computing, where the basic unit of information is the quantum bit (qubit), rather than the classical bit, and information is stored and manipulated using atomic-scale objects. Quantum computers can outperform any classical computer. The state of a qubit is determined by low-power microwave fields, which are measured by ultra-low noise amplification in superconducting integrated circuits. For communication networks, multicore fiber communication has demonstrated data rates of 172 Tb/s over a single long-haul transmission of 2,040 km, while research on 6G wireless networks has indicated delivery speeds of up to 8 Tb/s, and 5G networks are still being commercialized. Such high data rates have been driven by new services such as multimedia data streaming applications running on handheld devices or terminals in corporate offices and private homes. The availability of the large bandwidth in 5G networks has allowed various devices with communication capabilities in private homes, in workplaces, and in industrial and manufacturing facilities. These connected devices communicate with one another and constitute the Internet of Things (IoT), which has become a reality. The modern world has not been shaped by computers and communications alone; other, no less important applications of electronics include control, robotics, and instrumentation systems. These systems are also evolving very rapidly, with applications in medicine, biology, imaging, astronomy, and space technology.

As one might expect, electronic systems do not operate in isolation. They interact with other systems, with their environment, and their users in exchanging information through wireless and wireline communications and in collecting data from the environment through sensors. Information is received in the form of a physical signal, such as an electric signal, electromagnetic radiation, or an acoustic wave, whose energy can be very low and degraded because of the weakening incurred in the propagation medium. This is particularly important for wireless propagation, where the received power is inversely proportional to the square of the transmission distance. The signal is received with random variations from the transmission channel, referred to as noise. A very low signal can be drowned in the noise, in which case its reception is made impossible. Therefore, the signal can be distinguished from noise only if its power level is above the noise level at the input of the amplifier, referred to as the noise floor, and the ratio between the signal power and the noise power is better than a given value depending on the signal nature. However, because the signal level is still low, it must first be improved by an adequate amplifier before it can be processed. As most electronic components generate noise of their own, the intrinsic noise of the amplifier must be reduced to the possible minimum. Such an amplifier is called a low noise amplifier (LNA) due to its low intrinsic noise.

LNAs are found in several electronic applications that have shaped modern life, but high-performance LNAs are mostly used in wireless receiver systems as the first stage of a receiver chain, because they can amplify weak antenna signals with little noise degradation. They are found in mobile and fixed communication networks, such as 5G networks and wireless LANs, as well as in user terminals. For smartphones, several LNAs are generally required in one device, depending on frequency bands and applications available. Handheld user terminals for satellite-based radio navigation systems require LNAs for the optimum reception of satellite signals. Some automotive applications, such as collision-avoidance radars, and several defense applications, including radars and drones, require LNAs. In fact, before the explosion of communication capacity, microwave technology was almost exclusive to defense applications. LNAs enhance the received signal for further processing in radio astronomy and deep space applications. This enhancement is mostly required because signals from space can be extremely weak. In communications networks, signal enhancement is more important in today's networks because, as the data throughput increases and the transmission power decreases, the impact of noise on performance becomes more critical.

The noise generated by passive and active components of an LNA contributes to the total noise at the output, degrading the output signal-to-noise ratio. The output noise is particularly dependent on the noise performance of transistors and lossy passive components, such as resistors and inductors, to a lesser extent. Noise performance is measured by the noise figure (NF), which is a value in decibels given by the ratio of the signal-to-noise ratio at the input to the signal-to-noise ratio at the output. The minimum signal strength that an LNA can detect is measured by its sensitivity and is directly related to the NF. Therefore, to receive very weak signals, an LNA with a very low NF is required. This is particularly the case with radio astronomy receivers. Extremely low NFs are obtained at very low temperatures using MMIC processes, based on gallium arsenide (GaAs), gallium nitride (GaN), and indium phosphide (InP) semiconductors. LNAs for radio astronomy are generally based on InP high electron mobility transistors (HEMTs). Compared to silicon-based integrated transistors, GaAs, InP, and GaN transistors benefit from higher electron mobility, achieving lower noise and higher power gain. But GaAs and InP devices have a low degree of integration because of their low thermal conductivity. However, for applications demanding less sensitivity, silicon-based semiconductor processes are preferred because of their low cost, high scalability, and frequency of operation. Today, silicon-based semiconductors are implemented in millimeter wave circuits, operating up to 300 GHz.

Because silicon-based receivers can achieve sub-1 dB NFs, highly sensitive integrated front ends can be based on silicon processes, and the same chip can perform other receiver functional blocks, such as the intermediate frequency (IF) chain and the back end. The front end includes the LNA, filtering functions, and the mixer, which converts the signal amplified by the LNA into a signal at the operating frequency of the IF chain. The IF block implements further signal amplification and filtering, and the back-end block performs signal detection and limited processing.

The main performance criteria of LNAs, concerning the capability to improve the received signal, are the NF and the power gain. Although relevant to amplifiers in general, these criteria are less critical for amplifiers other than LNAs. Other performance criteria relevant to amplifiers in general, such as bandwidth, linearity, stability, dynamic range, and power consumption, are also applicable to integrated LNAs. These performance criteria depend on the semiconductor process, the LNA topology, the types of active and passive components, the matching networks, the biasing of active components, the biasing circuitry, the layout, and the packaging. To achieve the required performance, the designer must analyze all these performance areas and make the relevant choices and trade-offs for each of them.

This book focuses on integrated sub-1 dB NF LNAs at microwaves. It provides the theoretical understanding of the information required to design the chips around these performance areas to attain the required specifications. The book is organized in logical steps that present and discuss issues intended for researchers. The book first introduces ultra-low noise amplifiers and presents their state of the art. Because most sensitive LNAs have been designed for radio astronomy, some content of this book is derived from the author's PhD thesis (the late Dr Bimana) on highly sensitive square kilometer array receivers. This content represents less than 50% of the book text, as additional work is presented. The sections on LNA topologies, electronic noise, and semiconductor processes provide the background information for the readers. Further information is provided in the theoretical analysis section. Finally, typical examples of research on highly sensitive LNAs are provided. These examples discuss critical steps for research validation related to schematic simulation, layout, post-layout simulation before fabrication, and measurements.

AI-assisted language tools were used during manuscript preparation to improve grammar, readability, and consistency of presentation. These tools were not used to generate scientific content, figures, or original technical material. Responsibility for all content rests solely with the author.

Scholarly peer-review: Furthermore, the authors would like to recognize the support of technical reviewers, as well as language and graphic editors, who have all contributed to this process. We value the system of scholarly peer review and the approach that this adds to the production of research text, which adds to the body of scientific knowledge.

# About the Authors

**Wynand Lambrechts** (Senior Member, IEEE) received the BEng, MEng (with distinction), and PhD degrees in Electronic Engineering from the University of Pretoria, Pretoria, South Africa, in 2013. His major field of study was electronic engineering. He has worked as an electronic engineer and a radio frequency (RF) engineer and is currently a principal electronic engineer (semiconductors) based in South Africa. He also serves as a part-time research associate at the University of Johannesburg, Johannesburg, South Africa. He is the lead author of five technical books in the fields of sustainable energy and microelectronic engineering, all published by international publishers, and has co-authored six contributing chapters in books on green energy, technology, and the Fourth Industrial Revolution. He has authored four peer-reviewed journal publications and has presented at various local and international conferences. His current research interests include semiconductor design, sustainable energy systems, and microelectronic engineering. He is a senior member of the IEEE. He has been involved in IEEE activities and contributes to research and academic collaboration in the fields of analog and mixed-signal design.

**Saurabh Sinha** (Fellow, IEEE) received the BEng (with distinction), MEng (with distinction), and PhD degrees in Electronic Engineering from the University of Pretoria, Pretoria, South Africa. His major field of study is electronic engineering. He is currently Professor and Executive Dean of the Faculty of Engineering at the University of Canterbury, Christchurch, New Zealand. He also holds an appointment as Visiting Professor at the University of Johannesburg, Johannesburg, South Africa. Previously, he served at the University of Pretoria for over a decade, including as Director of the Carl and Emily Fuchs Institute for Microelectronics, and later as Executive Dean of the Faculty of Engineering and the Built Environment at the University of Johannesburg. He subsequently served as Deputy Vice-Chancellor: Research and Internationalisation at the University of Johannesburg. An established researcher rated by the National Research Foundation (NRF), he has authored or co-authored over 140 peer-reviewed publications in journals and at international conferences. His research interests include microelectronics, 5G/6G telecommunications, and artificial intelligence. Prof. Sinha is a fellow of the IEEE, the South African Institute of Electrical Engineers (SAIEE), and the South African Academy of Engineering

(SAAE). He is also a member of the Academy of Science of South Africa (ASSAf). He has served the IEEE as a member of the Board of Directors and as IEEE Vice President of Educational Activities.

**Abadahigwa Bimana**, SMIEEE, passed away in 2024. He received the "Diplôme d'Ingénieur" degree in Electronic and Computer Engineering from the University of Kinshasa in 1988, with distinction. He joined Telecel, the first African mobile operator in 1989. In 1993, he received the "Diplôme de Mastère" degree in Microelectronics (VLSI) from Télécom Paris Tech in Paris, France. From 1995 to 1999, he worked for the Council for Scientific and Industrial Research in South Africa where he was primarily engaged in the design and development of communication systems. He joined Alcatel-Lucent in 1999 as senior development engineer and has since held several positions within the company. He held an MBA and a PhD degree in Electronic Engineering (microelectronics) from the University of Pretoria. He is a senior member of the IEEE and Senior Research Associate at the Faculty of Engineering and the Built Environment, University of Johannesburg.

# Posthumous Recognition: Dr Abadahigwa ("Aba") Bimana

We would also like to acknowledge Dr Abadahigwa ("Aba") Bimana, whom we recognize posthumously as a key author of this work.

During the course of his doctorate, Dr Bimana undertook a wide range of intellectual explorations. While some of this work was secondary to his formal doctoral research, it naturally lent itself to a book-length treatment. Even while working in industry, Dr Bimana remained deeply committed to education and to explaining complex ideas in ways that empowered others.

Dr Bimana's formative education was profoundly influenced by the rigor and discipline of a well-structured, intellectually stimulating, and culturally enriched household, complemented by his early schooling in what was then Zaïre, now recognized as the Democratic Republic of Congo. This instilled in him a deep respect for first principles. He would question concepts down to the level of the electron, firmly believing that if one could explain a phenomenon at that level, true understanding would follow. For him, a strong grounding in fundamentals was essential.

Following the completion of his doctorate in 2016, Dr Bimana continued to write and collaborate actively, including on work that ultimately contributed to this publication. With coauthors based overseas, much of this collaboration took place via email and WhatsApp. Over time, we observed a gradual change in the consistency of correspondence, reflecting circumstances that were unfolding. Even during periods of hospitalization, Dr Bimana remained engaged, contributing thoughtfully whenever he was able. The full extent of the medical and personal challenges he was facing was not entirely apparent at the time, including to him in the earlier stages. With deep regret, the correspondence eventually fell silent, and it was only after his passing that we came to understand the profound difficulties he had been enduring.

These challenges extended beyond his health. Dr Bimana endured profound personal loss, including family members torn apart by the wars in the Democratic Republic of Congo, and the tragic passing of his elder brother, Berthuin. It was to Berthuin that he owed his early fascination with understanding electrical phenomena, radio reception techniques, demodulation and amplification processes, as well as his enduring love for music. In reflecting on his life, this context brings into sharp focus why peace was such a deeply held and important cause for him.

Dr Bimana left behind more than 170 pages of written material. Through careful and diligent work, particularly by Dr Lambrechts, a meaningful portion of this material could be integrated into the present publication. In recognition of his intellectual contribution, commitment, and authorship, we therefore acknowledge Dr Bimana posthumously as one of the key authors of this book.

Dr Bimana is a former doctoral candidate at the University of Pretoria, distinguished by remarkable dedication, perseverance, and warmth – an individual whose presence profoundly enriched all those around. Despite facing serious medical challenges in 2023 and further complications in 2024, he remained engaged, curious, and generous until the end. He passed away on 25 March 2024 at the age of 63.

When Aba joined the research group, his desire to contribute meaningfully to the body of knowledge was immediately evident. Holding two master's degrees, including an MBA, he had already achieved much, yet his thirst for learning and his wish to give back never waned. His generosity extended beyond words, often funding his own research-related expenses and travel.

His doctoral research, conducted in the context of the Square Kilometre Array project, reflected his innovative and unconventional thinking. His conviction that SiGe technology could be used to achieve receiver sensitivity without cryogenic cooling led to three peer-reviewed publications. This approach was affectionately referred to as "uncool," a reflection of both the technical idea and Aba's willingness to challenge orthodoxy.

Beyond academia, Aba had a deep love for music, analog record collections, and vacuum tube amplifiers. Engineering was not only his profession but also his hobby, although he remained steadfastly unimpressed by the wholesale digitalization of music.

Aba was marked by humility and kindness. He consistently sought feedback and struggled to understand the brevity and impatience sometimes found in modern academic reviewing, another reflection of his gentle and thoughtful nature. Even after completing his doctorate, he continued to collaborate actively, demonstrating a sustained commitment to advancing knowledge.

His message of 1 February 2024 carried the weight of a terminal diagnosis. Yet even then, his correspondence reflected grace, clarity, and care for others.

We are deeply grateful for Dr Bimana's journey of humility, his commitment to empowering others, and his lasting contributions to both the telecommunications industry and academia. May he rest in eternal peace.

# 1 Fundamentals of RF Receivers and Ultra-Low-Noise Amplifier Design

## RADIO WAVES

Several terms related to radio communications and integrated circuits are used in this section, and many of them are defined. Such terms include "noise figure," "electronic noise," and "noise floor." Terms that are not defined in the section are assumed to be mostly familiar to the reader. However, these terms are defined in other sections of the book, where their definition is more relevant. Signals in radio communications are mostly described by their power, frequency, and bandwidth. Power is generally expressed in relative power levels such as dBm or dBW, while amplitude measurements, such as μV and V, are only used for very low frequencies (e.g., audio frequencies). Input and output impedances are generally equal to 50 Ω. Typical power values range from a few pW to several kW. For low-noise amplifiers, signals and the input of receivers at the end of the transmission medium can be very weak. For example, the power of received global positioning system (GPS) signals can be lower than –150 dBm.

Wireless communication systems such as fixed wireless networks, cellular telephony, satellite communications, radio and television broadcast networks, and other means of wireless communication and control use electromagnetic waves to exchange signals between distant transmitting and receiving antennas. Electromagnetic waves consist of varying and interdependent electric and magnetic fields carrying the wave energy. In free space, the electric and magnetic field components are perpendicular and travel in a plane that is transverse to the propagation direction. Their velocity $c$ is 299,793 km/s, rounded to 300,000 km/s in most calculations. As electromagnetic waves propagate, the amplitudes of the electric and magnetic fields are attenuated, reducing the wave energy. The attenuation of electromagnetic waves during propagation depends essentially on two fundamental elements: their wavelength and the propagation medium. The analysis of the attenuation of electromagnetic waves through propagation is provided by the link budget. The link budget evaluates losses incurred by an electromagnetic signal through propagation from a transmitting antenna to a receiving antenna.

### The Link Budget

A transmitting antenna and a receiving antenna are placed in an isotropic medium. The power $P_R$ received by the receiving antenna can be determined from the transmitted

DOI: 10.1201/9781003459385-1

power $P_T$ transmitted by the transmitting antenna using the link budget. Assuming that the transmitting antenna is isotropic, and that the propagation medium is ideal and lossless, the transmit flux power density $P_D$ at a distance $D$ from the antenna is given by

$$P_D = \frac{P_T}{4\pi D^2}\left[\frac{W}{m^2}\right]. \tag{1.1}$$

This representation, derived from the inverse-square law, describes the uniform power flux density incident at the surface of a sphere with radius $D$, essentially an indication of the distance from the source. The denominator represents the surface area of a sphere (again, with radius $D$), indicating that the power transmitted from the source spreads uniformly over a spherical area, as it radiates outward. If the transmitting antenna (source) is not isotropic, with a dimensionless gain defined relative to an isotropic radiator equal to $G_T$ in the direction of the receiver, the power flux density at the same distance becomes

$$P_F = \frac{G_T P_T}{4\pi D^2}\left[\frac{W}{m^2}\right] \tag{1.2}$$

where the product in the numerator is defined as the effective radiated power (ERP) of the transmitting antenna. The power received at the receiving antenna is thus defined by

$$P_R = S_R \frac{G_T P_T}{4\pi D^2}[W] \tag{1.3}$$

where $S_R$ is the effective area or aperture of the receiving antenna. The effective area or aperture of the receiving antenna is a plane area next to the antenna and is perpendicular to the direction of the maximum gain of the radiation. The gain $G_R$ of the receiving antenna and its effective area are related and defined relative to an isotropic receiver element, such that

$$G_R = \frac{4\pi S_R}{\lambda^2} \tag{1.4}$$

where $\lambda$ is the wavelength in m. The ratio between the received power and the transmitted power is consequently defined by

$$\frac{P_R}{P_T} = G_T G_R \frac{\lambda^2}{4\pi^2 D^2} \tag{1.5}$$

which represents the Friis transmission equation in free-space communication, considering the gain of the antennas, the wavelength, and the distance between the transmitter and receiver. This ratio is often used to predict the received signal strength

(RSS) and optimize link performance. To fully develop the link budget, the free-space path loss should also be defined, with

$$A(dB) = 20\log_{10}\left(\frac{4\pi D}{\lambda}\right)[dB] \tag{1.6}$$

The total power at the receiving antenna is therefore determined by

$$P_R(dBm) = P_T(dBm) + G_T(dBi) + G_R(dBi) - A(dB) \tag{1.7}$$

where $G_T$ (dBi) is the gain of the transmitting antenna in the direction of the path and relative to an isotropic radiator element, and $G_R$ (dBi) is the gain of the receiving antenna, also in the direction of the path, relative to an isotropic radiator element. If it is assumed that the internal impedance of the receiving antenna is purely resistive (for simplicity) and equal to 50 Ω, the antenna and the receiver are matched, and the temperature is 290 K, the power of the thermal noise from the antenna, derived from the Nyquist–Johnson relation, is given by

$$N(dBm) = -174 + 10\log B \tag{1.8}$$

where $B$ is the bandwidth of the transmission channel. Assuming zero loss from the propagation medium, but accounting for free-space loss, the SNR at the output of the receiving antenna is given by

$$\frac{S}{N} = \frac{P_R}{N} \tag{1.9}$$

Additionally, the noise figure (NF) of a receiver is given by $10 \times \log_{10} F$ with F being its noise factor. The noise factor is the ratio between the SNR at the input of the receiver and the SNR at its output. This ratio is greater than unity due to the intrinsic noise of the receiver. A noisy receiver can be viewed as a noiseless two-port network where the intrinsic noise of the network is reported to the input. The total SNR at the input of the receiver gets degraded by the NF, such that

$$\frac{S}{N_i} = \frac{P_R}{N} - NF \tag{1.10}$$

which is used to determine the maximum range of transmission. Signal attenuation is also typically higher if the propagation medium losses are accounted for, typical in practical circumstances. In such cases, the environment is modeled, for example, an urban environment's free-space path attenuation is approximated as $A \approx (4\pi/\lambda)^2 D^n$, where $n$ is the path-loss exponent (or propagation factor), which describes how the signal attenuation increases with distance. The value of $n$ depends on the environment.

In free space (and typically used in rural or open areas), $n$ is typically 2, whereas in urban environments it can range between 2.7 and 3.5. In indoor environments, $n$ can be higher (e.g., 4 or more) due to obstructions such as walls or furniture. As a result, the total power received can be expanded with additional factors:

$$\begin{aligned} P_R(dBm) = {} & P_T(dBm) + G_T(dBi) + G_R(dBi) \\ & - A(dB) - L_{TX}(dB) - L_M(dB) - L_{RX}(dB) \end{aligned} \tag{1.11}$$

where $L_{TX}$ (dB) represents the transmitting feeder losses (coaxial cable, connectors), $L_M$ (dB) represents miscellaneous losses such as the fading margin, body loss, and polarization mismatch, and $L_{RX}$ (dB) accounts for the receiver feeder losses.

## The RF Spectrum

The first experiments on radio transmissions were done by Gugliemo Marconi in 1894 who, based on the scientific work by Oliver Lodge and Henry R. Hertz, invented the radio telegraph system. He created the Marconi Telegraph Company in London in 1899, two years before the first transoceanic wireless communication was carried out on December 12, 1901. More than 100 years after this milestone, wireless transmissions have experienced tremendous changes in range, capacity, and complexity and have become an integral part of our daily lives. Radiofrequency identification systems, short-range car/door openers, remote-controlled vehicles, and drones, remote sensing systems such as radar, and computer networks resort to wireless transmissions. Radio astronomy, space and satellite communications, navigation systems, and several other short and long-distance communication systems, such as cellular and broadcast networks, rely on the most recent wireless communication technologies.

Radio frequencies belong to the lower frequency region of the electromagnetic spectrum. The electromagnetic spectrum is commonly known to span from 3 Hz to approximately $3 \times 10^{26}$ Hz, where the highest frequencies correspond to Gamma rays (Salcedo and Martinez 2017). Gamma rays are the spectrum range above x-rays ($10^{17}$ Hz – $10^{20}$ Hz, 1 nm – 1 pm). Resulting from a radioactive process in the atomic nucleus, they radiate an extremely high energy and are ionizing. As a result, exposure to Gamma rays can cause damaging physiological effects on tissues and provoke genetic mutations and cancer. The typical frequency of Gamma rays is above $10^{20}$ Hz. The visible region of the electromagnetic spectrum, corresponding to the visible light between infrared rays ($10^{13}$ Hz – $4 \times 10^{14}$ Hz, 25 μm – 750 nm) and ultraviolet rays ($10^{15}$ Hz – $10^{17}$ Hz, 400 nm – 1 nm), is very small ($4 \times 10^{14}$ Hz – $7.5 \times 10^{14}$ Hz, 750 nm – 400 nm). Radio frequencies, as defined by the IEEE Standard 521-1984, are comprised between 300 kHz and 300 GHz, corresponding the wavelengths greater than 1 mm (IEEE Standard Letter Designations for Radar-Frequency Bands 2002). Although widely used, the IEEE system provides less granularity than the ITU classification shown in Table 1.1, where frequencies are further allocated by region; the world is divided into three regions for global spectrum management. However, the spectrum allocated to radio frequencies by the ITU goes beyond 300 GHz, including frequencies from $3 \times 10^{11}$ Hz to $10^{13}$ Hz, 1 mm–25 μm), which corresponds to tremendously high frequencies (THF) and terahertz frequencies, the spectrum evolving toward more bandwidth owing to next generations of wireless networks, to increased demand in wireless data transmission capacity and to new applications. Next wireless networks include 6G networks, and applications include the Internet of Things (IoT), autonomous driving, virtual/augmented reality (VR/AR), and wireless backhaul. It is reminded that the relative bandwidth of wireless signals, the size of transmit and receive antennas, as well as the effects of the transmission medium and the environment on the transmission channel, are related to the signal frequency. Consequently, wireless applications are also related to the signal frequency, as shown in Table 1.1. The span of the radio frequency spectrum, from extremely low frequencies (ELF) radio waves used by submarines to the lower limit of infrared frequencies (300 GHz), the allocation and designation of frequency bands by the ITU and the IEEE, as well as their applications are shown in Table 1.1 (Kraus and Marhefka 2001, McCulloch 2013, Salcedo and Martinez 2017).

Frequencies in ELF, SLF, ULF, and VLF bands are for long-distance transmissions and are the only ones that can penetrate deeply into oceans. The noise in these bands is very high and mostly of an impulsive type. MF transmissions, used mostly for AM broadcasts, have a range of only a few hundred kilometers but can achieve thousands of kilometers at night. Contrary to MF, HF transmissions are intended for very long distances, up to 10,000 km. They were, from the 1930s to 1970s, until the prevalence of satellite relays, the preferred frequencies for very long-distance voice, data, and image communications. However, they suffer from fading and dependence on daytime, seasonal, and Sun activity variations. VHF transmissions are intended for short-range line-of-sight communications. The characteristics of UHF and VHF transmissions are similar. Applications are also similar, but include communications between space vehicles and the Earth. Transmissions in the SHF band are very directional and, because of their small wavelengths, can use high-gain parabolic antennas. However, they suffer from impairment due to rain, fog, and solar outages.

Frequencies allocated to low-band fifth generation (5G) wireless systems and mid-band 5G are comprised between 600 MHz and 6 GHz (mainly 3.5–4.2 GHz), while frequency bands at 26, 28, 38, and 60 GHz are allocated to higher performance 5G networks (20 gigabits per second), the last two frequencies corresponding to mm-wave bands. Electromagnetic radiation at 60 GHz is absorbed by oxygen because oxygen molecules ($O_2$) resonate at 60 GHz. The 60 GHz band is used for satellite-to-satellite communications – they cannot be listened to by terrestrial stations – and by the intelligence community for high-security short-range communications, as risks of signal interception are considerably reduced. Another advantage of this band comes from easier frequency reuse because links in a relatively close region do not interfere. As a note for Table 1.1, mm-wave is generally used to refer to frequencies between 30 and 300 GHz, which includes the V- and W-bands and a portion of the Ka-band. Beyond 5G communications, sub-THz and THz (0.1–10 THz) frequencies are expected to be used for sixth generation (6G) wireless systems to attain much higher data rates (1 gigabit per second). Although frequencies between 275 GHz and 3 THz were not allocated by the ITU (2016) until the World Radio Communication

**TABLE 1.1**
**The RF spectrum**

| Band | Frequency | ITU Band | IEEE Sub- frequency | IEEE Designation | Applications |
|---|---|---|---|---|---|
| Extremely low frequency (ELF) | 3–30 Hz | 1 | | | Lowest frequency waves for wireless communications. Penetrates deeply in conducting materials. Used in underwater communication (submarines) and scientific applications (e.g., geological remote sensing). |
| Super low frequency (SLF) | 30–300 Hz | 2 | | | Submarine and electric grid communications |
| Ultra-low frequency (ULF) | 300 Hz–3kHz kHz | 3 | | | Mining communications and military applications |
| Very low frequency (VLF) | 3–30 kHz | 4 | | | It can penetrate through dirt and rock. Applications in geophysics, geophysics, and Earth monitoring. |
| Low frequency (LF) | 30–300 kHz | 5 | | | AM broadcasting, RFID, navigation (aviation beacons), amateur radio. |
| Medium frequency (MF) | 300–3,000 kHz | 6 | 300–3,000 kHz | MF | AM broadcasting, coast-to-sea communication, distress signals, etc. |
| High frequency (HF) | 3–30 MHz | 7 | 3–30 MHz | HF | Short waves AM broadcasting, aviation and marine communication, amateur radio, and weather broadcasting. |
| Very high frequency (VHF) | 30–300 MHz | 8 | 30–300 MHz | VHF | Analog television broadcasting, FM radio broadcasting, aircraft communications, medical equipment for magnetic resonance imaging, mobile land, and marine communication systems |

| | | | | | |
|---|---|---|---|---|---|
| Ultra-high frequency (UHF) | 300–3,000 MHz | 9 | 300–1,000 MHz | UHF | Satellite television, Wi-Fi, GPS, Bluetooth, television broadcasting, mobile communications (GSM), CDMA, and LTE services |
| | | | 1–2 GHz<br>2–4 GHz | L<br>S | Radars, DTH services, 5GHz Wi-Fi, radio astronomy, mobile networks, TV broadcasting satellites, microwave devices, broadcasting satellites, amateur radio. |
| Super high frequency (SHF) | 3–30 GHz | 10 | 4–8 GHz<br>8–12 GHz<br>12–18 GHz<br>18–27 GHz<br>27–40 GHz | C<br>X<br>Ku<br>K<br>Ka | Radars, DTH services, 5GHz Wi-Fi, radio astronomy, mobile networks, TV broadcasting satellites, microwave devices, broadcasting satellites, amateur radio:<br>C-Band: Wi-Fi, Satellite communications, 5G<br>X-Band: Satellite communications, Radar<br>Ku-Band: Satellite communications<br>K-Band: Radar, radio astronomy, Communications, 5G<br>Ka-Band: Radar, radio astronomy, satellite communications, 5G |
| Extremely high frequency (EHF) | 30–300 GHz | 11 | 40–75 GHz<br>75–110 GHz<br>110–300 GHz | V<br>W<br>mm or G | Radio astronomy, amateur radio, remote sensing, Earth Exploration, radio navigation, microwave relays, Inter-Satellite, 5G |
| Terahertz or tremendously high frequency (THF) | 300–3,000 GHz | 12 | | | Communications, 6G, Alternative to X-rays, imaging, spectroscopy, solid-state physics. |

Conference 2019 (WRC-19), frequencies between 275 GHz and 1 THz were identified for passive (receive-only) applications in radio astronomy, Earth exploration satellite services, and space research. Four THF bands were allocated by the WRC-19 to wireless and mobile communications for a total bandwidth of 137 GHz, following the approval of footnote RR no. 5.564A on radio regulations.

## RECEIVER ARCHITECTURES

A wireless receiver comprises several signal enhancement and processing functions implemented in back-to-back blocks referred to as the receiver chain. Hence, the receiver chain is the signal-receiving part of a radio frequency (RF) system. The receiver must reject undesired signals and select the desired one, amplify it, and recover from it the original information. These three functions correspond to selectivity, amplification, and demodulation tasks. To achieve them, the receiver should have a low noise figure, a low phase noise, a low intermodulation distortion, a high and flat gain across its bandwidth, a high dynamic range, a good selectivity, and an appropriate bit error rate (BER). The receiver is the most difficult element to design in a communication system.

A unidirectional RF system comprises a transmitter and a transmit antenna on the transmitting side, the transmission channel in the transmitting medium, the receive antenna, and the receiver on the receiving side. The transmitter processes and amplifies the RF signal to compensate for transmission losses and impairments in the transmission channel. Transmission losses are related to the transmission distance and the transmission medium. In most cases, this transmission medium is made up of the terrestrial atmosphere, but can also consist of solids, liquids, or space in the case of space communications. In free space, the power of the received electromagnetic signal is inversely proportional to the square of the transmission distance. Therefore, for large transmission distances, the received signal, which is already affected by the transmission medium (e.g., water vapor), signal reflections, and various sources of interference external to the RF system, is significantly weakened by the propagation distance. A key characteristic of the transmission channel is its transmission capacity: this is the theoretical limit of the maximum information flow that can be transmitted in real time in the channel. A unidirectional or simplex RF system is represented in Figure 1.1 (Luzzatto and Haridim 2016).

The front end of the receiver in Figure 1.1 comprises the low-noise amplifier (LNA), filtering functions, and the first mixer. The mixer is made of a nonlinear element that combines an incoming signal with a signal generated by a local oscillator. The output of the nonlinear element consists of several harmonic components, whose frequency is a linear combination of the input signal frequencies, from which the desired output is selected by filtering. In this way, the frequency of the incoming signal can be down-converted or upconverted. The strength of the signal at the input of the receiver depends on the antenna type and feed losses. Two types of feed are available: single-ended and differential feed. Differential feeds are used with differential LNAs for antennas with unbalanced outputs (e.g., dipoles) and allow reducing the common-mode interference, to improve the dynamic range and the linearity (Shrivastava and

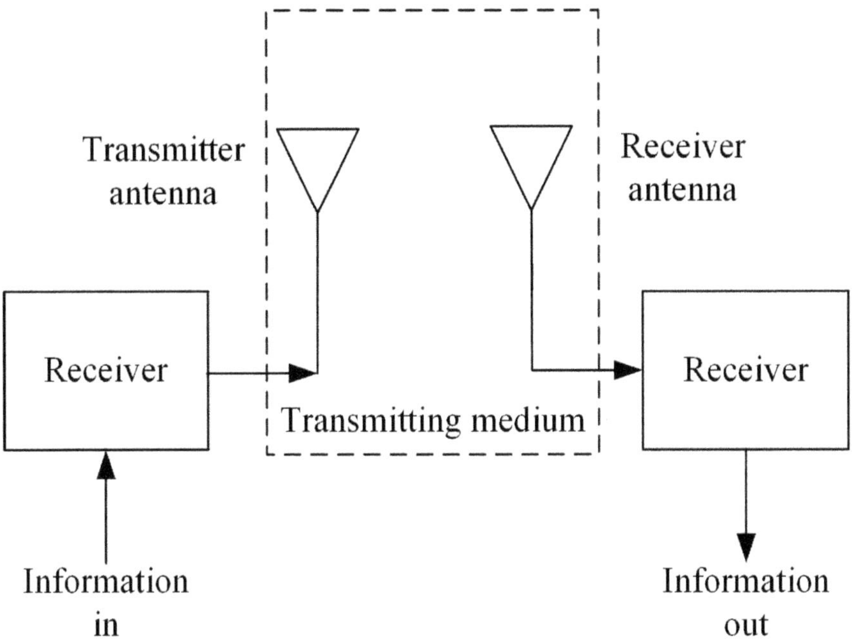


**FIGURE 1.1** Unidirectional RF communications system.

Khatri 2017). Even-order harmonics are canceled in differential LNAs, improving linearity. However, because several components are duplicated, differential LNAs have a higher NF, higher power consumption, and larger chip area when compared to similar single-ended LNAs. At microwave frequencies, losses in coaxial feeds can be significant for a single-ended LNA. To reduce losses, the feed line should be as short as possible, and semi-rigid coaxial cables can be used to reduce losses. Ideally, the LNA input can be directly coupled with the signal reception probe in the feed horn waveguide for a reflector antenna or an aperture antenna. The feed probe can also perform a waveguide to coaxial transition at the cost of coaxial losses, depending on the length of the coaxial line. Feed losses are also minimized when planar antennas are used, as the antenna can be implemented on the LNA printed circuit board (PCB) top metal layer. This book discusses only single-ended LNAs, as differential LNAs can be derived from the proposed single-ended architectures. A feed horn with a quarter-wave feed probe is presented in Figure 1.2. The reception probe is the quarter-wave element used inside a waveguide for most high-frequency feeds.

## Antenna Considerations

Antennas are devices that convert electrical energy into electromagnetic waves and vice versa. A transmitting antenna converts electrical signals into radio waves, while a receiving one converts radio waves into electrical signals. Because an antenna is the first element of a wireless receiver chain, the external noise entering first in the

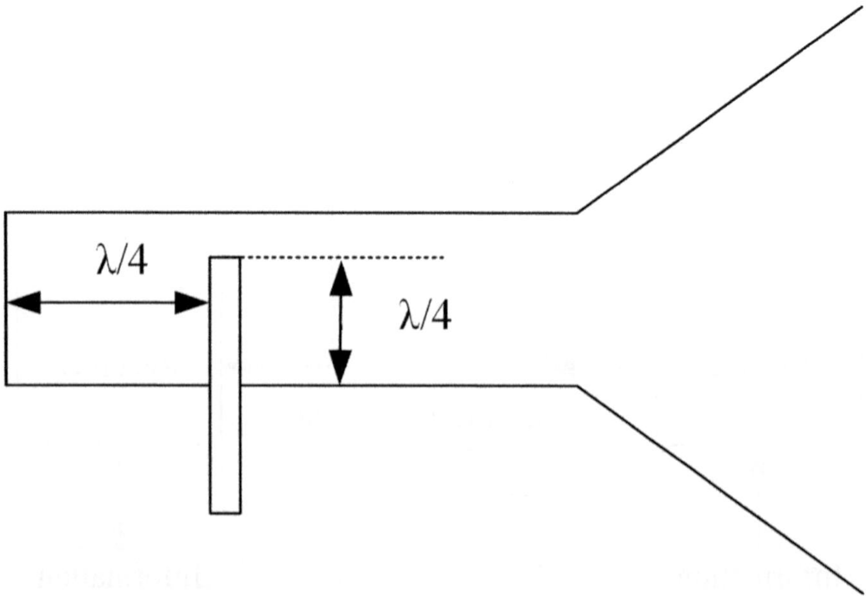


**FIGURE 1.2** Feed horn with a quarter wavelength probe.

receiver system first comes from the antenna and the feed line. This noise adds to internal noise from filters, couplers, and the LNA. This section discusses some basic considerations on antennas.

Antennas are defined by the IEEE as "*a means for radiating or receiving radio waves*." Several antenna properties are invariant in both the transmitting and receiving modes. They are therefore reciprocal and can be analyzed as receiving or transmitting devices, owing to the electromagnetic case for reciprocity (Burke *et al.* 2019). The electromagnetic case for reciprocity is stated in Burke *et al.* (2019) as follows:

> An antenna can be treated as a receiving device, gathering the incoming radiation field and conducting electrical signals to the output terminals, or as a transmitting system, launching electromagnetic waves outward. These two cases are equivalent because of time reversibility: the solutions of Maxwell's equations are valid when time is reversed.

Antenna reciprocity facilitates calculations and measurements because some parameters are easier to evaluate in one mode than in the other. For example, it is easier to evaluate the receiving power pattern of an antenna than its transmitting power pattern.

Based on the Thevenin circuit equivalent, a transmitting antenna appears as a circuit element with two terminals of impedance $Z_A = R_L + R_r + jX_A$ characterized by its radiation pattern. The same antenna is also characterized by its antenna noise temperature $T_A$ when used in receiving mode. $T_A$ is neither the physical temperature of the antenna nor a physical quantity, but rather a measure of the thermal noise received by the antenna from the region that the antenna is pointing at and from other radiation in its environment. The load resistance $R_L$ represents conduction and dielectric losses in the antenna structure, $R_r$, the radiation resistance, characterizes the power radiated by the antenna, hence its ability for power conversion, $X_A$, is the antenna reactance. The radiation resistance is the resistance that dissipates the same power as the power radiated by the antenna when the resistance and the antenna are crossed by the same current. For receiving antennas, the radiation resistance characterizes the power received from electromagnetic radiation. All energy received is transferred to the radiation resistance for ideal lossless antennas. Often, it is desirable to suppress the radiation of energy in certain directions and divert it to other ones, which is characterized by the radiation pattern and antenna gain $G$. The gain and the noise temperature define a common figure of merit for receiving antennas, $G/T_A$. To fulfill various antenna requirements related to performance, gain, radiation pattern (directivity), polarization, operating frequency, bandwidth, and size, several types of antennas are available. The types that are mainly used are wire antennas, aperture antennas, microstrip antennas, array antennas, reflector antennas, and lens antennas (Balanis 2016).

The equivalent circuit of a transmitting antenna based on the Thevenin model is shown in Figure 1.3.

The RF signal generator is represented by the voltage source $V_g$ and its impedance $Z_g$. The characteristic impedance of the transmission line is $Z_c$. The radiation resistance is $R_r$, the loss resistance related to conduction and dielectric losses is $R_L$. The reactance of the antenna is $jX_A$. Using the Thevenin equivalent circuit, the antenna impedance $Z_A$ is complex as indicated and can be written as

$$Z_A = R_A + jX_A \tag{1.12}$$

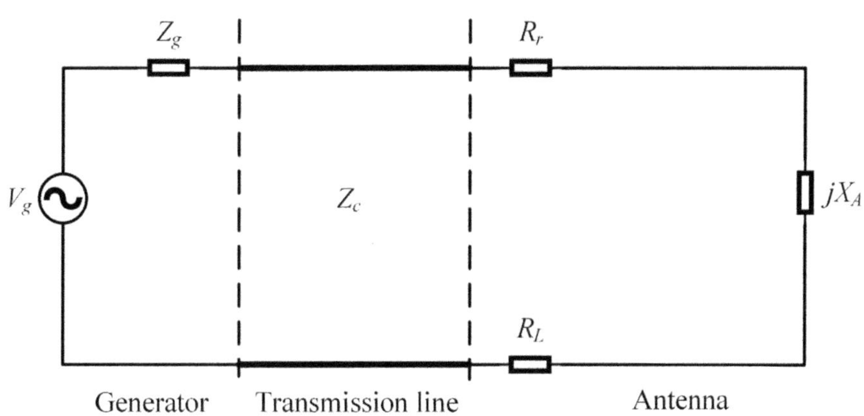


**FIGURE 1.3** Equivalent circuit of a transmit antenna.

where $R_A$ is the resistance of the antenna. At resonance, the reactive component of $Z_A$ is zero. $R_A$ comprises two parts, the radiation resistance $R_r$ and the loss resistance $R_L$, and its relationship is defined by

$$R_A = R_r + R_L \tag{1.13}$$

The radiation resistance $R_r$ is defined by the ratio of the total average power radiated by the antenna in all directions to the square of the RMS value of the current causing the radiation. A typical value of $R_r$ is 50 Ω. The definition of $R_r$ provided is related to the transmitting antenna; however, owing to reciprocity (Burke *et al.* 2019), all parameters presented apply to receiving antennas. The average power in a resistor $R$ when the current $I$ is time variant is given by

$$\langle P \rangle = \langle I^2 \rangle \times R[W] \tag{1.14}$$

where $<P>$ and $<I^2>$ are the averages of $P$ and $I^2$, respectively. If the current is sinusoidal, therefore in the form $I = I_0 \sin(\omega t)$, then $<I^2> = I^2_0/2$ and the average power is given by

$$P = \frac{I_0^2 R}{2} \tag{1.15}$$

The average power $P_r$ that is radiated from the antenna is therefore given by (Balanis 2016)

$$P_r = \frac{1}{2}\left[I_A\right]^2 R_r = \frac{\left[V_g\right]^2}{2}\left[\frac{R_r}{\left(R_r + R_L + R_g\right)^2 + \left(X_A + X_g\right)^2}\right] \tag{1.16}$$

and the power that is lost in $R_L$ is equal to

$$P_L = \frac{\left[V_g\right]^2}{2}\left[\frac{R_L}{\left(R_r + R_L + R_g\right)^2 + \left(X_A + X_g\right)^2}\right] \tag{1.17}$$

whereas $P_g$, the average power dissipated in the source resistance $R_g$, is given by

$$P_g = \frac{\left[V_g\right]^2}{2}\left[\frac{R_g}{\left(R_r + R_L + R_g\right)^2 + \left(X_A + X_g\right)^2}\right] \tag{1.18}$$

If $R_r + R_L = R_g$ and $X_A = -X_g$, a complex conjugate match occurs, and the maximum power received by the antenna is given by

$$P_r = \frac{\left[V_g\right]^2}{8}\left[\frac{R_r}{\left(R_r + R_L\right)^2}\right] \tag{1.19}$$

and the lost power in $R_L$ is also equal to

$$P_L = \frac{\left[V_g\right]^2}{8}\left[\frac{R_L}{\left(R_r + R_L\right)^2}\right] \tag{1.20}$$

where the average power $P_g$ dissipated in the source resistance $R_g$ becomes

$$P_g = \frac{\left[V_g\right]^2}{8R_g} \tag{1.21}$$

and the power dissipated by the signal source at complex conjugate matching is given by

$$P_S = \frac{\left[V_g\right]^2}{4R_g} \tag{1.22}$$

which shows that half of the source power is dissipated as heat within the source, with a part of the other half being radiated and the rest dissipated as heat in $R_L$. The maximum power that is available for radiation is only half of the source power and can only be achieved when $R_L$ (and its associated losses) is zero.

The efficiency of the antenna $e_0$ is related to reflections between the feedline and the antenna input, determined by the reflection coefficient Γ and by conductive and dielectric losses represented by $R_L$. Dielectric losses are caused by the variations of the electric field in dielectrics in the antenna structure when an electromagnetic wave passes through them. They also result from the conductivity of the dielectric. The antenna efficiency is given by

$$e_0 = e_{cd} \times e_r = \left(1 - |\Gamma|^2\right) \tag{1.23}$$

where $\Gamma$ is given by

$$\Gamma = \frac{Z_{in} - Z_0}{Z_{in} + Z_0} \tag{1.24}$$

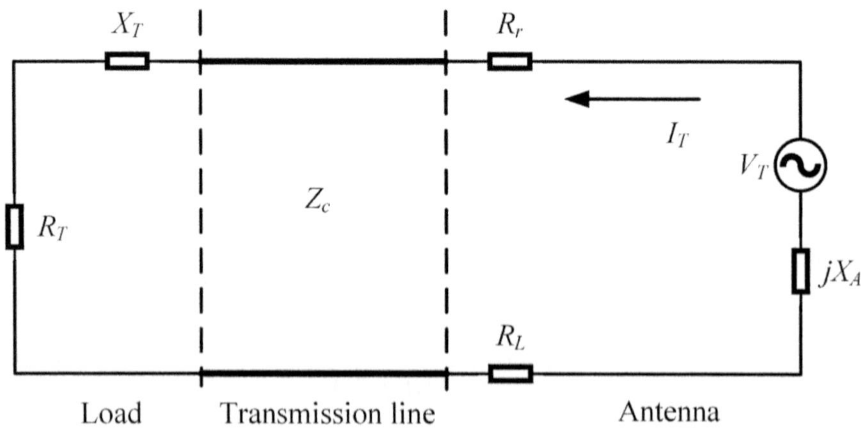


**FIGURE 1.4** Equivalent circuit of a receiving antenna.

and $Z_{in}$ is the antenna impedance at its terminal, and $Z_0$ is the characteristic impedance of the feed line. The noise at the input of a receiver due to the receiving antenna is reviewed in the discussion on receiving antennas in the following paragraphs. Three cascaded signal planes are considered: the antenna aperture, the antenna terminals, and the receiver input. At the antenna aperture, received noise is due to external noise sources only; at the antenna terminals, noise from antenna internal losses is added to external noise; at the receiver input, noise and losses from the feedline are added to previous noise sources. The discussion begins with the equivalent circuit of the receiving antenna. The Thevenin equivalent circuit of a receiving antenna is shown in Figure 1.4, where the electromagnetic wave received by the antenna induces the voltage $V_T$.

The equivalent circuit in Figure 1.4 shows that part of the captured power represented by the voltage source $V_T$ is dissipated in $R_L$, $R_r$ and $R_T$. The load is $R_T$, and only the power received by $R_T$ is sent to the receiver. The power received by $R_r$ is reradiated. For this reason, this power is referred to as scattered power. Compared to the transmitting antenna, $V_T$ can be viewed as the source, and $R_r$ as a fictive source resistance.

Complex conjugate matching is achieved when $R_r = R_L + R_T$ and $X_A = -X_T$. It is noted that the complex conjugate matching conditions differ from the same conditions for transmitting antennas. The power delivered to $R_T$, $R_r$, and $R_L$ at complex conjugate matching is given by

$$P_T = \frac{\left[V_T\right]^2}{8R_T} \tag{1.25}$$

$$P_r = \frac{\left[V_T\right]^2}{8}\left[\frac{R_r}{\left(R_r + R_L\right)^2}\right] \tag{1.26}$$

$$P_L = \frac{[V_T]^2}{8}\left[\frac{R_L}{(R_r + R_L)^2}\right] \tag{1.27}$$

If losses are assumed to be zero, the maximum power that can be received by $R_T$ under complex conjugate matching is equal to half of the captured power, and the other half is re-radiated and dissipated as heat in $R_L$. The power captured by the antenna is given by

$$P_c = \frac{[V_T]^2}{4}\left[\frac{1}{R_r + R_L}\right] \tag{1.28}$$

and is equal to the product of the received power density and the effective area of the antenna.

Electronic noise from the receiving antenna is due to two main sources: ohmic and dielectric losses in the antenna, corresponding to $R_{L,}$ and the noise picked up from remote sources, included in $V_T$. The region that the antenna is pointing toward and the antenna environment are sources of antenna noise because any object with a temperature superior to absolute zero (–273° C) and with finite conductivity radiates electromagnetic energy. The amount of noise picked by the antenna depends on its radiation pattern, namely, its beams and sidelobes. Hence, the antenna can pick thermal noise from celestial bodies such as the Sun, the Moon, the Earth, and galaxies, as well as noise due to atmospheric losses in gases (e.g., water vapor and oxygen) and rain absorption. Since the atmosphere and the rain attenuate the signal by absorbing electromagnetic radiations and generating heat, they also generate thermal noise because, when energy is absorbed, the absorbing body radiates the same energy in the form of noise, as shown by the black body emission theory, known also as Planck's law (Planck 1914). The contribution of various space and atmospheric regions crossed by a signal from space to the noise received by an Earth station is represented in Figure 1.5, where $T_{sky}$ is the thermal noise due to celestial bodies and galaxies, $T_{at}$ is the thermal noise due to atmospheric absorption, and $T_{rain}$ is the thermal noise resulting from the rain. The contribution to noise $T_{spill}$ due to ground radiation and due to conduction and dielectric losses in the antenna structure $T_{loss,}$ represented by $R_{L,}$ is not shown in Figure 1.5. All contributions to noise are analyzed at the aperture of the antenna.

At frequencies below 50 MHz, most noise comes from lightning strikes across the globe. The significant noise above 50 MHz comes from antenna losses, the Earth, atmospheric absorption, the Sun, and the galaxy. For frequencies below 1 GHz, the galactic noise is dominant. Above 1 GHz, reradiation after oxygen and water vapor absorption is dominant (Bokulic 1991).

In general, conduction and dielectric losses are accounted for by the antenna conduction–dielectric efficiency $e_{cd}$, because they are difficult to evaluate by calculation

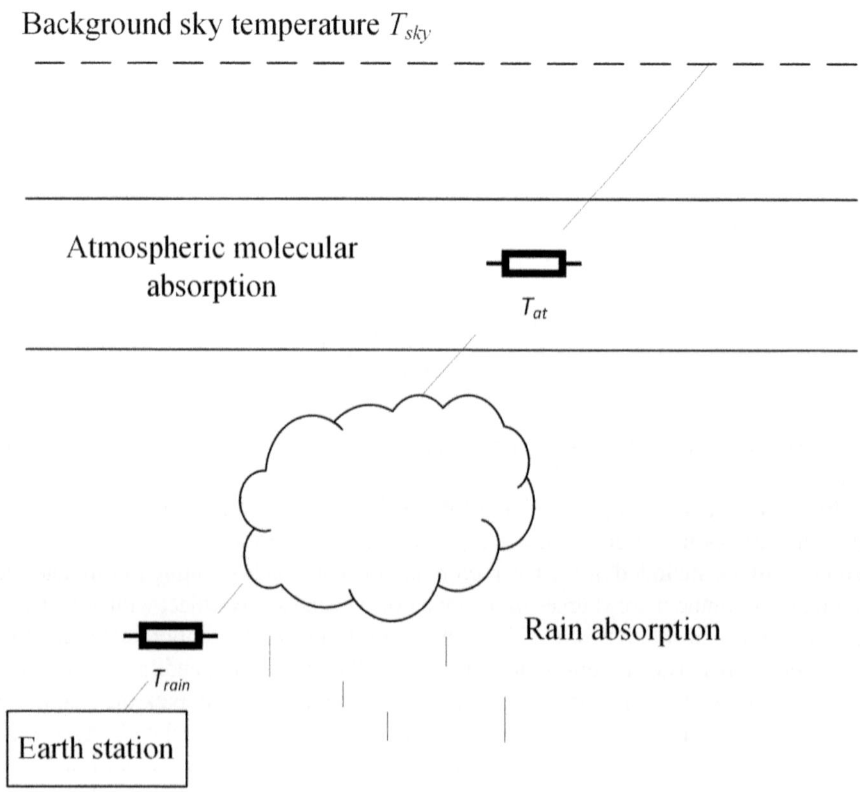


**FIGURE 1.5** Noise received by an Earth station.

or measurement. The antenna's total efficiency also accounts for reflections between the feed line and the antenna. The conduction-dielectric efficiency is given by

$$e_{cd} = \frac{R_r}{R_L + R_r} \tag{1.29}$$

where the skin effect must be considered in the calculation of $R_L$. The total noise temperature at the antenna aperture resulting from $T_{sky}$, $T_{at}$, $T_{rain}$, $T_{spill}$, and $T_{loss}$ can be represented by $T_A$, the antenna noise temperature, with $T_A = T_{sky} + T_{at} + T_{rain} + T_{spill} + T_{loss}$. Hence, antenna noise at the aperture signal plane can be evaluated by $T_A$.

It is seen that the antenna noise temperature includes not only external noise due to its environment but also antenna internal losses represented by $T_{loss}$ and related to the antenna's physical temperature. The determination of $T_{loss}$ is made by considering that the antenna is made of cascaded two-port components and by reporting the noise due to $R_L$ to the antenna aperture, the input of the cascade chain. Calculation details are provided in (Bokulic 1991), and $T_{loss}$ is given by

$$T_{loss} = \left(\frac{1}{e_A} - 1\right)T_p = \frac{R_L}{R_r}T_p \tag{1.30}$$

where $T_p$ is the physical temperature of the antenna and $e_A$ is its radiation efficiency with $0 \leq e_A \leq 1$. In Figure 1.4, $R_L$ can be seen as a two-port component at the physical temperature $T_0$. Losses in this component impair the radiation efficiency of the antenna, measured by $e_A$. For a two-port lossy component with $P_{in}$ as input power and $P_{out}$ as output power, the radiation efficiency is defined by the ratio

$$e_A = \frac{P_{out}}{P_{in}} \tag{1.31}$$

and by using equivalent circuit parameters, $e_A$ can be written as

$$e_A = \frac{R_r}{R_L + R_r} \tag{1.32}$$

where losses in the feed line also add noise to $T_A$ and should additionally be accounted for. At the antenna terminals signal plane, the noise $T_{L2}$ due to the feed line is given by

$$T_{L2} = \left(\frac{1}{e_L} - 1\right)T_L \tag{1.33}$$

where $T_L$ is the physical temperature of the feed line and $e_L$ is the thermal efficiency of the line, with $e_L = e^{-2\alpha l}$. The length of the feed line is $l$, and $\alpha$ is its attenuation in Nepers per unit length. To evaluate the contribution of $T_{L2}$ to noise temperature at the antenna aperture plane, $T_{L2}$ is reported to the aperture plane

$$T_{L1} = \frac{T_{L2}}{e_A} = \frac{1}{e_A}\left(\frac{1}{e_L} - 1\right)T_L \tag{1.34}$$

The noise from the antenna and the feed line enters the receiver at the receiver input signal plane. The system noise comprises the antenna noise, the feed line noise, and the receiver noise. The system noise $T_A^{sys}$ reported to the antenna aperture is given by

$$T_{sys}^A = T_A + \frac{1}{e_A}T_L\left(\frac{1}{e_L} - 1\right) + \frac{1}{e_A e_L}T_R \tag{1.35}$$

where $T_R$ is the receiver noise temperature reported at its input. The system noise reported to the receiver input is equal to

$$T_{sys}^{R} = e_A e_L T_A + T_L\left(1 - e_L\right) + T_R \tag{1.36}$$

and the antenna noise and the feed line noise at the input of the receiver are equal to

$$T_{AL}^{R} = e_A e_L T_A + T_L\left(1 - e_L\right) \tag{1.37}$$

The noise power $P_A$ of an antenna of noise temperature $T^R{}_{AL}$ at the input of the receiver with an input bandwidth $B$ is given by Planck's law of black body radiation or the equipartition theorem, approximated for radio frequencies as

$$P_A = kT_{AL}^{R}B \tag{1.38}$$

where $k$ is Boltzmann's constant ($k = 1.38 \times 10^{-23}$ J/K) and $B$ in Hz is the bandwidth under consideration. This representation is equivalent to the Nyquist–Johnson equation, as they are both derived from the equipartition theorem. $P_A$ can also be represented as the power generated by the thermal noise of a fictitious resistance of any value at $t^R{}_{AL}$ into a resistive load of the same value, as shown in Figure 1.6.

In Figure 1.6, the noise power density at $t^R{}_{AL}$ due to $R$ is given by the Nyquist–Johnson equation

$$v_{rms}^{2} = 4kT_{AL}^{R}R \tag{1.39}$$

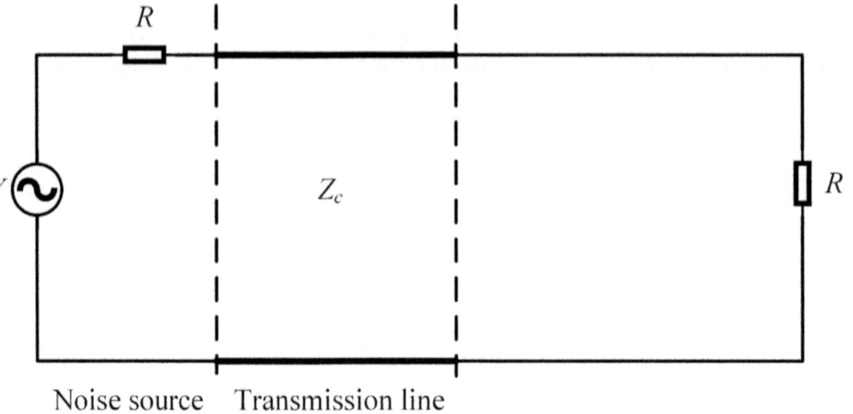


**FIGURE 1.6** Thermal noise due to the resistor $R$ into a matched load.

and the power in the resistive load due to $v$ is given by

$$P_R = \frac{v^2}{4R} \tag{1.40}$$

and the power contained in $B$ yields

$$P_R = kT_{AL}^R B \tag{1.41}$$

Since $P_A = P_R$ for any value of $R$ at the temperature $t^R{}_{AL}$, the antenna noise is equivalent to the thermal noise of a resistor of equal resistance as the input resistance of the LNA (at the input of the receiver and $t^R{}_{AL}$). Due to this complex conjugate matching, this resistance is equal to $R_T$, and the total noise power at the input of the LNA is equal to $P_R = kt^R{}_{AL}B$. This is noteworthy in the calculation of the noise of a matched antenna to an LNA. The antenna can be assumed to be a resistor whose value is equal to the LNA input resistance at the temperature $t^R{}_{AL}$ – the antenna noise temperature, with the noise temperature of a resistor being equal to the physical temperature of the resistor.

The impedance at the input of a transmitting antenna is real at the frequency of resonance and equal to $R_A$, as discussed in previous paragraphs. For a receiving antenna, the impedance at the output is real at resonance and equal to $R_r + R_L$. Maximum power transfer between the source and the antenna or the antenna and the receiver is achieved at complex conjugate matching. Generally, $R_L$ is small compared to $R_r$ except when the physical size of the antenna is small compared to the wavelength of the RF signal. If antenna losses are neglected, $R_A \approx R_r$. For a quarter-wave vertical over a ground plane, $R_r$ is about 50 Ω, for a half-wave dipole, $R_r$ is about 72 Ω, and 300 Ω for a folded half-wave dipole, all antennas being in free space. The noise at the output of the receiving antenna corresponds to the power of natural noise sources whose radiation is received by the antenna and to the noise due to antenna internal losses and feed line losses.

As indicated, antenna noise is caused by thermal bodies, atmospheric losses, cosmic radiation, ground noise, and antenna losses. It depends on the frequency and the time of day. It also varies with the elevation angle of the antenna. At night and for high elevation angles, the antenna points skyward and picks only atmospheric and galaxy noise; the noise temperature at the antenna aperture is very low and varies between 3 and 5 K for frequencies between 1 and 10 GHz. This frequency range is called the microwave or space window and is particularly beneficial for deep-space and satellite communications. The noise temperature increases considerably when the antenna is to the zenith, pointing to bright objects in the sky, such as the Sun. During low solar activity, the noise temperature of the Sun is in the range of 106 K between 50 and 200 MHz and 104 K around 10 GHz. The Sun's noise temperature increases by a factor of 100 to 10,000 during high solar activity. The Moon's noise temperature varies between 200 and 300 K, depending on its cycle, and is less dependent on frequency above 1 GHz. The antenna noise temperature varies between 100 and 150 K when the elevation angle is zero, the antenna pointing to the horizon. The noise temperature corresponding to the ground is close to 300 K, which is close to the standard ambient temperature.

In noise calculations at room temperature, the source is generally a resistive device, and both the source and the LNA are assumed to be at the same temperature. However, because the noise temperature of the antenna is generally different from room temperature, the calculated LNA total output noise will differ from the real output noise. However, the noise factor will not be impacted because the noise factor is a measure of the additional output noise caused by the LNA. It is assumed in this book that the noise temperature is room temperature (290 K) when the signal source is an antenna used for terrestrial applications (Kanar and Rebeiz 2014).

Important considerations on the equivalent circuit of receiving antennas, as presented in Balanis (2016) and Nikolova (2018) and discussed in this book, are required. It first appears that the equivalent circuit of the receiving antenna is derived from the reciprocity principle. However, it cannot be implied that the equivalent circuit of a receiving antenna is the same as the one for a transmitting antenna only because of the reciprocity principle. Second, it is admitted in the literature that only 50% at most of the source power can be radiated. Radiation happens through the radiation resistance $R_r$, which is in series with the resistance $R_d$ – representing antenna losses – and the antenna reactance. However, in calculations related to receiving antenna noise temperature, the hypothesis that the received noise can be reradiated through $R_r$ is not considered. Although $R_r$ is a fictive noiseless resistance, the reradiation of noise through $R_r$ and the effects of $R_r$ on total noise temperature should be considered if one must be coherent with calculations related to transmitting antennas. In noise temperature calculations, the total noise received by the antenna's external environment, the noise due to antenna losses, and the noise from the feedline are transferred in full to the receiver input (Nikolova 2018). In addition, according to the equivalent circuit of a receiving antenna, antenna internal losses due to conductivity in dielectric exist only when a load is connected to the antenna, which should not be the case. If calculations based on the model in Balanis (2016) and Nikolova (2018) are confirmed by measurements, the equivalent circuit of the receiving antenna is incorrect (incomplete) because it is based on incoherent assumptions.

The circuit model of antennas is based on Thévenin or Norton theorems. Although these models are found in most books, it is stated in Huang *et al.* (2021) that they are not valid for receiving antennas because they cannot be represented by a voltage or a current source, the radiated power received by the antenna being a product of the power density and the effective antenna area. According to Huang *et al.* (2021), a receiving antenna should rather be modeled as a constant power source, a device that is unavailable in circuit modeling. For a transmitting antenna, low efficiency results mostly from the power dissipated in the resistance of the source. For a receiving antenna, low efficiency is explained in most textbooks as resulting from the power reradiated in $R_r$. However, as indicated in the previous paragraph, the hypothesis that noise is reradiated in $R_r$ is not considered. Several papers have discussed antenna equivalent circuits, but no consensus has been found, and unresolved problems are

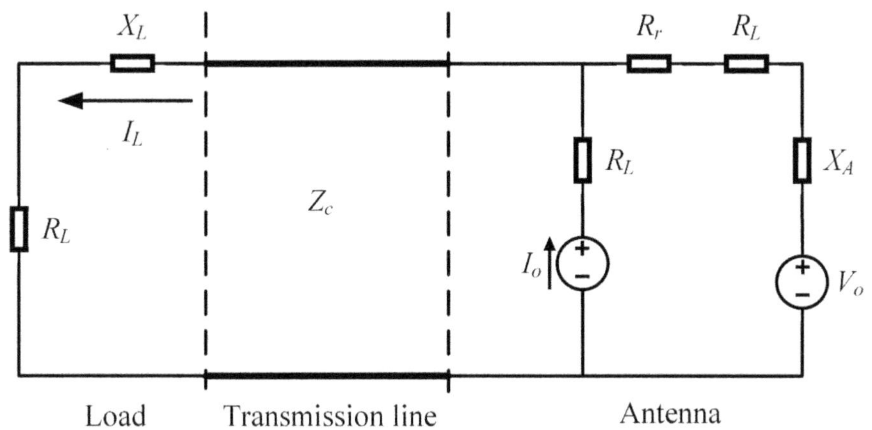


**FIGURE 1.7** Receiving antenna equivalent circuit model.

related to the circuit model and the efficiency of antennas. The model with both voltage and current sources proposed for receiving antennas in Huang *et al.* (2021) and validated by simulation and measurements is depicted in Figure 1.7.

In Figure 1.7, the voltage source $V_O$ and the current source $I_O$ are related to the electric field $E$ and the magnetic field $H$ by the following expressions:

$$V_O = E\sqrt{\frac{Z_A}{\eta}A_{eff}} \tag{1.42}$$

and

$$I_O = H\sqrt{\frac{\eta}{Z_A}A_{eff}} \tag{1.43}$$

where $A_{eff}$ is the effective aperture of the antenna and $\eta$ is the intrinsic impedance of the medium (377 Ω in free space), with $\eta = E/H$ and $V_O = I_O Z_A$. This calculation relates to the circuit model in Huang *et al.* (2021) whereas in the primary characteristics of this model, as given Figure 1.7, the following assumptions are made:

- The standard antenna input impedance $Z_A = R_r + R_L + jX_A$.
- The efficiency could effectively be higher than 50%, with a hypothetical loss-less 100%.
- Power is lost through $R_{loss}$ even if there is no load, reflecting the real-world scenario.
- Power is re-radiated through $R_r$, and losses are incurred for the second time in $R_{loss}$ when the load impedance and the antenna are mismatched.

It is noted that, for conjugate matching, the proposed model provides the same results for noise calculations as the common model. This can be justified by considering that an antenna in each environment is first analyzed using the commonly accepted model: the noise power at the aperture of the antenna is equal to $kTB$. In the same environment, the antenna is analyzed using the new model: the power of the noise from both voltage and current sources is equal to $kTB$. The antenna is then placed in an anechoic chamber where noise originating from the environment is not available. In this case, the common model can be viewed as fitted with a short circuit at the antenna aperture. In the same case, the proposed model can be viewed as a circuit where the voltage source is short-circuited, and the current source is opened. It is seen that the commonly used circuit model is equivalent to the new circuit model and that the noise due to $R_L$ ($R_{loss}$) is equal in both cases. Although both models yield the same results for noise calculations, the new model provides a better understanding of antennas and a better evaluation of the received power and antenna efficiency.

## Review of Receiver Architectures

The simplest method of transmitting a signal from one point to another consists of using baseband transmission. In baseband transmission, the frequency of the signal source, that is, the transducer where information is generated, is not altered during the transmission. Such a method is used when the transmission channel lends itself to it and when a dedicated physical transmission channel can be allocated to each communication. Baseband transmission is used in wired telephony, in-studio television on coaxial cables, and in many short-distance applications. In most transmission cases, the signal must be conditioned because of channel constraints, such as the bandwidth available, the attenuation, and distortion due to the distance and the medium (e.g., electric grid communications, optical fiber, submarine, Earth's atmosphere, or free space communications). The main signal conditioning operations include modulation and amplification. For wireless transmissions, these operations consist of modulating a signal carrier by the source baseband signal, upconverting the frequency of the modulated carrier, amplifying it, and radiating the amplified carrier from the transmitting antenna. These operations allow us to match the transmitted signal to the characteristics of the transmission channel. An example of changes in the transmitted signal to adapt to the transmission channel consists of frequent frequency changes encountered in amplitude-modulated short-wave broadcasts, aiming to benefit from the increase in the transmission range and quality caused by propagation variations in the ionosphere.

At reception, the received signal is first amplified before undergoing reverse conditioning in the receiver chain. Reverse conditioning includes down-converting the signal frequency and demodulating it. Frequency down-converting and up-converting are performed by frequency mixers. For receivers designed to demodulate RF signals of various frequencies in a frequency band, it is convenient for some sections of the receiver chain to operate at low and fixed frequencies, which facilitates the design of high-gain amplifiers and filters with high sharpness. Such frequencies are referred to as intermediate frequencies (IF). The use of IF improves the performance of receivers, increases their component count, but reduces their complexity, and can optimize their cost when standardized components are used. The sections of a receiver operating IF can suffer from spurious responses and interference resulting from image frequencies. An image frequency is an undesired frequency related to the desired one that falls in the IF band after frequency mixing and down-conversion. For example, mixing input signals of frequencies $f_i$ and $2f_{LO} - f_i$ with a local oscillator at $f_{LO}$ provides an output signal at the IF $f_{IF}$, where $f_{LO}$ is the frequency of the mixing oscillator, $f_{FI} = f_i - f_{LO,}$ and $f_i > f_{LO}$. If $f_i$ is the desired input frequency, $2f_{LO} - f_i$ is the undesired image frequency. Therefore, the difference between the desired frequency and its image is twice the IF.

The effects of the image frequency are reduced by filtering and by the receiver architecture. The architecture of receivers has considerably evolved with advances in digital communications and digital signal processing. Several analog functions are easier and better implemented in software, and issues related to alignment and maintenance are alleviated with softwarization. The first receiver systems were mostly analog radio sets converting RF signals modulated by Morse code, speech, and music to audio frequencies. Using heterodyne radio sets, an RF signal was directly converted to audio frequencies by a unique mixer owing to synchronizing a local oscillator to the RF input signal frequency. Regenerative and super-regenerative receivers (SRR), invented by Edwin Howard Armstrong in the 1920s, allowed for achieving very high RF gain with only a few components using positive feedback termed "regeneration." The performance of both heterodyne and SRR radio sets was improved by superheterodyne receivers, also invented by Edwin Howard Armstrong, which used an IF stage before the final conversion to audio. In this way, filtering and amplification could be performed at a fixed frequency, independently of the received frequency, which allowed for overcoming limitations in sensitivity and selectivity. However, because of their low component count, low consumption, and high energy efficiency, SSRs are still used in modern communications for ultra-low power and low data rate applications, such as home automation (e.g., gate opener), remote keyless entry devices, manufacturing, warehouse inventory control, robotics, biomedicine, administration in hospitals, wireless sensor networks (WSNs), computer peripherals (e.g., wireless mice) and radio-frequency identification (RFID).

SRRs can cause interference in radiating RFs from the receiver through the antenna. The RF amplifier ahead of the oscillator in SRR architectures aims to provide good antenna isolation between the antenna and the oscillator, but can also improve sensitivity. Despite their appearance of simplicity, the analysis of SRRs is complex because of their time-varying and nonlinear characteristics. According to Thomas H. Lee, the complexity and subtlety of SRRs "*has never been understood by more than a handful of people at a given time*" (Pala-Schonwalder *et al.* 2020). While research in SRRs remains a continuous effort, the current state of their knowledge is summarized in Pala-Schonwalder *et al.* (2020).

For the last few decades, driven by the increase in performance in application-specific integrated circuits, microprocessors, and inexpensive large memory, software-defined radio (SDR) has enabled digitalization of almost all functions performed in a receiver for frequencies below 1 GHz (Rodhe 2017). An SDR communication system can receive RF signals of various standards and modulation techniques in a large frequency spectrum owing to programmable hardware and software. Its RF front-end is not fully analog and consists of an antenna, several antennas or reconfigurable antennas, a gain variable low noise amplifier, and a programmable IF block. The input RF signal transducer may consist of an antenna, several antennas, or reconfigurable antennas. Because of its flexible architecture, SDR solves the need for communication between different types of equipment.

The previous paragraphs have introduced basic functions and types of receivers. An exhaustive review of these functions, as well as their theoretical analysis, is beyond the objectives of this section. A high-level view of the main functions of an analog receiver is provided in Figure 1.8, which is followed by their brief descriptions.

The antenna matching function consists of a preselector filter performing three functions: it matches the antenna impedance to the input of the next stage for maximum sensitivity, it attenuates any signal at the image frequency, and reduces overloading from strong signals. Overloading occurs when a powerful interfering signal overshadows the input signal. Its power spreads across the receiver bandwidth because of intermodulation, increasing the noise floor. This effect is also known as desensitization or blocking. However, the preselector filter introduces losses at the input, which impairs the noise figure. Gain and selectivity functions are performed by the LNA, the mixers, the IF block, and related filters. The gain of the LNA is variable.

Signal processing functions include automatic gain control (AGC), limiting/blanking, and adaptive equalization. AGC, initially developed to combat fading in HF receivers where this transmission impairment is common, optimizes the signal level within the receiver, allowing an acceptable SNR and maintaining the dynamic range for the receiver. The AGC is particularly required to maintain the signal in the dynamic range required by the analog-to-digital (A/D) converter when analog-to-digital conversion is used. The AGC can be distributed to more than one stage. The level of the received signal, referred to as RSS, varies over time. Three types of variation exist. Variations due to receiver mobility or changes in the communication channel (e.g., path loss) are slow when compared to the transmission rate in the case of digital receivers. The transmission rate, expressed in symbol rate, is related to the bit rate, with one symbol corresponding to several bits, depending on the modulation type. In digital transmissions, the physical signal at the input of the modulator – the output of the demodulator – is made up of a series of elementary waveforms like a pulse. Each elementary signal constitutes a symbol, and each symbol corresponds to

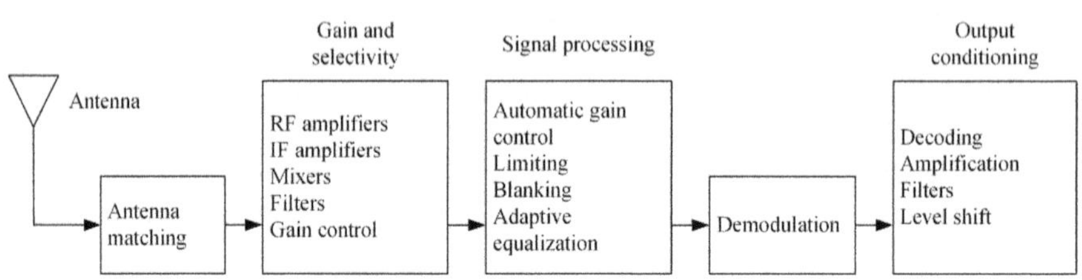


**FIGURE 1.8** Receiver block diagram and functions.

a series of bits whose length depends on the type of modulation. The AGC allows for successfully reducing the effects of slow RSS variations. For rapid channel variations such as Rayleigh fading, corresponding to the second type of RSS changes, the AGC is unsuccessful. They can be addressed by channel equalization or forward error correction (FEC). Finally, the last types of RSS variations are due to signal blocking, interference, or jamming.

Short-duration pulses can saturate the receiver in creating desensitization when the AGC response is slower than the duration of pulses. Such pulses may come from the ignition system of heat engines, from radar, corona discharge, lightning discharge, etc. The limiting function in Figure 1.8 reduces the level of such pulses, while the blanking function tries to detect pulses at an early stage to cancel the gain of the amplifier. Adaptive equalization attempts to reduce inter-symbol interference between symbols. Because of perturbations in the transmission channel, such as multipath, symbols transmitted one after the other on the transmitter side may undergo different delays, causing them not to fit together correctly at reception. As a result of inter-symbol interference, the receiver's resistance to noise is reduced.

Operations performed by the modulator in the transmitter are reversed by the demodulator. Traditionally, analog modulation has been used for speech, music, and television. Analog demodulation tries to reproduce as faithfully as possible the waveform at the input of the transmitter to the receiver output. Amplitude modulation (AM), frequency modulation (FM), and phase modulation (PM) are the three main types of analog modulation from which hybrid forms can be derived. However, FM can be viewed as a form of PM. Some hybrid modulations are the double-sideband with a suppressed carrier (DSB-SC) and the single-sideband (SSB), vestigial-sideband (VSB). Unlike analog demodulation, digital demodulation tries to reproduce exactly and at a discrete time, a discrete level from a sequence of discrete levels at the transmitter input (Rodhe 2017). In this way, impairment related to linearity, distortion, and external disturbances in analog transmission can be alleviated. But the RF signal out of the transmitter and at the input of the receiver front-end remains analog. It reminds us that the physical world, the world that we live in, is analog. This RF signal is the modulator output amplified, with the signal entering the modulator being symbol waveforms, each symbol corresponding to a series of bits. AM, FM, PM, or their hybrids can be used. Their equivalents in the digital domain are amplitude shift keying (ASK), frequency shift keying (FSK), and phase shift keying (PSK) modulations. In these discrete analog modulations, amplitude, frequency, and phase are shifted between discrete values set by the bits to transmit. Typical examples found in digital satellite television (DVB-S, DVB-S2) are quadrature phase-shift keying (QPSK), 8PSK, 16APSK, and 32APSK, which are phase shift keying modulations and combinations of ASK and PSK modulations to increase the symbol set. At demodulation, received symbols are discriminated against and converted into bits. Bits encoded within user information for error detection and correction are extracted by decoding using baseband processing functions. On the transmitting side, the shape of the signal can be changed by baseband processing functions to improve the transmission. Such a change is reversed at reception by the baseband processing, as shown in Figure1.8.

Reference was made to terms related to digital transmission in previous paragraphs, terms not in Figure 1.8: A/D conversion, bits, symbols, and FEC. This is justified because the output signal from the baseband processing block can be supplied to a data demodulator for digital output from the baseband signal. But, although the receiver output in Figure 1.8 can be supplied to a data demodulator and digitized to provide a digital output, the functions depicted correspond to a classic analog receiver. Modern digital receivers tend to perform digital conversion closer to the antenna, with most signal processing done in the digital domain. This is the case of discrete analog M-*ary* demodulation, such as QPSK, 8PSK, 16APSK, and 32APSK for the DVB-S2 standard for satellite communications, where the signal in the IF band at 70 MHz is digitized and down-converted again by a digital down-converter (DDC) to provide I/Q signals for processing and demodulation. Modern digital receivers are much more complex than basic analog systems: in addition to performing RF signal filtering and amplification, like the analog front-end in analog systems, more operations, such as digital clock extraction, A/D conversion, digital processing, data decoding, error correction, complex demodulation schemes, and other examples, are done. As a result, digital receivers are more flexible in using multiple communication standards and multiple frequency bands. In addition, their performance is easily improved using digital filters, and they can require less chip area, consume less power, and achieve cost savings.

A typical digital receiver system, other than a software-defined radio system, can be viewed as consisting of an antenna, an analog front-end, an analog IF and baseband, a data conversion subsystem, a frequency generation subsystem, and a digital baseband subsystem (Rouphael 2014). The analog front end determines the sensitivity and the linearity of the receiver. It is made of filters, low-noise amplifiers (LNAs), mixers, and possibly switches. The IF subsystem, also made of filters and amplifiers, is key for selectivity and interference mitigation. The conversion of the analog signal from the IF subsystem to digital is performed by the data conversion subsystem. This conversion, done by analog-to-digital conversion architectures, determines jitter performance.

Several receiver architectures have been developed over a few decades. The most common architectures are the superheterodyne architecture, the direct conversion or zero-IF architecture, the low-IF architecture, the superheterodyne and direct conversion, the direct RF sampling architecture, the tuned RF architecture, and the image reject receiver (Hartley and Weaver) architecture. The most used of these architectures, as reported in the literature, are briefly discussed in the following paragraphs.

## The Heterodyne Architecture

The heterodyne architecture is one of the cornerstone methods in radio receiver design, introducing a flexible and robust solution for frequency conversion in communication systems. A typical heterodyne receiver includes several distinct stages designed to isolate, amplify, and down-convert a selected signal, as illustrated in Figure 1.9.

Initially, the received RF signal is processed through a preselection filter, or band selection filter, which narrows the incoming signal spectrum to the desired band, mitigating interference from out-of-band signals. This initial filtering ensures that the

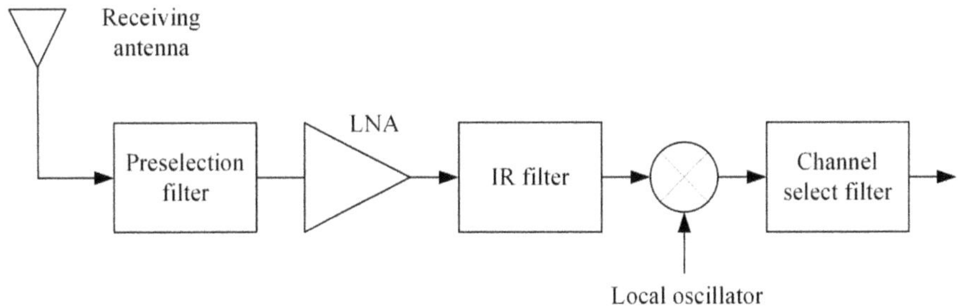


**FIGURE 1.9** The heterodyne architecture.

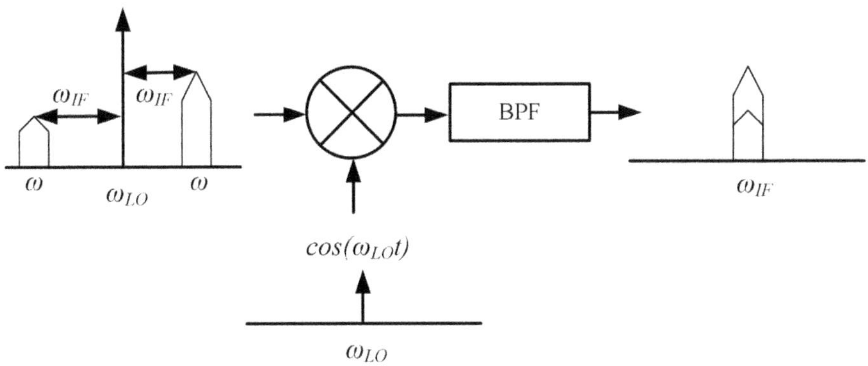


**FIGURE 1.10** Signal and image frequency through the mixer.

LNA, which immediately follows, can operate with reduced noise contributions from unwanted signals, thereby maximizing the SNR of the selected band. The amplified signal then encounters an image rejection (IR) filter, an essential component to prevent signals from the undesired image frequency from being down-converted alongside the desired signal, which would otherwise cause interference and degradation in receiver performance.

The next critical component is the mixer, which leverages a tunable local oscillator (LO) to shift the frequency of the incoming RF signal to a lower IF, facilitating easier processing. This conversion from RF to IF is essential because it simplifies the requirements for subsequent filtering stages, allowing for fixed-frequency, high-selectivity channel filters to perform efficient signal isolation. In this architecture, frequency conversions occur in two stages: the first shifts the RF to an IF, and the second moves it from IF to a baseband signal. Importantly, setting a low IF frequency can significantly ease the design constraints of the IF bandpass filter. However, a lower IF also narrows the separation between the desired signal and the image frequency, increasing the burden on the channel selectivity filter to effectively isolate the signal.

The presence of image frequency in the heterodyne design introduces a unique challenge; signals symmetrically spaced around the LO frequency are down-converted to the same IF, as shown in Figure 1.10.

This behavior necessitates that selectivity requirements become stricter as the IF decreases because the image frequency moves closer to the desired RF signal, leading to potential overlap. The relationship between these frequencies can be expressed mathematically, with the image frequency defined by $f' = 2f_{LO} - f_i$ and the desired signal frequency by $f_i = f_{LO} + f_{IF}$, where $f_{LO}$ is the LO frequency and $f_{IF}$ is the IF. This interdependence highlights a compromise in the heterodyne design: while a lower IF frequency simplifies filtering requirements in some respects, it simultaneously demands greater precision in channel filtering to prevent image interference.

## The Superheterodyne Architecture

The superheterodyne architecture remains a dominant and highly adaptable choice in the realm of RF receivers, particularly valued in complementary metal-oxide semiconductor (CMOS) designs for its versatility with various modulation schemes and its capacity to meet rigorous performance standards in both narrowband and broadband applications. This architecture's core function – double-frequency conversion – is key to its success, efficiently down-converting a modulated RF signal into baseband in-phase and quadrature (I/Q) signals. The output I/Q signals are then digitized and directed to a digital signal processor (DSP), which performs critical post-processing, including AGC through a programmable gain amplifier (PGA). By operating across two stages, superheterodyne receivers eliminate the risk of image frequency fold-over, ensuring robust selectivity and reduced interference. In this regard, the architecture has traditionally excelled; however, emerging alternatives like direct conversion and undersampling architectures present competition due to their simplified structures and fewer components.

In contrast to superheterodyne, undersampling achieves frequency down-conversion directly, often avoiding the need for an IF gain stage, high-frequency local oscillators, mixers, or IF filters. This approach notably reduces receiver complexity but comes with trade-offs in terms of flexibility and potential noise limitations, depending on the application. Superheterodyne, on the other hand, leverages double conversion stages and phase-locked loops (PLLs) to provide more controlled, stable signal tuning. As illustrated in Figure 1.11, the architecture incorporates a variable first local oscillator for flexible tuning and down-conversion, followed by a second local oscillator with a fixed frequency to maintain consistent IF processing.

The superheterodyne approach boasts multiple advantages, beginning with its ability to achieve high gain without the instability issues often caused by RF interstage feedback. Additionally, its IF stage offers good selectivity, as the fractional bandwidth at IF is greater than that at RF, making high selectivity filtering simpler at this intermediate stage. This selectivity enables efficient signal isolation and allows the receiver to tackle issues like the "half-IF" problem, which may arise when an RF input frequency aligns with half of the IF frequency, producing undesired interference due to second-order nonlinearities. The problem can be effectively managed by increasing the first IF frequency and incorporating differential IF stages, thus preserving signal integrity.

Beyond these benefits, a high first IF also expands the receiver's tuning range, granting it versatility in a wider spectrum. The use of a low second IF, conversely,

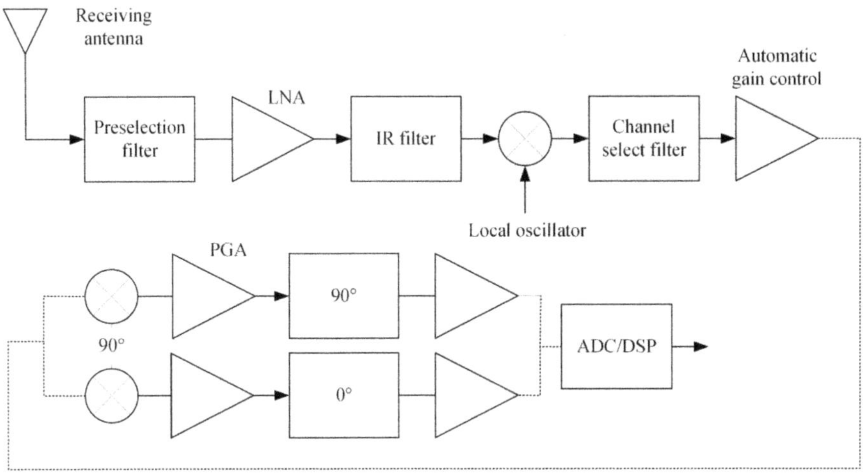


**FIGURE 1.11** The superheterodyne receiver.

enhances selectivity by increasing the fractional bandwidth. A further advantage is the immunity to $1/f$ noise and avoidance of DC-offset, both of which are critical for maintaining signal fidelity in lower frequencies and minimizing baseline drift. However, these performance gains come at a cost: the architecture is relatively complex, requiring more components and, consequently, higher power consumption, particularly from the local oscillators. This increase in complexity also necessitates a larger chip area, driving up production costs, which can be a disadvantage in resource-constrained designs.

Various adaptations of the superheterodyne architecture help address specific use cases. For example, some designs opt for a local oscillator frequency higher than the desired input frequency to improve selectivity further. In applications requiring even greater precision, a third mixer stage may be introduced. Additionally, for receivers with extensive input frequency tuning ranges – such as high-frequency (HF) receivers or spectrum analyzers – a higher first IF followed by a lower second frequency is utilized, accommodating input frequencies over a wide range, from 3 to 30 MHz, while maintaining reliable down-conversion.

The following approach eliminates IF, directly converting RF signals to baseband. This simplification offers advantages in terms of reduced component requirements and smaller chip area, though it also brings its own set of challenges.

## The Direct Conversion Architecture

The direct conversion architecture, also known as the homodyne or zero-IF architecture, directly down-converts an RF signal to baseband using quadrature down-conversion to produce in-phase (I) and quadrature (Q) signals, which are subsequently digitized. This design, illustrated in Figure 1.12, omits the IF stage entirely, a distinguishing feature that makes it "zero-IF." The architecture introduces both

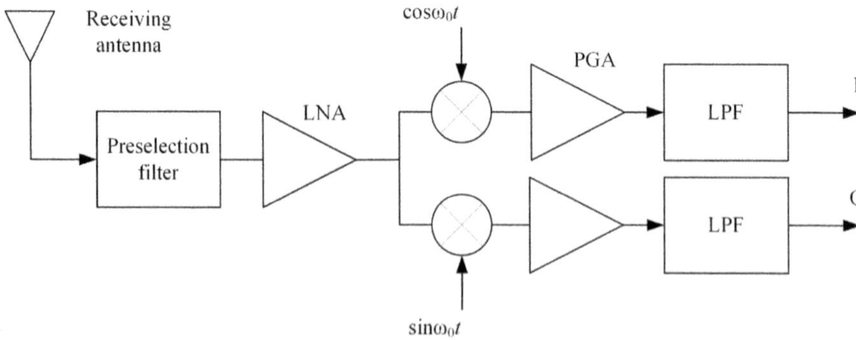


**FIGURE 1.12** The direct conversion (homodyne) receiver.

advantages and unique challenges due to the absence of an IF stage and the direct nature of its signal path.

Since there is no IF stage, the image frequency remains an issue in direct conversion because the image signal, unlike in superheterodyne, is identical to the desired signal itself. This results in mirrored lower and upper sidebands that overlap in the baseband, making separation impossible through traditional filtering. Instead, the architecture utilizes down-conversion with both sine and cosine local oscillators, producing quadrature I/Q signals that facilitate the demodulation process. This is achieved by measuring the phase difference between I and Q branches, often through algorithms like the COordinate Rotation DIgital Computer (CORDIC) in a DSP. The primary benefits of the direct conversion architecture are as follows:

- No image rejection filter required: Direct down-conversion eliminates the need for an image rejection filter since there is no IF stage, simplifying the receiver structure.
- High tuning range: The architecture can handle a wide frequency range without needing multiple stages.
- Low power consumption and cost: Fewer components and simplified structure contribute to reduced power usage and manufacturing costs.
- Absence of off-chip components: The design is fully integrable, lending itself well to compact, on-chip implementations in CMOS technology.
- Flexibility for multi-standard applications: Its broad tuning range and low-cost implementation make it suitable for multi-standard, low-power applications, including cellular and Bluetooth devices.

However, these advantages come with trade-offs, particularly related to interference and noise challenges, including the following limitations:

- RF leakage and interference: The LO in direct conversion operates at the signal frequency, making it susceptible to RF leakage back through the antenna, where it may inadvertently transmit and interfere with the signal.

This is primarily due to the crosstalk between the RF input and LO input of the mixer.

- DC offset: The LO can re-enter the mixer, creating a DC offset. The resulting near-DC component must be removed with high-pass filtering, which requires long time constants (~0.1 seconds). This filtering often removes part of the desired signal, leading to output distortion that worsens as time constants increase. Moreover, every frequency change necessitates receiver re-settling, further slowing response times. Modern DSPs, as shown in Figure 1.13, dynamically calculate and correct the DC offset using feedback.
- Dependence on quadrature generator precision: Any imbalance in gain or phase between I/Q signals adversely affects performance, requiring continuous recalibration for accurate signal reproduction, particularly in applications demanding wide bandwidths.
- Susceptibility to low-frequency interference: Processing at baseband exposes signals to low-frequency noise from the power supply and digital circuitry. This architecture is especially vulnerable to $1/f$ noise, which can degrade performance in low-frequency applications.
- High IP2 linearity requirement: Baseband amplification within the I/Q stage demands extremely high second-order intercept points (IP2) to avoid nonlinearities, which can severely limit the architecture's utility in applications requiring extensive signal amplification.

Despite these limitations, the direct conversion architecture has seen increased usage due to advancements in integrated circuits, which allow for better I/Q balance, especially in CMOS technologies. Implementing a differential design mitigates

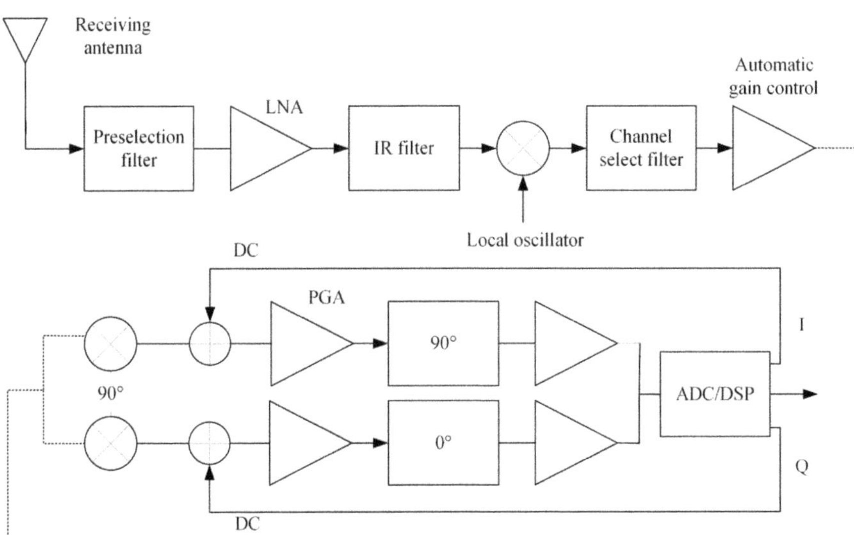


**FIGURE 1.13** Zero-IF with DC compensation feedback.

common-mode interference, enhances linearity by canceling even-order harmonics, and reduces baseband noise sensitivity, improving IP2 linearity.

Near-zero IF is an extension of direct conversion that addresses DC offset and low-frequency noise by offsetting the LO frequency by half the RF input signal bandwidth. This configuration – often employed in receivers for Bluetooth, cellular base stations, and mobile phones – shifts the signal slightly from zero frequency, avoiding the issues centered at DC. Unlike direct conversion, near-zero IF requires additional down-conversion after filtering and amplification, typically managed by a DSP.

The low-IF or non-zero IF architecture reintroduces the image frequency problem. This architecture is equivalent to the zero-IF architecture except that the RF frequency is offset from the LO by half of its frequency. The baseband signal is just higher than DC. The final shift of the signal to zero DC is performed by the DSP. This architecture provides a practical compromise by positioning the IF near, but not at, baseband, thus reducing many of the challenges seen in direct conversion while maintaining high selectivity and image rejection.

#### Superheterodyne and Direct Conversion

The superheterodyne and direct conversion architecture combines the advantages of the superheterodyne and the direct conversion architectures while mitigating several of their disadvantages. In this architecture, only one frequency conversion stage is implemented in the superheterodyne. Any number of conversions can be done. This architecture is found in several applications, such as mobile cellular handsets and wireless LANs.

#### Direct RF Sampling and Tuned IF Architectures

In this architecture, the RF signal undergoes preselection filtering, followed by amplification, and is then digitized without requiring any IF conversion. This approach relies heavily on a DSP for further signal analysis and processing, allowing for a highly compact and power-efficient design. However, to ensure optimal performance, several stages of amplification and filtering are usually needed to achieve the substantial gain (typically between 50 and 110 dB) required by the analog-to-digital converter (ADC).

#### Tuned IF Architecture

The analog counterpart to this architecture is known as the tuned IF configuration. This setup also amplifies and processes the RF signal without intermediate conversion, yet suffers from several distinct limitations:

- Poor stability: Interstage feedback, often through the power supply, impacts the stability of the signal, making it challenging to maintain signal integrity throughout the stages.
- Gain limitations due to stability concerns: Achieving the optimal gain in this architecture is often difficult. High gain levels may exacerbate stability issues, necessitating a delicate balance between gain and stability, especially in applications requiring consistent performance.

- Narrow fractional bandwidth requirements: The architecture requires filters with exceptionally narrow fractional bandwidths to ensure signal fidelity. Designing filters that meet these stringent specifications can be complex and costly.
- High-frequency detection challenges: As detection occurs at the receiver signal frequency, which can be very high, there are limitations in performance. Most detectors are optimized for lower frequencies, making high-frequency detection less efficient and potentially less accurate.

The simplicity and cost-effectiveness of the tuned IF configuration make it suitable for low-cost applications where performance requirements are moderate. Typical applications include RFID tags and keyless entry devices, where functionality outweighs the need for high precision or extensive range.

Following this discussion, the image reject receiver architecture will be introduced. This architecture addresses a critical issue faced in other designs – the presence of image frequencies that can degrade signal quality. By implementing specialized techniques to reject these unwanted image signals, the image reject receiver offers a more refined approach to handling RF signals, especially in environments with significant interference.

## Image-Reject Receiver Architecture

The image-reject receiver architecture is designed to address the challenges related to image rejection, a common issue in both the superheterodyne and low-IF architectures. The fundamental problem is that any unwanted signals at the image frequency can degrade the quality of the desired signal. In general, effective image rejection is achievable when there is a large frequency separation between the desired input signal and its image, requiring a relatively high IF. In contrast, the near-zero IF architecture, with a lower IF, places the desired RF carrier and the IF signal closer together, making effective image rejection challenging. Additionally, achieving both effective image rejection and good interference suppression with a low IF is particularly difficult, given that the image-reject filter's passband must be sufficiently wide to encompass the entire desired frequency band.

To mitigate the limitations of traditional image rejection techniques, Hartley and Weaver architectures provide a robust solution. These designs use phase manipulation and signal cancellation to remove unwanted image signals without relying heavily on high-performance filters, making them effective in cases where interference suppression is crucial.

The Hartley image-reject architecture achieves image rejection by splitting the incoming RF signal from the LNA into two paths. In each path, signal processing occurs in such a way that, upon recombination, the signals at the image frequencies cancel out, leaving only the desired signal. Specifically, a 180° phase difference between the signals in the in-phase (I) and quadrature (Q) branches ensures that unwanted image signals in the two paths are out of phase when they reach the combiner, resulting in effective image cancellation. Figure 1.14 illustrates the Hartley

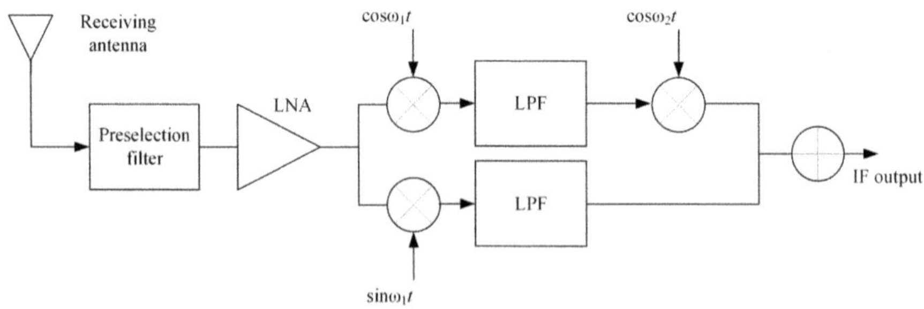


**FIGURE 1.14** Hartley image-reject architecture.

image-reject configuration, where the architecture's phase manipulation mechanism enhances image rejection capabilities.

The Hartley architecture's approach to image rejection provides several benefits:

- Effective image rejection without requiring complex or high cost filtering stages, enabling a more compact design.
- Adaptability across various RF bands, as the architecture relies on phase cancellation rather than physical filters to eliminate image signals.
- Minimal filter design constraints, allowing flexibility in applications where small filter sizes or specific filter bandwidths are essential.

The Weaver image-reject architecture is like the Hartley design but introduces a quadrature down-converter instead of a 90° phase shifter. This design choice results in improved quadrature matching, especially over wideband frequencies. By replacing the phase shifter with a quadrature down-converter, the Weaver architecture achieves better performance in applications requiring precise quadrature matching across a broad spectrum. However, this comes at the cost of added complexity, as the Weaver architecture requires additional IF mixers and oscillators to achieve the desired level of image rejection. Figure 1.15 presents the Weaver image-reject configuration.

The Weaver architecture offers:

- Enhanced wideband quadrature matching, making it suitable for wideband applications where precision is essential.
- Higher image rejection capability than standard low-IF architectures, providing improved signal integrity in challenging RF environments.
- Scalability for various applications, including scenarios where wideband performance and adaptability to multiple RF bands are necessary.

The image-reject receiver architecture, through Hartley and Weaver implementations, addresses the limitations inherent to near-zero and low-IF configurations, particularly the challenge of effectively rejecting image frequencies close to the desired signal. Both architectures present innovative approaches to image

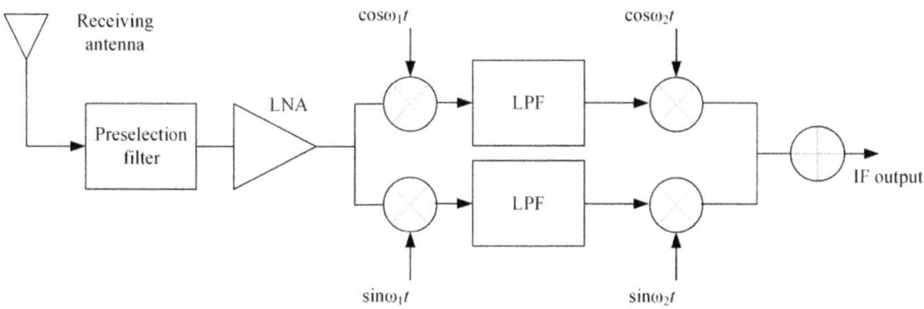


**FIGURE 1.15** Weaver image-reject architecture.

rejection, balancing the trade-offs between complexity, component count, and signal purity.

## THE RF FRONT END

A detailed discussion on RF front ends is required to close this review of receiver architectures. The definition of the front end differs among different authors. Some authors define the front end as the circuitry between the antenna and mixer (Rouphael 2014), while other authors refer to the circuitry between the antenna and the demodulator (Luzzatto and Haridim 2016). In this book, the front end is defined as all hardware and software modules operating at the final RF frequency. This includes RF front filters, LNAs, RF frequency signal processing modules, RF sampling/undersampling ADC, and high-frequency mixers (Luzzatto and Haridim 2016). Similarly, the backend is defined as all hardware and software modules that operate below the first IF frequency or at a frequency other than the final RF frequency. The backend includes the baseband processing module and the detector. A receiver may have several IF frequencies, as implied in this definition. The IF chain includes all hardware and software modules that operate at a nonzero IF (Luzzatto and Haridim 2016).s

The RF front end is interfaced, on the channel side, to the antenna, and, on the detector side, to the IF chain (Luzzatto and Haridim 2016) or the demodulator, depending on the front end definition. The signal from the antenna to the detector determines the boundary conditions on the front end. The designer first selects the receiver architecture and converts these boundary conditions, expressed in terms of SNR, into circuit requirements for the front end, such as gain, NF, etc. [18]. From these requirements, boundary conditions related to the sub-blocks of the RF front end are derived. Finally, the architecture of sub-blocks as well as the values of their passive and active individual components are derived from their boundary conditions. These derivations are beyond the scope of this book, but related information can be found in [18].

The RF signal from the antenna suffers generally from several impairments, including a low signal level (in the order of microvolts), in-band and out-of-band noise, adjacent channel interference, and co-channel interference. Co-channel interference appears when the same frequency is used in a distant channel, as is possibly found in mobile cellular networks. The RF front end allows improving the signal level without major distortion and deterioration of its SNR, as well as reducing the interference. This can be done in several ways, depending on the architecture of the receiver. The input signal must be amplified to increase its level before demodulation. But it is remembered that in increasing the signal level, both the noise and the interference are amplified. Hence, a need for an LNA, an amplifier that limits the possible minimum, the additional noise inherent to it, and suppresses the noise contributed by subsequent blocks. Interference at the input of the LNA can saturate the amplifier and generate intermodulation products of the same frequency as the input signal because of nonlinearities in the LNA signal response. Such intermodulation products will distort the desired input signal. A band or preselection filter at the input of the front end can reduce or filter out the out-of-band interference. However, such a filter should have a narrowband and be of high order, which is difficult to achieve at high frequencies. Still, the approach mostly used consists of using out-of-band interference filtering at the input of the front end, aiming to avoid saturation and intermodulation effects. However, this suppression cannot be complete because the filter cannot be ideal, as it must remain practical at RF. Because a high amplification gain is required to achieve the signal level required by the demodulator, interfering signals will still appear at the output of the amplifier for a high-gain LNA. The solution to this problem consists of a partial amplification (lower gain) in the first stage, followed by high-Q IR filtering and down-conversion of the RF signal frequency to an IF where further filtering for channel selection and amplification are achieved. This process is repeated until the desired signal level and SNR are achieved at the input of the demodulator. It is noted that low noise amplification, as will be shown, is mostly required only for the first stage of amplification. The desired signal frequency is $f_R$, the IF is $f_{IF,}$ and the frequency of the local oscillator is $f_{LO}$. The first and second bandpass filters are the preselection and IR filters. The IR filter is a high-Q off-chip filter. The IF filter is the channel selection filter. This filter is a high-Q off-chip filter that can be based on ceramic crystals of surface acoustic wave (SAW) technologies. The selected signal level is adjusted in the IF chain before quadrature down-conversion in the backend. The backend is IQ-based, and the baseband outputs $I(t)$ and $Q(t)$ can be digitally sampled for processing and detection by a DSP. The boundary conditions on the detection side are the SNR of the signals $I(t)$ and $Q(t)$.

## APPLICATIONS IN MODERN COMMUNICATIONS

The rapid advancements in integrated ultra-low noise amplifier design are at the core of enabling modern wireless communication standards. Integrated amplifiers today are pivotal to supporting the high sensitivity and low noise performance required by an expanding variety of wireless technologies. These technologies include cellular, local-area, and personal-area communications, each demanding different performance characteristics for optimized data transmission, error reduction, and power efficiency. As integrated circuit design progresses, modern ultra-low noise amplifiers continue to evolve, supporting increasingly complex standards while simultaneously meeting the growing requirements for low-power operation and compact form factors.

The main standards in mobile wireless communications are the Global System for Mobile Communications (GSM), the Digital Communication System (DCS), and the personal communications services (PCS). These standards are very similar, with similar protocol layers but differences in the physical layers since they use different frequency bands. The modulation is the Gaussian minimum shift keying (GMSK), with time multiplexing and a 200 kHz channel bandwidth that can deliver a data rate of 270 kb/s.

While the core technology supporting these standards remains largely the same, the continued optimization of integrated ultra-low noise amplifiers has enabled these communication systems to scale their performance without sacrificing sensitivity. As mobile devices incorporate multiple bands and modes within a single platform, integrated amplifiers with high linearity and wide dynamic range are essential for handling the wide variety of signals and maintaining robust communication quality across different network types.

Superheterodyne, zero-IF, and low-IF architectures have been implemented for these standards with success. WCDMA and UMTS systems, known as 3G wireless systems, can deliver data rates of up to 3.84 Mb/s, using QPSK modulation. Because WCDMA channels are wideband with a spread spectrum nature, the direct conversion (Zero-IF) architecture is suitable for receivers if issues related to dc-offset and $1/f$ noise are resolved. This can be done through a servo loop and AC coupling.

In these architectures, the role of integrated ultra-low noise amplifiers is crucial. By reducing noise at the input, these amplifiers prevent degradation of signal quality, enhancing the reliability of communication even in crowded frequency environments. As data rates continue to rise with new standards, the requirements on noise performance become even more stringent, especially in wideband systems. In such systems, the LNA must maintain performance across a broad spectrum of frequencies, further underscoring the importance of precision design in integrated amplifier technology.

To achieve very high data rates and very low latency, carrier aggregation is used in 4G and 5G mobile networks. A receiver is required for each carrier. The conventional zero-IF architecture is used. The zero-IF architecture's reliance on ultra-low noise amplifiers is especially critical here. LNAs in such architectures not only improve signal reception under challenging conditions but also enhance the system's overall energy efficiency, an essential aspect for high-speed data networks such as 4G and 5G. Modern integrated amplifiers used in these networks are designed to provide precise control over gain while keeping distortion to a minimum, thereby optimizing signal clarity even when handling multiple carriers simultaneously.

In 5G applications, amplifiers must often contend with high-frequency millimeter-wave signals, which introduce unique noise challenges. To address these, integrated amplifiers are engineered to exhibit minimal NF and to maintain stability across wide frequency ranges. This level of sophistication is essential in supporting carrier aggregation and multiple-input multiple-output (MIMO) configurations that define 5G's advanced connectivity capabilities.

The 802.11g standard for wireless LANs (WLAN) can deliver up to 54 Mb/s using a 64-quadrature amplitude modulation (64-QAM) in the 5 GHz band and 11 Mb/s in the 2.4 GHz band using complementary code keying (CCK) modulation. Both the zero-IF and superheterodyne architectures are possible. For the superheterodyne architecture, the IF must be high, in the range of 3 GHz, to allow on-chip image rejection.

As WLAN standards evolved to include higher data rates and multiple bands, integrated LNAs adapted to handle these requirements efficiently. The compact form of ultra-low noise amplifiers makes it possible to support both 2.4 GHz and 5 GHz bands within a single WLAN chip, accommodating multiple frequency bands while reducing signal interference. In applications like 802.11g, where data throughput and latency are paramount, amplifier design emphasizes the minimization of harmonic distortion and signal loss, even in high-speed environments.

Standards for wireless personal-area networks (WPAN) include Bluetooth (802.15.4), ZigBee (802.15.1), and ultra-wideband (802.15.3). Bluetooth uses Gaussian frequency shift keying (GFSK) modulation and 1 MHz channel bandwidth to deliver a data rate of up to 1 Mb/s in the 2.4 GHz ISM band. The Zigbee standard was developed for wireless machine-to-machine (M2M) and IoT networks. It achieves 250 kb/s in the 2.4 GHZ ISM band using GFSK modulation.

With IoT applications on the rise, ultra-low noise amplifiers play a foundational role in WPAN standards, where power efficiency and noise control are critical. In Bluetooth and ZigBee implementations, integrated amplifiers are required to be low-power yet capable of maintaining signal integrity over short ranges. Modern IoT devices are increasingly compact and energy-conscious, necessitating LNAs that deliver consistent performance even with minimal power consumption.

The ultra-wideband standard uses very low power for short-range and high-bandwidth communications. It achieves up to 480 Mb/s using 500 MHz channels in the 3.1-10.6 GHz band and shaped-pulse (SP) or frequency-switched orthogonal frequency division multiplexing (FS-OFDM) modulation. Bluetooth and ZigBee use low-IF receivers, while ultra-wideband receivers are based on zero-IF and low-IF architectures. Low-IF architectures are also used in mobile phones, FM radio, and televisions.

The unique requirements of ultra-wideband (UWB) communications place stringent demands on LNA design, particularly for high signal integrity over a broad frequency range. Integrated ultra-low noise amplifiers in UWB devices must manage extremely low noise levels, allowing them to operate effectively in both short-range high-bandwidth applications and in crowded environments where multiple devices share the same frequencies. This combination of high selectivity, low noise, and efficiency makes integrated LNAs essential for UWB and other low-IF architectures, supporting applications from mobile phones to television broadcasting.

The most used architecture for mobile communications in the mid-2000s was the superheterodyne receiver. However, the trend for lower cost, lower power, and higher integration was ongoing. At the same time, the evolution of wireless terminals into universal multi-service devices was promoting multi-standards (GSM, WCDMA,

IEEE 802.11, Bluetooth, ultra-wideband, GPS) and specifications architectures. Building blocks in these architectures were not duplicated but rather reconfigured, which allowed for to reduction in manufacturing costs. In 2016, the zero-IF and low-IF architectures were the most used in low-power receivers. The trend in multi-service device implementation is to duplicate RF paths, RF ports, LNAs, filters, and switches.

In summary, integrated ultra-low noise amplifiers have become indispensable in modern communications, evolving alongside the growth of standards and network architectures. As the industry moves toward universal, multi-standard devices, LNAs must continue to deliver unprecedented noise reduction and power efficiency while remaining adaptable across different architectures and frequency bands.

## CONCLUSION

In concluding this chapter on antenna efficiency, noise, and impedance matching, we see how fundamental concepts converge in determining antenna performance, particularly within communication systems requiring high fidelity and sensitivity, such as satellite and deep-space applications. Antenna efficiency ($e_0$), calculated from both conductive-dielectric losses ($e_{cd}$) and impedance mismatches ($\Gamma$), is a critical measure of the antenna's ability to convert captured electromagnetic waves into usable signal power. The efficient transfer of this signal to the receiver depends on both minimizing dielectric and conductive losses and a careful match between the antenna and the feedline impedances, where any reflection is minimized. This efficiency reflects not only the material and design constraints but also the antenna's operational environment and the spectrum used.

Noise analysis further deepens our understanding by considering the unavoidable internal losses within the antenna structure and external environmental factors. External noise from celestial and terrestrial sources – such as galactic noise, thermal noise from the Sun and Moon, and atmospheric attenuation effects – interacts with the antenna's design, particularly regarding radiation patterns and elevation angle. At higher frequencies, where absorption in atmospheric gases and rain can be significant, a rise in noise temperature impacts system sensitivity, necessitating optimal placement, orientation, and material selection for antennas in these applications. Understanding this noise's spectral characteristics and spatial behavior enables engineers to design antennas that exploit "quiet" regions, like the 1–10 GHz range, or the so-called "microwave window," ideal for space communications.

The interaction of antenna impedance, characterized at resonance, and its matching with the receiver is also shown to be vital for maximizing power transfer and reducing losses. By achieving complex conjugate matching, we ensure that maximum signal power is delivered to the receiver while minimizing power loss in the form of reflected waves. This becomes especially crucial in low-noise amplification stages, where the receiver's LNA operates most effectively when the antenna's noise temperature is aligned with that of the LNA's input impedance. Thus, antenna noise is often considered equivalent to that of a resistor at the same noise temperature, a concept that allows simplified calculations for noise matching and system sensitivity.

Through the discussion, achieving low system noise temperature and high antenna efficiency requires precise management of both intrinsic antenna losses and extrinsic noise sources, particularly in applications where signal integrity over long distances is paramount. The interplay of antenna efficiency, impedance matching, and noise temperature provides a complete picture of what engineers must consider when designing antennas that perform reliably in complex and often hostile electromagnetic environments. By mastering these principles, we advance the capabilities of communication systems, particularly in fields where precise signal reception and minimal signal degradation are essential. This holistic understanding underpins the effectiveness of antennas in modern communication networks, laying the groundwork for future advancements in antenna technology.

## REFERENCES


Balanis, C.A., 2016. *Antenna Theory: Analysis and Design*. 4th ed. Wiley.

Bokulic, R.S., 1991. Use Basic Concepts to Determine Antenna Noise Temperature. *Microwaves & RF*, 30, 107–108.

Burke, B.F., Graham-Smith, F., and Wilkinson, P.N., 2019. *An Introduction to Radio Astronomy*. Cambridge University Press.

Huang, Y., Alieldin, A., and Song, C., 2021. Equivalent Circuits and Analysis of a Generalized Antenna System [Antenna Applications Corner]. *IEEE Antennas and Propagation Magazine*, 63 (2), 53–62.

IEEE Standard Letter Designations for Radar-Frequency Bands, 2002. The Institute of Electrical and Electronics Engineers, Inc. (IEEE).

ITU, 2016. Radio Regulations, Part 1 Chapter II Article 5 (Frequency Allocations) and Part 3 Resolution 767 (Studies towards an Identification for Use by Administrations for Land-Mobile and Fixed Services Applications Operating in the Frequency Range 275-450GHz) [online]. Available from: www.itu.int/pub/R-REG-RR-2016 [Accessed 23 August 2024].

Kanar, T., and Rebeiz, G.M., 2014. X- and K-Band SiGe HBT LNAs With 1.2- and 2.2-dB Mean Noise Figures. *IEEE Transactions on Microwave Theory and Techniques*, 62 (10), 2381–2389.

Kraus, J.D., and Marhefka, R.J., 2001. *Antennas for All Applications*. 3rd ed. McGraw-Hill.

Luzzatto, A., and Haridim, M., 2016. *Wireless Transceiver Design: Mastering the Design of Modern Wireless Equipment and Systems*. 2nd ed. Wiley.

McCulloch, M., 2013. *Enhancing the Noise Performance of Low Noise Amplifiers-with Applications for Future Cosmic Microwave Background Observatories*. University of Manchester.

Nikolova, N.K., 2018. Notes on Antenna Engineering. McMaster University.

Pala-Schonwalder, P., Bonet-Dalmau, J., del Aguila-Lopez, F., Moncunill-Geniz, F.X., and Giralt-Mas, R., 2020. Superregeneration Revisited: From Principles to Current Applications. *IEEE Microwave Magazine*, 21 (2), 35–47.

Planck, M., 1914. *The Theory of Heat Radiation*. Blakiston.

Rodhe, U.L., 2017. *Communication Receivers, Principles and Design*. 4th ed. McGraw-Hill.

Rouphael, T.J., 2014. Receiver Architectures. In: *Wireless Receiver Architectures and Design*. Elsevier, 439–480.

Salcedo, A., and Martinez, E., 2017. Analysis of the Electromagnetic Spectrum under the Extremely Low Frequency Band: Frequency Sub-Bands Classification. In: *2017 International Conference on Mechatronics, Electronics and Automotive Engineering (ICMEAE)*. IEEE, 157–162.

Shrivastava, N., and Khatri, R., 2017. Design of a 2.4-GHz Differential Low Noise Amplifier Using 180nm Technology. In: *2017 International Conference on Recent Innovations in Signal processing and Embedded Systems (RISE)*. IEEE, 494–497.

# 2 Low-Noise Amplifiers and Their Role in the Square Kilometer Array

## THE ROLE OF THE LNA

The role of the LNA is to improve the input signal power level without adding significant noise and distortion and suppress the noise contributed by subsequent blocks, as indicated. As the first and most sensitive element of a receiver chain, the LNA determines the NF and the linearity of the receiver. The main four LNA characteristics are the NF, the gain, the bandwidth, and the linearity of the dynamic range. Performance requirements for LNAs depend on their applications. For applications with a low signal level and large bandwidth, the most important requirements are the NF, the input matching over the bandwidth, and the linearity. In the case of the square kilometer array (SKA), linearity specifications are less stringent than noise and matching requirements owing to the selection of sites that are free from radio interference, as poor linearity reduces the immunity to interference.

A survey to compile LNA data from 570 designs is provided in Belostotski (2023). LNA data, such as gain, bandwidth, NF, $S_{11}$, P1dB, IIP3, power consumption, area, and the manufacturing process, are presented, per technology, for many designs published in peer-reviewed research papers. The lowest and highest performance parameters per technology resulting from this survey are presented in Table 2.1.

In Table 2.1, the values in rows do not necessarily correspond to the same design; they indicate the highest and lowest survey data for all published designs for the technology. $f_o$ is the center frequency, *BW* is the bandwidth, $S_{11}$ is the input return, *P1dB* is the 1 dB compression point, *IIP3* is the input third-order intercept point, and *Tech* is the technology node, also known as process node or process technology. The technology node, expressed generally in μm or nm, refers to the manufacturing process and design rules of a semiconductor fabrication process. Table 2.1 shows that the lowest NF is achieved by cryogenic cooling at 10 K, corresponding to 1 K or 0.01 dB. The 1 K NF LNA has been designed for bands 2–5 of the SKA (0.94–13.8 GHz) and is based on 50 μm HEMT (Schleeh *et al.* 2017). The table also shows that sub-1 dB NFs can be achieved at room temperature by CMOS and SiGe processes. InP, GaAs, and GaN processes have been used traditionally on low-noise designs and subsequently can achieve sub-1 dB NFs.


DOI: 10.1201/9781003459385-2

**TABLE 2.1**
**LNA survey**

| Process | | *NF* *[dB]* | $f_0$ *[GHz]* | *BW* *[GHz]* | *Gain* *[dB]* | $S_{11}$ *[dB]* | *P1dB* *[dBm]* | *IIP3* *[dBm]* | *Tech* *[μm]* |
|---|---|---|---|---|---|---|---|---|---|
| CMOS | H | 11 | 211 | 45 | 32 | –2.5 | –43.8 | –24 | 0.8 |
| | L | 0.2 | 0.1 | 0.045 | 5.2 | –45 | 3 | 22 | 0.016 |
| SiGe | H | 11 | 5.95 | 8.1 | 50 | –3 | 12.9 | 20 | 0.8 |
| | L | 0.75 | 0.6 | 0.008 | 5 | –27.6 | –44 | –33 | 0.055 |
| GaAs | H | 6.5 | 587 | 78 | 43 | –2 | 1 | 20.1 | 0.6 |
| | L | 0.26 | 0.6 | 0.1 | 4.3 | –18 | –41 | –25 | 0.02 |
| GaN | H | 7.6 | 140 | 60 | 31 | –2.3 | 19.5 | 49.2 | 0.25 |
| | L | 0.4 | 1 | 0.8 | 10.9 | –15 | –4 | 9 | 0.04 |
| InP | H | 13 | 828,5 | 65 | 50 | –2 | – | – | 0.25 |
| | L | 0.15 | 0.75 | 0.6 | 5 | –12 | – | – | 0.025 |
| Cryogenic | H | 191 K | 210 | 0.4 | 57 | –1.5 | –10 | 0 | 0.2 |
| | L | 1 K | 0.3 | 60 | 15.75 | –15 | –53 | –47 | 0.028 |

## TRENDS IN LNA PERFORMANCE

The NF of active components increases with frequency, for example, the LNA in Leong *et al.* (2015) achieves an NF of 11.1 dB at 0.8 THz while the LNA in Schleeh *et al.* (2016) achieves an NF of 0.15 dB at 0.75 GHz. Both LNAs are based on InP HEMTs and are at room temperature. However, the NF of active components decreases with the increase of the unity gain frequency $f_T$ and with scaling. Hence, the NF is expected to increase less with frequency as the size of transistors decreases. In Andersson *et al.* (2017), trend curves based on reported LNAs predict the evolution of the NF versus the operating frequency for several technology nodes. The trend curves are depicted in Figure 2.1 and show that by 2022, NFs better than 1.5 dB were achieved at 20 GHz and that by 2027, NFs better than 1 dB can be attained at 20 GHz.

For the interest of the reader, it can be noted that the typical NF for an LNA for a GPS receiver is 1.5 dB at 1.575 GHz. In fact, for a global navigation satellite system (GNSS) receiver, the received satellite signal can be as low as –130 dBm. For LTE, a typical NF at 882 MHz is 0.89 dB and 0.95 dB at 3.5 GHz. In general, the NF for LTE LNAs is comprised of between 0.77 and 1.6 dB for frequencies between 2.3 and 4 GHz (Maalik 2013). For 5G, the NF is expected to be between 1 and 3 dB, for frequencies up to 28 GHz. In Andersson *et al.* (2017), NFs of 5, 7.8, 8.7, and 10.1 dB are expected at 2, 30, 45, and 70 GHz, respectively, for 5G receivers. For digital video broadcasting terrestrial (DVB-T), the desired NF is about 1.3 dB (Tsaraklimanis and Karagianni 2011). For radar, the typical NF varies from 1.3 to 1.7 dB at 5 GHz (Suana 2017). For radio astronomy, the NF can be extremely low, especially at cryogenic temperatures (e.g., 0.1 dB or less).

New trends in LNA research and development are characteristics of the Fourth Industrial Revolution (Industry 4.0). The Fourth Industrial Revolution is characterized by the convergence of advanced technologies leading to the automation of activities requiring human interaction. Such activities include the production of goods and services that are pattern-based or routine-based, as well as emerging applications, such as

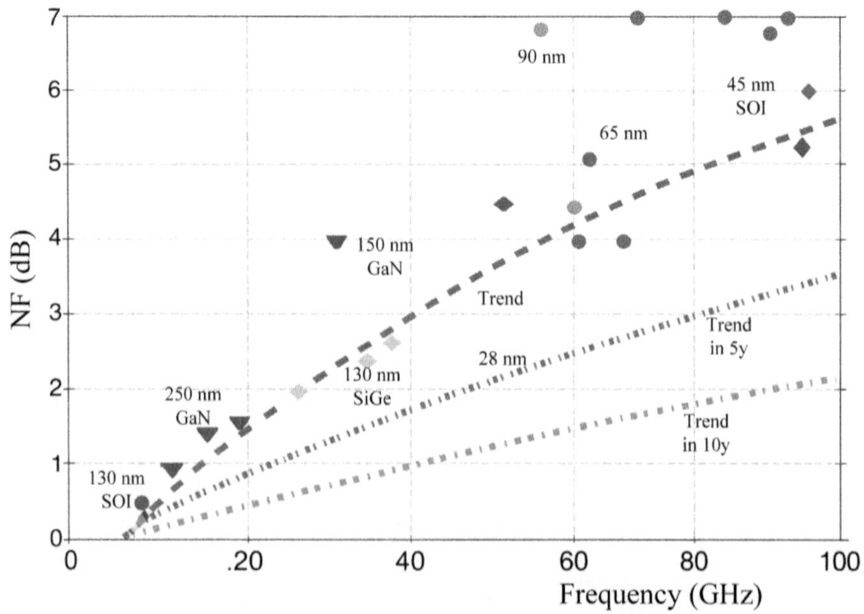


**FIGURE 2.1** Plot of NFs for reported LNAs and a predicted trend (Andersson *et al.* 2017).

virtual reality (VR) and autonomous vehicles. Their automation relies on the learning capacity of machines, hence artificial intelligence, and on a massive communication infrastructure demanding low-latency because of its real-time nature. This infrastructure will comprise countless Internet of Things (IoT) sensors that will communicate through 6G networks. It is considered that these high-speed and low-latency data networks will operate at millimeter-wave (mm-wave) or near-THz-based communications.

In this book, ultra-low noise LNAs are defined as LNAs that achieve sub-1 dB NFs and a power gain of more than 10 dB. Such a definition can also be found in the literature. It is shown in Chapter 1 that sub-1 dB NFs are expected to be achieved only at frequencies of a few GHz using current CMOS, SiGe, and GaN processes, at room temperature. Therefore, this book does not cover mm-wave LNAs and is limited to LNAs that are based on lumped parameter models, in opposition to distributed parameters. The typical gain is comprised of between 10 and 20 dB, the lower gain limit allowing a significant reduction in the input-referred noise from subsequent stages of the receiver.

## LNA PERFORMANCE VARIABLES

The five main LNA parameters are related to several variables that affect performance. Because some of these variables are competing, a trade-off between them is required to attain the required specifications, which is a source of complexity in LNA design. These variables are depicted in Figure 2.2.

The structure of this book is based on the five main LNA performance criteria (NF, gain, bandwidth, linearity, and dynamic range), but with a focus on the NF. The book

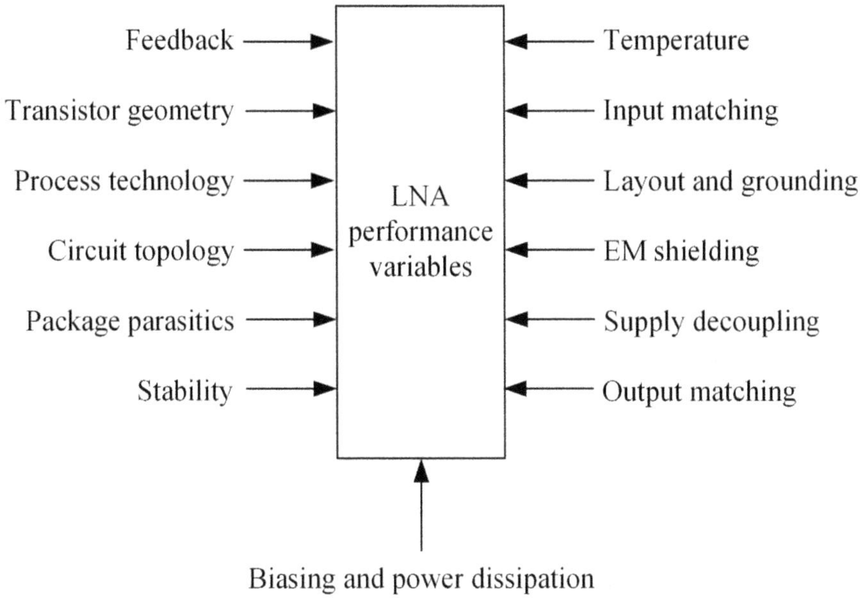


**FIGURE 2.2** LNA performance variables.

presents and analyzes dependencies between performance parameters and variables in Figure 2.2. The following sections of this chapter introduce these variables. The dependencies between these variables and the performance parameters are developed further in subsequent chapters.

## Sensitivity

Sensitivity and NF are strongly related, and it is desired to devote a paragraph on sensitivity before presenting and discussing the elements of Figure 2.2. The NF of an amplifier determines its sensitivity. This section presents the basic concepts related to sensitivity, the minimum possible signal power that the LNA can receive, and the noise floor. At the input of an amplifier, the signal emerges from the background noise as shown in Figure 2.3, and its quality, relative to the background noise, can be evaluated by the signal-to-noise ratio (SNR) at the input, which is given by $S_i/N_i$. The background noise at the input can be analyzed as resulting from the source noise and the amplifier noise reported to the input. Reporting the noise of a noisy two-port network to its input is a technique used in noise analysis, a technique that will be presented in subsequent chapters. In this technique, a noisy two-port network is replaced by a noiseless two-port network, and two noise sources are reported to its input. Hence, the amplifier is assumed to be noiseless, and its noise is represented by equivalent noise sources at its input. Because sensitivity is limited by the total noise at the input, its lower limit corresponds to the power of the source noise $N_i$, when the noise due to the LNA is very small and can be neglected. The lower sensitivity limit

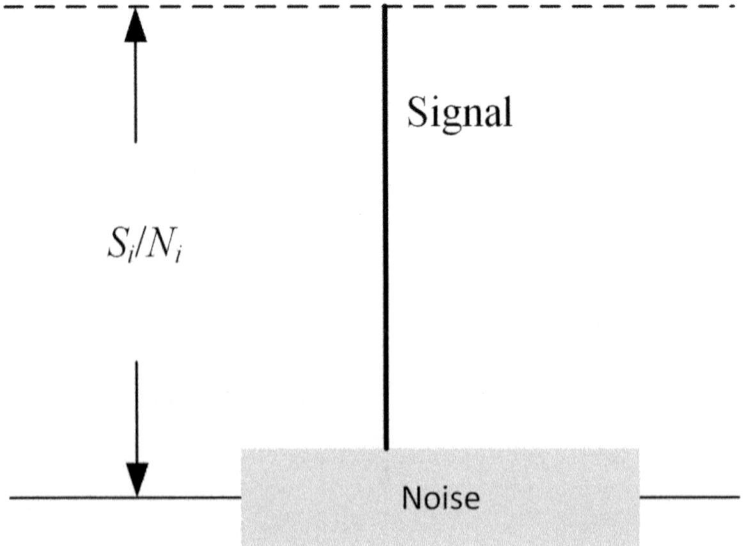


**FIGURE 2.3** Noise at the input of an amplifier.

is then equal to the noise power at the input due to the source thermal noise. This power is given by the expression $kTB$ – where $k$ is Boltzmann's constant, $T$ is the temperature, and $B$ is the bandwidth – when the LNA input impedance is matched to the source, corresponding to $N_i = -174$ dBm/Hz. The temperature is 290 K, which is the IEEE standard for room temperature.

When the noise of the two-port network and its equivalent noise sources are taken into consideration, the input signal must be greater than $N_i$ to emerge from white noise. The minimum discernible signal for an LNA with a bandwidth $B$ and a noise figure NF is given by the noise floor:

$$Noise\ floor_{dBm} = 10\log_{10}\left(kT \times 1000\right) + NF + 10\log_{10} B \tag{2.1}$$

and shows that the noise floor decreases when NF decreases, increasing the sensitivity of the LNA. The SNR at the output of the LNA decreases with the increase of the noise due to the LNA, as shown in Figure 2.4, where the output SNR decreases by 10 dB when the gain of the LNA is 20 dB and NF is 10 dB.

Because the noise factor $F$ is given by the expression $F = (SNR)_{in}/(SNR)_{out}$, a low noise factor is desired for a high-output SNR and a high sensitivity.

## Process Technology

Possible transistor technologies for ultra-low noise receivers are from low-noise foundries based on GaAs, on mHEMT, pHEMT, SiGe HBTs, CMOS, and InP devices. Over the past few decades, HEMT has been the primary choice for low-noise technology. HEMT transistors have high transconductance ($g_m$) and high unity

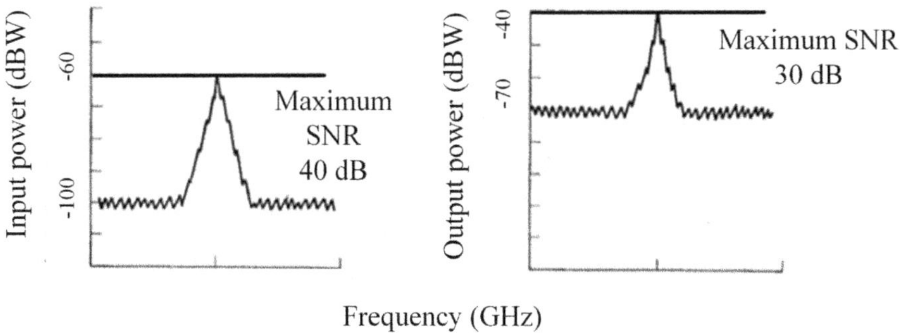


**FIGURE 2.4** LNA input and output signals.

gain frequency ($f_T$). They also have extraordinarily low NF and low parasitic resistance. The NF of HEMT LNAs is dramatically reduced with InP pHEMT and GaAs mHEMT transistors. InP and GaAs are the most widely used technologies for LNAs in radio astronomy. At cryogenic temperatures, a noise factor of less than 0.1 dB is often achieved with InP transistors (Bij de Vaate *et al.* 2010). Despite offering a high $f_T$ coupled with high noise performance, InP and GaAs processes have a low integration level, high power consumption, and remain expensive. They are not suitable for the integration of complex digital logic.

Devices based on gallium nitride (GaN) have a higher breakdown voltage than silicon and can achieve higher power as well as higher electron mobility. They have been dominant in lasers and photonics but are being commercialized rapidly into HEMTs, power amplifiers, and THz devices, owing to their unique properties.

The SiGe HBT process combines the performance of transistors in the III-V technologies (e.g., GaAs and InP) with the maturity, integration level, and cost of Si fabrication. SiGe HBT devices of earlier generations could achieve an $f_T$ of more than 50 GHz and an NF of less than 0.7 dB at 2 GHz (Cressler 1998). Today, devices operating up to near-Tera Hz speeds are available. A combination of SiGe HBT and BiCMOS (SiGe BiCMOS) can be used for on-chip systems. SiGe BiCMOS can integrate passives, a low-noise amplification function, and baseband processing on the same chip.

A reduction in the receiver cost can be achieved by integrating an entire receiver chain into a single chip. Such a solution has been proposed with a Si CMOS process (Jackson 2005). However, compared to SiGe technology, a Si CMOS process has low $f_T$ and poor RF performance. Moreover, the process has high substrate losses and poor-quality passives. Nevertheless, the technology is mature and has attracted important investments to improve the speed of transistors and their integration. As the size of transistors decreases and $f_T$ increases, the CMOS process can be used in ultra-low noise receivers.

A qualitative evaluation of various technologies is provided in De Vaate (1999). InP and GaAs HEMT technologies have the best noise performance, but they are very costly. The NFs of SiGe and GaAs LNAs are comparable, but SiGe devices consume less power and are cheaper, about one-third of the cost of GaAs devices and 5% of

**TABLE 2.2**
**Technology performance criteria of front-end circuits**

| Technology | RF noise | 1/*f* noise | Power consumption | Linearity | Integration density | Cost |
|---|---|---|---|---|---|---|
| InP | Very good | Poor | Fair | Good | Low | High |
| GaAs PsHEMT | Very good | Poor | High | Good | Low | High |
| GaAs HBT | Good | Good | Fair | Very good | Low | High |

the cost of InP devices. The qualitative evaluation, as well as the criteria used for the evaluation of the most relevant technologies, is provided in Table 2.2 (De Vaate 1999).

Passive components such as resistors, inductors, capacitors, transformers, and baluns are essential in electronic systems. They are used for matching networks, filtering, signal attenuation, decoupling, LC tank circuits, and biasing circuits (Tilmans *et al.* 2003). However, integrated passive components fabricated in standard Si processes have failed to achieve high $Q$-factors, mostly because of substrate losses. A typical value of the $Q$-factor of an integrated inductor in a BiCMOS process or bipolar processes is about 10, in contrast to external inductors having a $Q$-factor that can reach 150 or more. Double metal-layer integrated inductors or complex fabrication processes can be used to improve performance, achieving $Q$ factors up to 30. To achieve ultra-low noise figures, it may be required to use external inductors.

## Transistor Geometry

Electronic circuits are made of five basic electronic components: resistors, transistors, capacitors, inductors, and diodes. For integrated circuits, these components are implemented in a semiconductor substrate and are connected between them and to input or output pads by metal tracks. All electronic components generate inherent noise, and, in integrated circuits, the main sources of noise are transistors, resistors, and inductors. An appropriate selection of their characteristics can significantly affect the noise performance of integrated LNAs. Integrated circuits are three-dimensional devices where the third dimension results from substrate doping, etching, and successive implantation of metal and insulators in different layers. Front-end-of-line (FEOL) refers to the bottom layers of the integrated circuit, where the basic electronic devices are patterned in semiconductor material. Back-end-of-line (BEOL) refers to the top layers where interconnections between devices are implemented.

Semiconductor technologies are characterized by the size of their transistors. In terms of the size of active devices, as most technologies support Si CMOS processes, the length of the gate of the associated Si CMOS process, implemented on the same die with BJT-based devices, is used to characterize BJT-based technologies. Other technologies are characterized similarly. The length of the gate of transistors and the width of the base for BJT-based processes are fixed for a given technology (e.g., 0.08 μm (80 nm) and 0.130 μm (130 nm)). However, the emitter length for BJTs and the channel width for MOS devices are not fixed and are determined by the designer. The base resistance in a BJT decreases with the increase in the length of the emitter. In addition, the capacitances of the base-emitter junction and the base-collector junction are related to the geometry of the transistor.

For MOS devices, the gate-source capacitance is related to geometry. The transconductance $g_m$ for BJTs and MOS transistors is also dependent on transistor geometry. For both BJTs and MOS devices, the unity frequency gain $f_T$ is a function of $g_m$ and parasitic capacitances. For BJTs, the noise factor is related to base and emitter parasitic resistances, to the DC gain $\beta$, to gm, to $f_T$, and the emitter current density. The noise factor of MOS transistors varies with the gate-source capacitance, the transconductance, the gate voltage, and the drain current. Therefore, the noise parameters of BJT and MOS devices are dependent on transistor geometry and bias.

## Circuit Topology and Feedback

The performance, size, and power consumption of an LNA are strongly related to its input type. The first part of this section discusses single-ended LNAs versus differential LNAs, while the second part presents LNA topologies.

## Single-Ended and Differential LNAs

Differential circuits have been mostly used in low-frequency analog circuits and digital circuits rather than in RF applications. They provide immunity against external noise sources that may come from couplings with adjacent circuits. These couplings generally operate in common mode and are reduced in differential mode. In addition, differential circuits improve linearity since even-order harmonics are canceled. For an antenna with an unbalanced output, a balun is required when a single-ended LNA is used. The contribution of a passive balun to noise can be higher than 16 K (Bhaumik and George 2011). Therefore, the balun noise will decrease the receiver noise performance.

Single-ended LNAs have one signal input and output and a common ground reference. A differential LNA has two signal inputs and one signal output. The ground is the common signal reference, but the LNA amplifiers only amplify the difference between the input signals. The diagram of a single-ended and a differential amplifier is presented in Figure 2.5 (Maalik 2013).

Because a common signal may appear at the input of the LNA due to interference and parasitic couplings, the LNA has two gains: the common-mode gain and the differential-mode gain. The common-mode rejection ratio, known as CMRR, is the ratio between the differential gain and the common-mode gain. This ratio is expected to be very high, ideally infinite, to suppress common-mode interference. The major difficulty in the design of differential LNAs consists of attaining a very high CMRR.

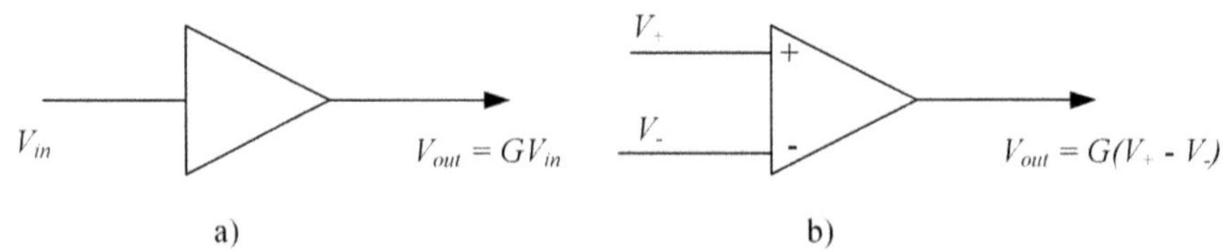


**FIGURE 2.5** A (a) single-ended and (b) differential amplifier.

Differential LNAs are duplications of single-ended LNAs and can be analyzed as such (Cassan and Long 2003, Hazenboom *et al.* 2006, Maalik 2013). Therefore, single-ended LNAs achieve better NFs than differential LNAs. In addition, the credibility and acceptability of differential amplification for very low noise amplifiers have not been established (Bhaumik and George 2011).

## LNA Topologies

The performance of LNAs is strongly related to their topologies, as the selected topology determines performance criteria such as NF, gain, and the matching technique and the required biasing. Topologies can be classified into one-stage and two-stage topologies. They can also be grouped into narrow-band and wide-band topologies. Because the input matching is determined by the first stage, only topologies for a one-stage LNA are presented in this introductory chapter. The high-level classification based on one-stage LNAs is as follows (Xin 2004):

- Wideband LNAs:
  - Resistive termination
  - Common base/gate
  - Resistive shunt-feedback
- Narrowband LNA
  - Inductive degeneration

Resistive termination refers to the resistive input of the first transistor. An LNA with resistive termination is presented in Figure 2.6. The input impedance of a common-source/common-emitter transistor is capacitive. However, when a resistor is placed at the input, the input impedance of the transistor may become resistive at low frequencies and be made to match the source impedance $R_S$ in a wide bandwidth.

The NF of this topology is high, typically superior to 3 dB for CMOS processes, because of the noise contribution from $R_m$. In addition, the gain of the input transistor decreases by 3 dB because of the voltage division resulting from resistive termination. Therefore, the LNA with resistive termination topology is not applicable for ultra-low noise applications.

For the common-gate/common emitter topology in Figure 2.7, the input impedance can be approximated by $1/g_m$ for $g_m >> \omega C_{gs}$, where $\omega$ is the angular frequency and $C_{gs}$ is the gate-source capacitance. Because the source is resistive (50 Ω), perfect

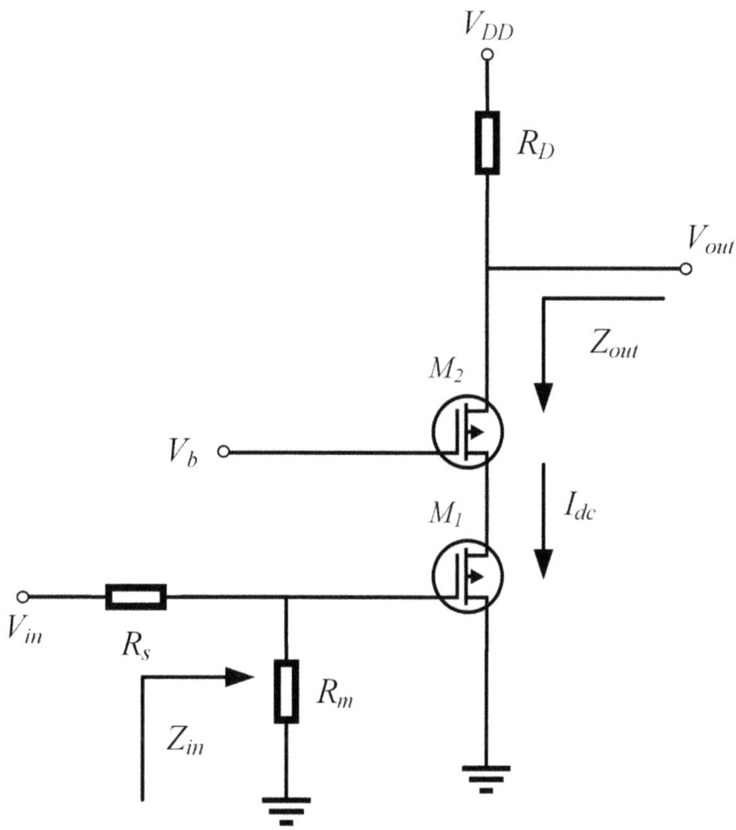


**FIGURE 2.6** LNA with resistive termination.

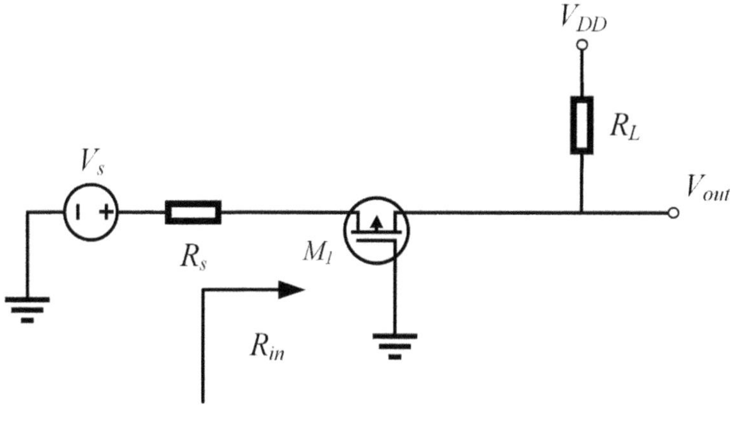


**FIGURE 2.7** Common-gate amplifier.

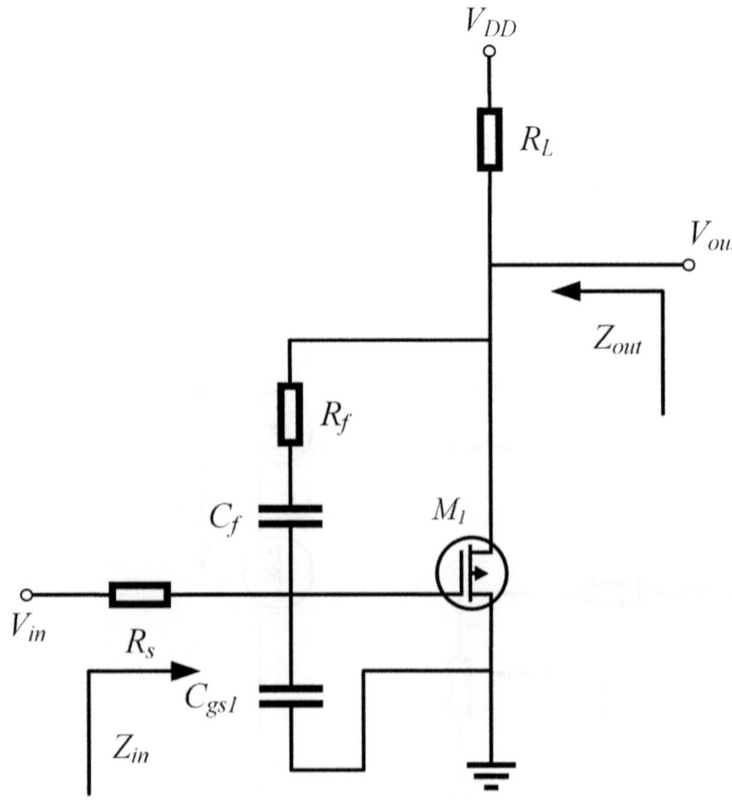


**FIGURE 2.8** LNA with resistive shunt feedback.

impedance matching is achieved in a large bandwidth, a bandwidth that is only limited by the condition $g_m >> \omega C_{gs}$.

The noise factor of the common gate is 2.2 dB for long-channel and 4.8 dB for short-channel transistors. The channel length of short-channel MOSFET devices is typically less than 0.1 μm, in contrast to long-channel transistors, which have a channel length greater than 0.1 μm. Second-order effects such as velocity saturation, threshold voltage variations, and hot carrier effects, not present in long channel devices, appear in MOSFETs as channel lengths decrease below 0.1 μm.

Figure 2.8 presents an LNA with shunt resistive feedback, where impedance matching is achieved by the resistive feedback due to $R_f$. The capacitor $C_f$ is used to decouple the DC voltage component between the gate and the drain. Because of the resistive feedback, the noise performance of this topology is moderate.

The inductively degenerated common-source/common-emitter LNA is presented in Figure 2.9. As shown in Chapter 4, this topology achieves narrowband impedance matching without impairing the noise factor of the input transistor.

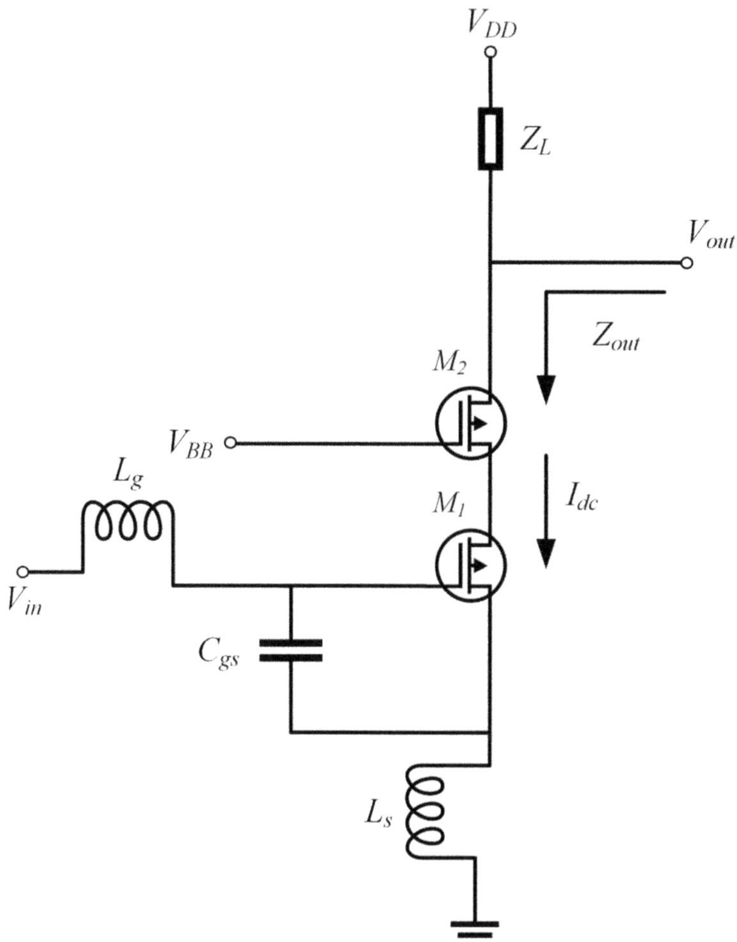


**FIGURE 2.9** LNA with inductive degeneration.

This topology is largely used in LNA design. A comparison between the four LNA topologies is provided in Table 2.3.

## The Effect of Temperature

The density of carriers in a doped semiconductor increases with temperature as semiconductor (e.g., Si) and dopant atoms get ionized with thermal energy. Because the number of semiconductor atoms is much larger than the number of dopant atoms, the impact of dopant atoms on conduction becomes insignificant, and the flow of carriers is no longer controlled by the potential barrier in p-n and n-p junctions. The junctions become simple resistors. The maximum temperature of operation in Kelvin

**TABLE 2.3**
**Comparison between transistor topologies**

| Parameters | Resistive Termination | Common Gate | Shunt Feedback | Inductive Degeneration |
|---|---|---|---|---|
| NF | > 6 dB | 3–5 dB | 2.8–5 dB | ~2 dB |
| Gain | 10–20 dB | 10–20 dB | 10–20 dB | 15–25 dB |
| Sensitivity to parasitics | Low | Low | Low | Large |
| Input matching | Simple | Simple | Simple | Complex |
| Linearity | –10 to 20 dBm | –5 to 5 dBm | –5 to 5 dBm | –10 to 0 dBm |
| Power | 1–50 mW | 5 mW | > 15 mW | > 10 mW |
| Highlight | Simple matching | Simple broadband matching | Simple broadband matching | Narrowband matching, small NF |
| Drawback | Large NF | Large NF | Stability | Large area |

is proportional to the bandgap energy of the semiconductor (e.g., 1.12 eV for Si) by approximately a factor of 500 (e.g., $T = 500 \times 1.12 \sim 560$ K). For very low-temperature operation, there is also a lower temperature limit below which the ionization energy for dopants is insufficient for weakly doped semiconductors. The number of carriers becomes insufficient when the doped semiconductor is in this state. The state is referred to as "freeze-out." Si attains freeze-out at 40 K, Ge at 20 K. The temperature corresponding to freeze-out decreases with doping concentration. As the temperature decreases for BJTs, the current gain ß decreases. The dependence of the current gain on temperature is given by

$$\beta \propto e^{\frac{\Delta E_g}{kT}} \tag{2.2}$$

with $\Delta E_g = E_{gE} - E_{gB} < 0$, and $\Delta E_g$ is the difference in bandgap between the emitter and base regions because of doping. A typical example is the case of the commonly used discrete BJT 2N2222: at room temperature, $\beta$ is equal to 225, but it drops to 35 at 77 K.

Carrier mobilities and saturation velocities of most semiconductors increase at low temperatures due to phonon scattering (Balestra and Ghibaudo 2001). Junction capacitances and track resistances also decrease. However, changes in transistor performance parameters related to temperature can be very unequal, depending on semiconductor technology. This is the case of the gain of Si-based BJTs, their speed and frequency response degrade significantly with cooling unless fitted with a special profile, while CMOS transistors exhibit improved performances at cryogenic temperatures (Balestra and Ghibaudo 2001). Ordinary Si-based BJTs cease operation below 100 K, not because of freeze-out but due to low emitter-base injection efficiency and $\beta$ temperature dependence. The change of the bandgap difference between the emitter and the base through bandgap engineering allows for improving $\beta$ temperature dependence. This is the case of SiGe HBTs, where the increase of base doping and the change of base material (SiGe) decrease the freeze-out temperature and improve $\beta$ temperature dependence. SiGe HBTs can still operate at 4.2 K, and $\beta$ increases at low temperatures.

The changes in the performance of transistors with temperature are summarized here. FETs:

- Performance improves with temperature, but high-frequency noise increases below 100 K.
- Lower temperature of operation depending on material: Si JFETs 40 K due to freeze-out, Ge JFETs like Si JFETs but achieve lower temperatures. GaAs n-channel JFETs can operate at very low temperatures.
- For enhancement MOSFETs and CMOS, freeze-out is limited by channel ionization caused by the gate. They can operate at very low cryogenic temperatures.
- Heterostructure FETs (e.g., HEMTs) based on III-V semiconductors can operate at all cryogenic temperatures (close to 0 K) as thermal ionization of dopant is not required.

BJTs:

- Band gap engineering allows operation and performance improvement at very low temperatures. SiGe HBTs can operate at very low temperatures. Ge and GaAs homojunction transistors can also operate at very low temperatures.

Flicker noise and burst noise are frequency-dependent and can be neglected at RF. Therefore, the noise voltage $\overline{v_i^2}$ and noise current $\overline{i_i^2}$ at the input of a BJT can be $i$ $i$ approximated by (Gray *et al.* 2009).

$$\overline{v_i^2} = 4kTR_S + 4kT\left(r_b + \frac{1}{2g_m}\right) + R_S^2 2q\left[I_B + \frac{I_C}{|\beta(jf)|^2}\right] \tag{2.3}$$

where $R_S$ is the source resistance, $r_b$ is the base parasitic resistance, $I_B$ is the base bias current, and $I_C$ is the collector bias current. For the MOS transistor, the total noise current at its input is given by

$$\overline{i_i^2} = 2qI_G + \frac{\omega^2 C_{gs}^2}{g_m^2}\left(4kT\frac{2}{3}g_m\right) \tag{2.4}$$

where $I_G$ is the gate leakage current, $\omega$ is the angular frequency, and $C_{gs}$ is the gate-source capacitance.

## Input Impedance and Noise Matching

The NF of a transistor decreases with the increase of its gain. The maximum gain is achieved when the source impedance is equal to the conjugate of the input impedance of the transistor. Noise matching is achieved when the noise impedance of the transistor is equal to the source impedance, which is expected to be real. The transistor noise from the input stage is not reflected toward the input and is amplified when noise matching is achieved.

## Output Matching

The optimization of the transfer of RF power from the output transistor of the LNA to the next stage is achieved by an output matching network. The matching network should not degrade the gain, and, generally, a trade-off between the gain and the IP3 is required.

## Packaging

In several cases, the LNA is a stand-alone circuit that is close to the received signal terminal and is separated from digital circuits to reduce possible disturbances resulting from switching circuits. The LNA must then be packaged. For packaging, bond pads on the chip and bond wires are required. Bond pads add parasitic capacitances on the chip, and bond wires add parasitic inductances, which must be considered by the designer. Signal pins on the package must be isolated enough to avoid couplings that can create feedback and affect the gain and stability. Several pins for the ground can be used to avoid performance impairment. Typical packages include DFN, QFN, and flip-chip.

## Biasing and Power Dissipation

Transistor biasing sets the DC operating voltages and currents, which determine how well the input RF signal is amplified, as well as the power dissipation of the transistor. LNAs operate in Class A for the best linearity, and RF voltages and currents are sufficiently small compared to the bias point: they do not shift the bias point. The transconductance $g_m$, $f_T$, $f_{MAX}$, and noise parameters of BJT-based devices and of MOS transistors are related to the bias point. Biasing the transistor is the third most important step in LNA design, after the selection of the topology and the input transistor.

## Layout and Grounding

The relative positioning of active and passive components on the chip is a critical design phase. This positioning of components, referred to as floor planning, aims to determine the preferred locations of the components of the chip while optimizing its area and limiting possible undesired coupling between RF signals. Special attention is required for on-chip inductors. During chip layout, the size, the metal type, the layer level, and the relative positioning and configuration of tracks allow for reducing losses and mutual couplings between tracks that can impair performance. Ground planes can be used for DC power distribution and shielding. Although the layout of digital ICs can be automated, the layout of RF ICs is manual, iterative, and very intensive.

### EMI Shielding and Supply Decoupling

External electromagnetic interference (EMI) can cause a chip to malfunction or damage it. The chip can also generate electromagnetic emissions that can disturb other circuits on a PCB. For LNAs, the shielding of the chip from external emissions and the decoupling between on-chip RF signal tracks are required to ensure good operation. The propagation of interfering signals between different blocks of the chip can be carried by internal power supply tracks within the chip. This propagation is avoided by several decoupling capacitors at selected locations within the chip layout circuit. EMI can be suppressed using a shielded package or an absorbing material between the chip and the package.

## MEASUREMENTS

For sub-1 dB NFs, measurement uncertainties can be comparable to the NF that is being measured. In general, factors contributing to the uncertainties should be identified based on the device under test (DUT) and used to provide corrections. It is recommended to consult the documentation provided by the test equipment manufacturer for the correction process. Techniques used to correct deviations from measurements caused by the measurement setup are presented in subsequent chapters. Although this book focuses on the impact of electronic noise on sensitivity, other characteristics of receivers are affected by noise:

- Electronic noise limits the gain because an amplifier may saturate due to the amplified noise, setting the dynamic range.
- Electronic noise limits the accuracy of measurements, as referred to earlier.
- The accuracy of systems based on pulse-echo techniques, such as radars, sonar, and medical imaging systems, can be affected by noise, as the phase of the received signal can be misread by the system.
- Noise can cause errors in digital systems, especially as supply voltages and the size of devices decrease, leading to increased coupling capacitances and crosstalk.

## APPLICATIONS OF ULTRA-LOW NOISE RECEIVERS: THE SKA

Astronomers need to look more than 10 billion years back to learn about the early universe. They need to study radiation emitted by objects that are some billion light-years away from the Earth, such as quasars at the beginning of the universe. Such radiation is extremely weak, and its detection requires highly sensitive radio telescopes. A highly sensitive radio telescope requires a large signal-receiving area.

The SKA, still under construction, will become the radio telescope with the largest signal-collecting area.

### The Scientific Case of the SKA

Four research areas in radio astronomy are limited by the performance of existing radio telescopes. These areas (the Cradle of Life, strong-field gravity tests using pulsars and black holes, the origin and evolution of cosmic magnetism, and galaxy evolution) have provided research direction for the SKA. Phase 1 of the SKA was expected to start in 2016 and was driven by the following subset of research directions (Schilizzi 2011):

- The history of hydrogen from the Dark Ages.
- The test of gravity theories by the detection of pulsars, the study of gravitational waves, and the investigation of supra-nuclear matter.

Phase 2 of the SKA will allow improvement of the sensitivity, the timing resolution, and the sky observable area of Phase 1. The increased resolution and observable area will allow further investigations into the origin of the magnetic field. With improved sensitivity, the SKA is expected to be able to probe all standard and millisecond pulsars in our galaxy, increasing the probability of finding the first pulsar-black hole system. The high-frequency end of the SKA in Phase 2 will allow scientists to probe the distribution of organic molecules and to study the origin of life.

### Requirements of the SKA

The technical requirements of the SKA have been derived from its science drivers. The frequency bands of the SKA are mapped to frequencies of signals that science will study to achieve the five SKA science drivers and satisfy the technological requirements of SKA receivers. Three frequency bands will be used:

- Low band: 70–450 MHz
- Mid-band: 0.3–10 GHz
- High band: 5–25 GHz

The specifications of the SKA have been evolving. Initially, low-frequency aperture arrays were expected for the low band, while for the mid-band, aperture arrays from 0.3 to 1.4 GHz and dish-based arrays from 1.2 to 10 GHz were to be used. Following the revision in Garret *et al.* (2010), it appears that to achieve the major science goals of the SKA set up in Phase 1 (i.e., the role of neutral hydrogen in the universe and detection of pulsars to test fundamental physics), a low-frequency sparse aperture array will be used from 70 to 450 MHz and 15-meter antennas using single-pixel feeds are expected from 0.45 to 3 GHz. Some variations are, however, still found in the literature. In York *et al.* (2016), reference is made to 70–300 MHz for the low band, 0.3–3 GHz for the midband, and 3–25 GHz for the high band.

The sensitivity of a radio telescope is determined by the ratio of the total effective collecting area ($A_e$) to the noise temperature of the system ($T_{sys}$) (Dewdney *et al.* 2009; Bij de Vaate *et al.* 2010). The strength of the weakest signal that the SKA can detect each time is proportional to $T_{sys}/A_e$. $A_e/T_{sys}$, is expected to be close to 20 000 $m^2/K$. The time of observation required to detect a source within the SKA field of view (FoV) is proportional to $(T_{sys}/A_e)^2$. The FoV is the solid angle where the sensitivity of the antennas is greater than half of the maximum sensitivity. The SKA will achieve an FoV of 1 $deg^2$ at 1.4 GHz with an angular resolution of 0.1 arcsec (Bhaumik and George 2011).

The low-frequency aperture array subsystem will achieve sensitivity ($A_e/T_{sys}$) of up to 2,000 $m^2/K$. Some of the collecting elements of the array will be located up to 100 km from the core, which will condense most of the elements (Belostotski *et al.* n.d.). Each collecting element of the array is an antenna that will require an LNA as the first element of the receiver chain. LNAs with differential inputs will be required, and differential input impedances of 50, 150, and 300 Ω are expected.

## The Impact of Sensitivity on Noise Figure

The sensitivity of a radio telescope is determined by the minimum flux, $\Delta S$, that the radio telescope can detect. For a radio telescope with a single antenna, $\Delta S$ is proportional to $T_{sys}/A_e\sqrt{B_\tau}$, where $T_{sys}$ is the noise temperature of the single antenna, $A_e$ is its effective area, $B$ is the bandwidth of the system, and $\tau$ is the duration of the observation (Burke *et al.* 2019; Dewdney *et al.* 2009). For antenna arrays, $\Delta S$ remains proportional to $T_{sys}/A_e\sqrt{B_\tau}$, for many $N$ antennas in the array. In such a case, $A_e$ is the sum of the effective areas of each antenna in the array, and is the effective area of the array. $T_{sys}$ is the total noise temperature of the system. The sensitivity of a radio telescope, such as the SKA, increases when $\Delta S$ decreases and is proportional to $A_e/T_{sys}$. Therefore, there are two ways of improving the sensitivity of radio telescopes: by decreasing $T_{sys}$ and/or by increasing $A_e$. In today's radio telescopes, the decrease of $T_{sys}$ has almost reached the limit for centimeter wavelengths. Therefore, a major improvement in sensitivity at an effective cost can only be achieved by increasing the radio telescope's signal-collecting area and by achieving a system with optimal noise cost-effectively.

The system noise of a radio telescope system is the sum of various noise sources, as expressed by

$$T_{sys} = T_{rx} + T_{cal} + T_{loss} + T_{spill} + T_{sky} + T_{bg} \tag{2.5}$$

where $T_{rx}$ is the receiver noise temperature, $T_{cal}$ is the contribution of calibration signals, $T_{loss}$ is the noise contribution due to losses in the feed, $T_{spill}$ is the contribution of the earth radiation, $T_{sky}$ is the noise contribution from the atmosphere, and $T_{bg}$ is the noise contribution from microwave and galactic backgrounds. In practical cases, the contribution of the noise temperature of the radio source that is being observed, $T_a$, is neglected because $T_{sys} >> T_a$.

To achieve the scientific requirements of the SKA, a value of $2 \times 10^4$ m²/K is required for $A_e/T_{sys}$ at 1.4 GHz in addition to a frequency coverage of 0.2–20 GHz (Schilizzi *et al.* 2024). In Chippendale (2002), a basic model suggests a value of 50 K for $T_{sys}$. The model assumes a collecting area of 1 km² and a single type of antenna and receiver. Noise models for $T_{sky}$ and $T_{bg}$ are based on Penzias and Wilson (1966). When calculating $T_{rx}$ by subtracting all other noise contributions from $T_{sys}$, it is found that the upper limit of $T_{rx}$ at 1.4 GHz is 19 K (0.28 dB). The main contribution to the receiver noise temperature, $T_{rx}$, comes from the LNA. Typically, 30% of the total noise budget of the SKA is provided by the LNA (Bhaumik and George 2011). The approximation of the SKA receiver noise temperature from the basic model is close to the 16 K noise temperature allotted to the LNA (Bhaumik and George 2011). The target of a noise temperature of 16 K (0.23 dB) for the LNA is derived from the SKA noise budget in (Bij de Vaate *et al.* (2010), corresponding to a system noise temperature of 40 K (0.55 dB). More stringent requirements are found in Dewdney *et al.* (2009). In Dewdney *et al.* (2009), the typical NF of the LNA is expected to be lower than 0.2 dB, which is equivalent to 14 K noise temperature. However, such a target is difficult to reach. Recent specifications that have been emerging propose 20 K (0.29 dB) for noise temperature (Bhaumik *et al.* 2011) of the LNA of the SKA. According to Bhaumik *et al.* (2011), the expected gain of the LNA is 30 dB, and the required power consumption is 30 mW.

## LNA FOR THE SKA

LNAs for radio astronomy require the lowest noise figure. The research on SKA receivers is ongoing, and several architectures and technology processes have been proposed. This paragraph presents the actual design used on the MeerKAT. The MeerKAT is a radio telescope that was built in South Africa as the precursor of the SKA. The radio telescope consists of an array of 64 dishes, each with a diameter of 13.5 m. Dishes are fitted with L-band LNAs (900–2100 MHz) and with cryogenic cooling at 15 K. The LNA is a single-ended three-stage hybrid amplifier based on InP HEMTs and high-Q discrete passive components. The hybrid architecture was selected because MMIC processes are limited in implementing passive components of large values. Noise matching is performed not only at the input of the LNA but also between amplification stages, which increases the complexity of the design. The LNA achieves a noise temperature of 2 K and more than 40 dB of gain, $S_{11}$ and $S_{22}$ are better than –11 dB and –15 dB, respectively (Jiang *et al.* 2014).

## CONCLUSION

This chapter explored the critical role of LNAs and other low-noise technologies in ultra-sensitive radio astronomy applications, such as the SKA and its precursor, the MeerKAT telescope. The drive toward achieving sub-1 dB noise figures has highlighted the intricate design and measurement challenges in reducing noise, improving gain, and maintaining accuracy across wide frequency bands. With stringent requirements for noise reduction, precise measurement, and environmental stability, LNAs for radio telescopes demand advanced architectures and cutting-edge materials like InP HEMTs, coupled with cryogenic cooling.

The SKA's ambitious scientific goals require achieving unprecedented sensitivity by minimizing receiver noise and maximizing the effective collecting area of its array. The use of cryogenically cooled hybrid amplifiers on the MeerKAT demonstrates the effectiveness of careful noise matching between amplification stages and the application of discrete high-Q components. This design achieves remarkably low noise temperatures and high gain, setting a benchmark for SKA's future receiver requirements.

Ultimately, this chapter underscores that while LNAs are essential to radio astronomy, meeting these noise and sensitivity goals necessitates continual advancements in amplifier technology, sophisticated measurement techniques, and the close integration of hardware and environmental considerations. These efforts will enable SKA and similar projects to push the boundaries of observational astronomy, advancing our understanding of the universe.

## REFERENCES


Andersson, S., Sundstrom, L., and Mattisson, S., 2017. Design Considerations for 5G mm-Wave Receivers. In: *2017 Fifth International Workshop on Cloud Technologies and Energy Efficiency in Mobile Communication Networks (CLEEN)*. IEEE, 1–5.

Balestra, F., and Ghibaudo, G., 2001. *Device and Circuit Cryogenic Operation for Low Temperature Electronics*. Springer US.

Belostotski, L., 2023. Low-noise-amplifier (LNA) performance survey [online]. *Online*. Available from: https://schulich.ucalgary.ca/contacts/leo-belostotski [Accessed 30 October 2023].

Belostotski, L., Haslett, J.W., and Veidt, B., n.d. Wide-Band CMOS Low Noise Amplifier for Applications in Radio Astronomy. In: *2006 IEEE International Symposium on Circuits and Systems*. IEEE, 4.

Bhaumik, S., and George, D., 2011. Review of Square Kilometre Array LNA Technologies and Topologies. In: *Proceedings of Wide Field Astronomy & Technology for the Square Kilometre Array — PoS(SKADS 2009)*. Sissa Medialab, 046.

Bhaumik, S., Roy, M., and George, D., 2011. Comparative Study of HEMTs for LNAs in Square Kilometer Array telescope. In: *2011 IEEE MTT-S International Microwave Symposium*. IEEE, 1.

Bij de Vaate, J.G., Witvers, R.H., and Woestenburg, E.E.M., 2010. Front-End Integration Requirements for the Square Kilometre Array Radio Telescope. In: *2010 IEEE Compound Semiconductor Integrated Circuit Symposium (CSICS)*. IEEE, 1–3.

Burke, B.F., Graham-Smith, F., and Wilkinson, P.N., 2019. *An Introduction to Radio Astronomy*. Cambridge University Press.

Cassan, D.J., and Long, J.R., 2003. A 1-V Transformer-Feedback Low-Noise Amplifier for 5-GHz Wireless LAN in 0.18-μm CMOS. *IEEE Journal of Solid-State Circuits*, 38 (3), 427–435.

Chippendale, A.P., 2002. Technology Issues for Square Kilometre Array Receiver Design. In: *Proceedings of the Workshop on Applications of Radio Science (WARS 2002)*. Leura.

Cressler, J.D., 1998. SiGe HBT Technology: A New Contender for Si-Based RF and Microwave Circuit Applications. *IEEE Transactions on Microwave Theory and Techniques*, 46 (5), 572–589.

De Vaate, B.G.B., 1999. RF-IC Developments for Wide Band Phased Array Systems. In: *Perspectives on Radio Astronomy: Technologies for Large Antenna Arrays*, edited by A. B. Smolders and M. P. van Haarlem. Netherlands Foundation for Research in Astronomy (ASTRON), 143–147.

Dewdney, P.E., Hall, P.J., Schilizzi, R.T., and Lazio, T.J.L.W., 2009. The Square Kilometre Array. *Proceedings of the IEEE*, 97 (8), 1482–1496.

Erturk, M., 2008. *The Bias Dependence of CMOS 1/F Noise Statistics*. University of Vermont.

Garret, M.A., Cordes, J.M., Deboer, D.R., Jonas, J.L., and Schilizzi, R.T., 2010. Square Kilometre Array: A Concept Design for Phase 1. In: *International SKA Forum 2010*. Hof van Saksen, 1–8.

Gray, P.R., Hurst, P.J., and Meyer, R.G., 2009. *Analysis and Design of Analog Integrated Circuits*. 5th ed. Wiley.

Hazenboom, S., Fiez, T.S., and Mayaram, K., 2006. A Comparison of Substrate Noise Coupling in Lightly and Heavily Doped CMOS Processes for 2.4-GHz LNAs. *IEEE Journal of Solid-State Circuits*, 41 (3), 574–587.

Jackson, S.A., 2005. RF Design of a Wideband CMOS Integrated Receiver for Phased Array Applications. In: *The Square Kilometre Array: An Engineering Perspective*, edited by Peter J. Hall. Springer Netherlands, 201–210.

Jiang, F., Claude, S., and Garcia, D., 2014. Hybrid Cryogenic Low Noise Amplifier for the MeetKAT Array. In: *2014 16th International Symposium on Antenna Technology and Applied Electromagnetics (ANTEM)*. IEEE, 1–2.

Leong, K.M.K.H., Mei, X., Yoshida, W., Liu, P.-H., Zhou, Z., Lange, M., Lee, L.-S., Padilla, J.G., Zamora, A., Gorospe, B.S., Nguyen, K., and Deal, W.R., 2015. A 0.85 THz Low Noise Amplifier Using InP HEMT Transistors. *IEEE Microwave and Wireless Components Letters*, 25 (6), 397–399.

Maalik, S., 2013. *Structured LNA Design for Next Generation Mobile Communication*. Master of Science Thesis in the Master Degree Program Radio and Space Science Engineering. Chalmers University of Technology, Goteborg.

Penzias, A.A., and Wilson, R.W., 1966. Determination of the Microwave Spectrum of Galactic Radiation. *Astrophysical Journal*, 146, 666.

Schilizzi, R.T., 2011. *SKA Project Execution Plan, Pre-Construction Phase for the Square Kilometre Array*. SKA Program Development Office (SPDO), The University of Manchester.

Schilizzi, R.T., Ekers, R.D., Dewdney, P.E., and Crosby, P., 2024. *The Square Kilometre Array*. Springer International Publishing.

Schleeh, J., Moschetti, G., Wadefalk, N., Cha, E., Pourkabirian, A., Alestig, G., Halonen, J., Nilsson, B., Nilsson, P.A., and Grahn, J., 2017. Cryogenic LNAs for SKA band 2 to 5. In: *2017 IEEE MTT-S International Microwave Symposium (IMS)*. IEEE, 164–167.

Schleeh, J., Wadefalk, N., Nilsson, P.A., and Grahn, J., 2016. 10 K Room Temperature LNA for SKA Band 1. In: *2016 IEEE MTT-S International Microwave Symposium (IMS)*. IEEE, 1–4.

Suana, J.A.R., 2017. *Design of a Low-Noise Amplifier for Radar Application in the 5 GHz Frequency Band*. University of Gävle.

Tilmans, H.A.C., Raedt, W. De, and Beyne, E., 2003. MEMS for Wireless Communications: From RF-MEMS Components to RF-MEMS-SiP. *Journal of Micromechanics and Microengineering*, 13 (4), S139–S163.

Tsaraklimanis, A., and Karagianni, E., 2011. Low Noise Amplifier Design for Digital Television Applications. *Journal of Electromagnetic Analysis and Applications*, 3 (7), 291–296.

Vasilescu, G., 2006. *Electronic Noise and Interfering Signals: Principles and Applications*. Illustrated. Springer Science & Business Media.

Xin, C., 2004. *Radio Frequency Circuits for Wireless Receiver Front-Ends*. Doctor of Philosophy. Nankai University, Nankai.

York, D.G., Gingerich, S.N., and Zhang, S.N., 2016. *The Astronomy Revolution*. CRC Press.

# 3 Electronic Noise Theory and Its Role in Low-Noise Amplifier Performance

## SIGNALS AND NOISE

In the context of electronics, a signal represents a measurable form of electrical energy as it propagates through a circuit. Signals carry information, and depending on how they are described, they may be deterministic or stochastic. A deterministic signal can be written explicitly as a function of time (e.g., a sinusoidal input at a known frequency), while a stochastic signal takes on random values that cannot be predicted. Electronic noise falls into this latter category: its instantaneous voltage or current fluctuates randomly in time, but its statistical properties remain stable when averaged over longer intervals. This stationarity ensures that noise can be analyzed even if its exact behavior cannot be predicted for every sample.

Electronic noise is any unwanted stochastic disturbance that degrades an information signal. The fact that its generation is inherently random means that noise must be treated statistically. In practice, this requires us to describe noise not by its instantaneous values, but through metrics such as the mean value, the mean-square value, the root-mean-square (RMS, or rms is also used throughout this book) value, or correlation measures that describe how two noisy signals relate to one another.

One particularly important quantity is the correlation coefficient between two noise voltages, denoted here as $v_a$ and $v_b$. It is defined as:

$$c = \overline{v_a v_b} / (v_a)_{rms} (v_b)_{rms} \tag{3.1}$$

where the numerator represents the mean value of the product $v_a v_b$, and the terms $(v_a)_{rms}$ and $(v_b)_{rms}$ are, respectively, the rms values of $v_a$ and $v_b$. The coefficient $c$ thus provides a dimensionless measure of the statistical relationship between two noise processes. A value of $c = 1$ indicates perfect correlation (the signals scale together exactly), $c = 0$ indicates complete independence, and negative values indicate anticorrelation. In (a), the horizontal line drawn above the product term $v_a v_b$ indicates a statistical averaging operation, sometimes referred to as the expectation or ensemble average. This notation is used consistently throughout the chapter whenever random variables are involved. Since noise voltages and currents fluctuate unpredictably in time, their


DOI: 10.1201/9781003459385-3

instantaneous values are not particularly meaningful on their own. Instead, what matters are the averaged properties that remain stable across repeated observations of the random process. For example, while the instantaneous product $v_a v_b$ may swing between positive and negative values, its time-averaged or ensemble-averaged value converges to a quantity that characterizes the degree of correlation between the two noise signals.

It is important to emphasize that this averaging does not imply a short-term or local smoothing of the waveform, but rather a fundamental statistical operation that extracts the long-term behavior of the random process. In practice, one can think of this as an infinite-time average of a stationary process, or, equivalently, as an average across an ensemble of independent realizations of the same process. In most noise analysis, these two interpretations are treated as interchangeable because of the assumption of ergodicity, which allows time averages to be equated to ensemble averages.

The importance of this definition in (a) lies in the fact that electronic circuits do not differentiate between signals and noise in how they process them. Both undergo the same transformations outlined by circuit laws, Fourier analysis, and transfer functions. This similarity is why the same mathematics used for deterministic signal analysis, impedance and admittance representations, network functions, and frequency-domain transforms can be applied directly to noise analysis.

It is worth noting that the field has developed multiple definitions for noise parameters, often tailored to the needs of different disciplines such as device physicists, circuit designers, or system engineers. For consistency, the definitions presented in Vasilescu (2019) will be followed throughout this book. However, where there is ambiguity, additional context will be provided to show how alternative definitions fit within the broader theoretical framework.

## NOISE FUNDAMENTALS

Before analyzing how the noise figure (NF) of a transistor amplifier depends on circuit design choices, it is essential to revisit some fundamental aspects of electronic noise. Noise in an electronic system can be broadly defined as everything except the desired signal (Lee 2004). In practice, this means that noise can degrade, mask, or entirely obscure the useful information carried by a signal.

Noise can be divided into two broad categories: artificial noise and fundamental (or intrinsic) noise. Artificial noise arises from external sources such as unwanted radio transmissions that interfere with a receiver or voltage spikes introduced by power electronics switching circuits. These types of noise can often be mitigated with well-known engineering techniques such as filtering, grounding, and electromagnetic shielding. Fundamental noise, on the other hand, is unavoidable. It originates from the inherent random behavior of charge carriers in a material and cannot be eliminated with filtering or shielding. The most important types of intrinsic noise are thermal (Johnson) noise, shot noise, flicker or $1/f$ noise, and generation–recombination (G-R) noise. Each of these has distinct physical origins, but all are statistical in nature.

Because noise originates from the random motion of electrons, it is inherently a stochastic process, meaning that its instantaneous value is unpredictable, but its long-term behavior can be described using probability and statistics. This requires a formal mathematical framework to characterize noise voltages and currents in a way that is compatible with circuit analysis.

Let us consider a random voltage $v(t)$. At any instant $t$, the $k$th sample of this process is denoted by $v_k(t)$, which has an associated probability $p(v_k(t))$. The statistical behavior of $v(t)$ can be described by a probability density function (PDF). For many physical processes, the PDF is Gaussian, which is given by

$$p\left(v_k(t)\right) = \frac{1}{\sqrt{2\pi}\sigma} \exp\left(-\frac{\left(v_k(t)-\mu\right)^2}{2\sigma^2}\right) \tag{3.2}$$

where $\mu$ is the mean, $\sigma$ is the standard deviation, and $\sigma^2$ is the variance of $v(t)$ at time $t$. From the PDF, one can define various statistical averages. For example, the mean value of $v(t)$, which can be estimated from $n$ samples, is

$$\mu(t) = \lim_{n\to\infty} \sum_{k=1}^{k=n} v_k(t)\, p\left(v_k(t)\right) \tag{3.3}$$

where $p(x)$ is the PDF of $v(t)$ at time $t$. The mean-square value of $v(t)$, used to quantify the average power, is expressed by the equation

$$\sigma^2(t) = \int_{-\infty}^{\infty} v^2(t)\, p(x)\, dx \tag{3.4}$$

In (3.3) and (3.4), the emphasis is on ensemble averages: the quantities are computed by considering many possible realizations of the process at a fixed instant in time. However, in practical electronic systems, we typically measure long-time records of a single realization. If time averages produce the same results as ensemble averages, the process is said to be ergodic. This assumption is central in noise analysis, since it allows laboratory measurements taken over time to be interpreted as statistical quantities.

An ergodic Gaussian random process $v(t)$ has several useful properties. Using the notation <> to denote time averaging, these include:

- The average/mean value $<v(t)>$ represents the DC component of the process.
- The square of the average value provides the DC power of the distribution, $<v(t)>^2$.
- The mean square of the process provides the total average power $<v^2(t)>$.
- The variance of the process $\sigma^2$ represents its average AC power, which fluctuates around the mean value.

- The RMS value is defined as the square root of the mean-square, and for a zero-mean process, it equals the standard deviation $\sigma$.

The RMS voltage is the square root of the expected value of the square of the voltage and is defined as:

$$RMS = \sqrt{< v^2(t) >} \tag{3.5}$$

The DC value of an electronic intrinsic noise signal has been zero, $<v(t)> = 0$, and $<v^2(t)> = \sigma^2$. Therefore, the RMS value of a zero-mean signal is equal to its standard deviation. The average $<v(t)>$ of $v(t)$ is expressed by

$$\langle v(t) \rangle = \lim_{T \to \infty} \frac{1}{T} \int_{-T/2}^{T/2} v(t)\,dt \tag{3.6}$$

and the average AC power of the random signal $<v(t)>$ is given by

$$\langle v^2(t) \rangle = \lim_{T \to \infty} \frac{1}{T} \int_{-T/2}^{T/2} v^2(t)\,dt \tag{3.7}$$

A related and very important concept is autocorrelation, which measures the similarity of a signal with a delayed version of itself. For a random function $v(t)$, the autocorrelation function is

$$\langle v(t)v(t+\tau) \rangle = \lim_{T \to \infty} \frac{1}{T} \int_{-T/2}^{T/2} v(t)v(t+\tau)\,dt \tag{3.8}$$

The correlation between two signals is a measure of their similarity. The similarity is measured by the correlation coefficient $c$, which is comprised between –1 (uncorrelated) and 1 (perfectly correlated). The coefficient is equal to –1 for total similarity but opposite signals, nil for two dissimilar signals, and 1 for completely similar signals. In other words, two deterministic waveforms are correlated if their mathematical expressions and phases are identical, with equal or different amplitudes. The waveforms are partially correlated when their phases differ slightly. The autocorrelation of a signal, expressed by the correlation function, is the comparison between the signal and itself, but delayed in time. A special case of the auto-relation function is when the delay time is zero; in this case, the autocorrelation function represents the mean of the square value of the signal, which is equal to the mean of the average power in the signal. A similar comparison between two separate signals is referred to as cross-correlation.

The correlation function provides insight into the power and bandwidth of a random signal, a property that is used in the analysis of electronic noise. Correlation techniques allow the detection of weak periodic signals buried in noise and are used in several applications, including telecommunications, radar, sonar, astronomy, and medical electronics.

The power spectral density of a stationary random signal is expressed by the Fourier transform of the autocorrelation function as formulated by the Wiener–Khinchin theorem:

The power spectral density $S_{vv*}$ of a stationary random signal $v(t)$ is expressed by the Fourier transform of the autocorrelation function as formulated by the Wiener–Khinchin theorem, where $f$ represents the frequency and $T$ is a time variable:

$$S_{vv^*}(f) = \int_{\tau=-\infty}^{\tau=\infty} \left[ \lim_{T\to\infty} \frac{1}{T} \int_{-T/2}^{T/2} v(t)v(t+\tau)\,dt \right] e^{-j2\pi f\tau} d\tau \tag{3.9}$$

such that

$$S_{vv^*}(f) = \lim_{T\to\infty} \frac{1}{T} \int_{-T/2}^{T/2} \left[ \int_{-\infty}^{\infty} v(t) e^{-j2\pi f\tau} dt \right] v(t+\tau)\, d\tau \tag{3.10}$$

and by using the change of variable $x = t + \tau$, yields

$$S_{vv^*}(f) = \int_{\tau=-\infty}^{\tau=\infty} \left[ \lim_{T\to\infty} \frac{1}{T} \int_{-T/2}^{T/2} v(t) e^{-j2\pi f\tau} dt \right] v(x) e^{-j2\pi fx} dx \tag{3.11}$$

and since the right-hand side term after the "]" can be written as $v(f)$, and $v(f)$ is the Fourier transform of $v(t)$, the equation becomes

$$S_{vv^*}(f) = \left[ \lim_{T\to\infty} \frac{1}{T} \int_{-T/2}^{T/2} v(t) e^{-j2\pi f\tau} dt \right] v(f) \tag{3.12}$$

and now, since the term inside the brackets is equal to $v^*(f)$, which is the complex conjugate of $v(f)$, the power spectral density becomes

$$S_{vv^*}(f) = \lim_{T\to\infty} \frac{v(f)v^*(f)}{T} \tag{3.13}$$

such that

$$S_{vv^*}(f) = \lim_{T\to\infty} \frac{|v(f)|^2}{T} = \overline{v(f)v^*(f)} = \overline{vv^*} \tag{3.14a}$$

where the operator $\overline{\phantom{x}}$ is the mathematical operator representing the average of $v(f)v^*(f)$, which is the power spectral density of an ergodic signal. The Fourier transform $v(f)$, which is truncated, is calculated over the range (-*T*/2 + *T*/2). The following properties of the operator will be used throughout this book:

If *a* and *b* are the Fourier transforms of two uncorrelated noise sources with zero mean value, then

$$\overline{ab^*} = 0 \tag{3.14b}$$

and

$$\overline{ab^*} = \left(\overline{ba^*}\right)^* \tag{3.14c}$$

The time average of the square of $v(t)$ over a long period of time $T$ is proportional to the power dissipated by $v(t)$. The average is determined by (3.15).

$$\overline{v^2(t)} = \lim_{T\to\infty} \frac{1}{2T} v(t)^2\, dt \tag{3.15}$$

At this stage, the mathematical foundation needed for noise analysis has been established. We can now turn to the distinct noise sources themselves, each of which has different physical origins but can be treated within the same statistical framework outlined above.

## OVERVIEW – NOISE SOURCES IN ELECTRONIC SYSTEMS

Electronic systems are invariably affected by noise, a fundamental limitation arising from the very physics of charge transport. Noise cannot be eliminated, but it can be understood, modeled, and minimized. Among the many types of noise encountered in circuits, four stand out as the most fundamental: thermal (or Johnson) noise, shot noise, flicker (1/*f*) noise, and generation–recombination (G-R) noise. Each originates from a different physical mechanism and manifests differently in circuit behavior. For the design of low-noise amplifiers (LNAs), which are highly sensitive to even minute fluctuations at the input, the first three – thermal, shot, and flicker noise – play the most significant roles. G–R noise is also important in some semiconductor devices, but has less immediate impact on RF amplifier design and will not be emphasized here.

## JOHNSON (THERMAL) NOISE

Thermal noise, sometimes referred to as Johnson noise after J. B. Johnson, who first measured it in 1926, arises fundamentally from the random motion of charge carriers in a conductor. At any finite temperature, electrons in a material are not stationary but in constant thermal agitation, colliding with the atomic lattice and with each other. This microscopic motion gives rise to fluctuating voltages and currents at the terminals of the conductor, even in the absence of any applied signal.

The underlying phenomenon is rooted in Brownian motion, a concept observed as early as 1827 by Robert Brown when he studied the erratic movement of pollen grains suspended in water. While Brown initially speculated that the motion was biological, it was later shown that the randomness arose from collisions with surrounding water molecules. By analogy, electrons in a conductor undergo similar agitation, colliding randomly with the lattice. This microscopic disorder translates macroscopically into measurable electrical noise. Albert Einstein, in 1906, provided a theoretical explanation of Brownian motion, extending its principles to particles at the microscopic scale, including free electrons. His work laid the foundation for connecting statistical mechanics with electrical phenomena.

The thermal noise power frequency spectrum is independent of the material and measurement frequency at RF and varies only with temperature and electrical resistance. Because its spectral density is uniform, thermal noise is white noise. The dependence of noise from resistors on temperature was found experimentally by John Johnson in 1926, and a theoretical model based on fundamental physics was developed by Harry Nyquist, both were from Bell Labs.

The probability distribution of thermal noise is a Gaussian or normal distribution. Because the distribution of thermal noise is Gaussian, the instantaneous amplitude of the noise is within $\pm\sigma$ 68% of the time. Similarly, the instantaneous amplitude is within $\pm 2\sigma$ 95.4% of the time, $\pm 3\sigma$ 99.7 % of the time, and $\pm 3.4\sigma$ 99.94% of the time, respectively. If the instantaneous amplitude of the noise is $v(t)$, then $v^2\left(t\right)= \sigma 2$.

For a resistor of value $R$ and with a noise voltage $v(t)$ on its terminals, the power spectral density is equal to $\overline{vv^*} = 2kTR$, an expression that is valid up to several THz, typically $6 \times 10^{12}$ Hz. The open circuit voltage, $v_{rms}$ equivalent to the total power due to random noise resulting from the integration of the power spectral density from $-f$ to $f$, with $f$ approaching infinity. The voltage $v_{rms}$ is given by $v_{rms} = \sqrt{4kTRB}$, where $B$ is the bandwidth under consideration.

From a practical standpoint, thermal noise sets the ultimate sensitivity limit of any receiver or sensor. The concept of NF in amplifiers, for example, is defined relative to the unavoidable thermal noise floor $kTB$. No circuit, no matter how well designed, can reduce this intrinsic limit; it can only add to it. Thus, in LNA design, thermal noise is not just an abstract physical effect but the benchmark against which all performance is measured.

Thermal noise in a resistor of value $R$ is measured by the mean square value of the noise voltage $v_d$ on its terminals or by the mean square value of the noise current $i_d$ through the resistor:

$$\overline{v_d^2} = 4kTR\Delta f \tag{3.16}$$

$$\overline{i_d^2} = 4kT\frac{\Delta f}{R} \tag{3.17}$$

where $T$ is the absolute temperature, $k$ is Boltzmann's constant, and $\Delta f$ is a small bandwidth at frequency $f$, in Hz, under measurement. The thermal noise of two series resistors, $R_1$ and $R_2$, is given by

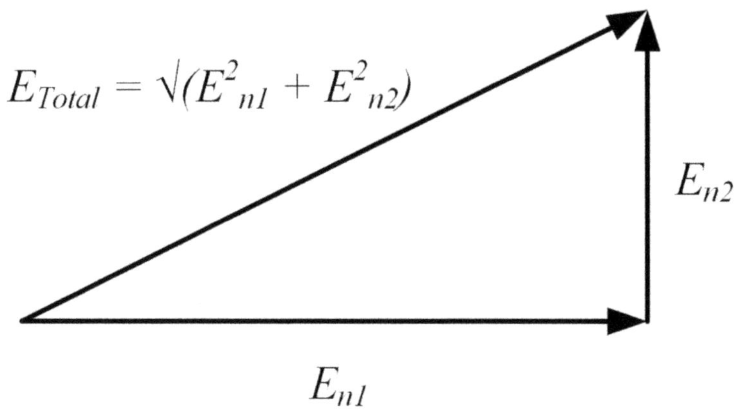


**FIGURE 3.1** Noise RMS voltage of two uncorrelated noise sources.

$$\overline{E^2_{Total}} = 4kTR_1\Delta f + 4kTR_2\Delta f = \overline{E^2_{n_1}} + \overline{E^2_{n_2}} \tag{3.18}$$

The expression of the total noise power shows that the noise RMS voltages of two resistors in series can be added as two orthogonal vectors, as shown in Figure 3.1, because the noise voltages are uncorrelated.

In summary, thermal noise represents an unavoidable baseline present in every resistive element, independent of biasing or device structure. Its Gaussian nature and statistical independence across uncorrelated sources allow noise contributions from different resistors to be combined as RMS values rather than simple arithmetic sums. This property greatly simplifies noise analysis in larger circuits, since each element's contribution can be treated independently and then combined. For LNA design, this principle is particularly important: it defines how resistive elements within matching networks, feedback paths, or biasing circuits accumulate to set the minimum achievable noise floor.

## PLANCK'S LAW

The electronic noise generated by a resistor affects the sensitivity of receivers and has played a critical role in their development. A short digression on Planck's Law from which (3.17) is derived may be desired before discussing other noise types. The laws describing the spectral density of electromagnetic radiation from a black body at thermodynamic equilibrium were derived by Rayleigh-Jeans and Planck for a three-dimensional system. The thermal noise of a resistor is derived from the application of these laws to a one-dimensional system. A black body is a body that absorbs all radiations at all frequencies. According to Kirchhoff's law, a blackbody should emit radiation at all frequencies as well. A resistor absorbs all electrical energy supplied to it, as a black body. The thermal noise generated by a resistor cannot be distinguished from the noise received by an antenna pointed toward a black body, as presented in

Chapter 1. A box with four sides in conductive materials and enclosing electromagnetic waves and having, on one side, a small hole allowing some radiation out, was used to investigate radiation out of a black body. The Rayleigh-Jeans law, derived from this investigation in 1900, showed that, for a given frequency $f$, the energy radiation per unit frequency and per volume is given by:

$$u(f) = \frac{8\pi f^2 kT}{C^3} \tag{3.19}$$

where $C$ is the speed of light (299,792,458 m/s). However, the energy density radiated from a black body and given in (3.19) diverges at high frequencies: this is referred to as the Ultraviolet Catastrophe. The equation is validated by experimentation at low frequencies but not at high frequencies. The solution to the Ultraviolet Catastrophe was found from Einstein's infamous 1905 paper: he proposed that, under certain conditions, light can be considered as a flow of particles, each with the energy $hf$, where $h$ is Planck's constant. This hypothesis, validated 9 years later, allowed for explaining some features of the photoelectric effect. Considering that each vibration mode in the black box corresponds to a flux of particles, Planck's distribution function, derived from (3.19), is given by (3.20).

$$u(f) = \frac{8\pi h f^3}{C^3} \frac{1}{e^{hf/kT} - 1} \tag{3.20}$$

where $h = 6.626 \times 10^{-34}$ Js is Planck's constant. This distribution recovers the Rayleigh-Jeans result at low frequencies (where $hf << kT$) and suppresses the divergence at high frequencies, thereby resolving the Ultraviolet Catastrophe. When applied to electrical noise, Planck's formulation reduces to the familiar $4kTR$ expression in the classical limit but also provides the necessary corrections at terahertz and optical frequencies.

In practice, this connection between resistor noise and black-body radiation is important for understanding why the simple Johnson model begins to fail above a few terahertz and why the quantum-corrected Nyquist expression must be used in such cases. For LNA design in RF and microwave bands, the classical expression remains sufficient, but the Planck-based formulation provides a rigorous foundation and helps merge the treatment of thermal noise with the broader physics of radiation and quantum electronics.

## NYQUIST DERIVATION

Several methods have been used to prove Johnson's findings. The original proof by Nyquist is presented in this section. A one-dimensional system is represented in Figure 3.2, where two resistors of value $R$ each and held at temperature $T$ are connected to the extremities of a transmission line of length $L$. Each extremity of the line is excited by a fluctuating voltage of amplitude $V_o$ and resulting from the random movements of agitated free electrons in the resistors. Assuming that the characteristic impedance of the transmission line is $R$, there is no signal reflection at the end of each

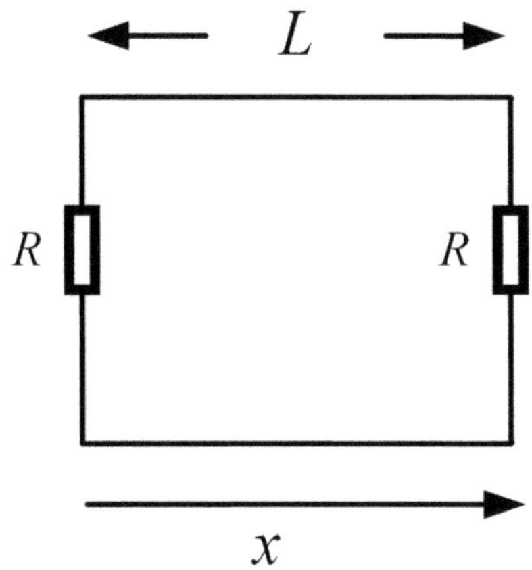


**FIGURE 3.2** Resistor and transmission line.

line, and the power of each incoming signal is absorbed and transformed into heat. However, no heat can be generated when the system is at equilibrium, according to the laws of thermodynamics. This means that the transmission line can only support transmission modes with zero voltages at its extremities.

The original derivation by Nyquist studies only TEM modes and considers shorting the resistors, which is a source for unanswered questions related to thermal equilibrium (Abbott *et al.* 1996). However, because voltages are nil at the end of the transmission line, the propagation on the line can be viewed as the result of two waves of equal amplitudes traveling in opposite directions. The first wave travels from the left to the right and is given by:

$$V_R = V_0 e^{j(\alpha x - \omega t)} \tag{3.21}$$

while the second wave travels from the right to the left and is given by:

$$V_R = V_0 e^{j(-\alpha x - \omega t)} \tag{3.22}$$

where $\omega=2\pi f$, $f$ is the frequency, $a=\omega/c'$, and $c'$ is the velocity of the wave on the transmission line. In general, $c'$ is different from $C$, the speed of light in vacuum.

The equilibrium condition implies that the line length $L$ is a multiple of $\lambda/2$, where $\lambda$ is the wavelength of the wave, and yields the expression:

$$a = \frac{\pi m}{L} \tag{3.23}$$

where $m = 0,1,2,3, \ldots$ The number of propagation modes as a function of frequency is given by:

$$m = \frac{2L}{C'} f \tag{3.24}$$

and the number of transmission modes per unit length per unit frequency is given in:

$$p_f = \frac{2}{C'} . \tag{3.25}$$

To explain the Ultraviolet Catastrophe, the wave-particle duality is introduced for the one-dimensional system. It is assumed that a transmission mode of energy $E_N$ consists of $N$ photons, each of energy $hf$. This is expressed in

$$E_N = Nhf \tag{3.26}$$

where the probability for the energy level $E_N$ to be populated when the system is in equilibrium at temperature $T$ is given by:

$$p\left(E_N\right) = be^{-Nhf/kT} \tag{3.27}$$

where $b$ is a positive coefficient. The value of $b$ can be derived from (3.27) because $\Sigma p(E_N) = 1$. Therefore, this equation yields

$$p\left(E_N\right) = \frac{e^{-Nhf/kT}}{\sum_{N=0}^{\infty} e^{-Nhf/kT}} \tag{3.28a}$$

and the denominator is of type

$$\sum_{i=0}^{\infty} y^i = \frac{1}{(1-y)} \tag{3.28b}$$

when $|y| < 1$ and therefore becomes

$$p\left(E_N\right) = \left(1 - e^{-hf/kT}\right) e^{-Nhf/kT} \tag{3.29}$$

which provides the probability distribution of photons for the transmission mode of energy $E_N$ at frequency $f$. The average number of photons $\bar{N}$ is then given by:

$$\bar{N} = \sum_{N=0}^{\infty} Np\left(E_N\right) \tag{3.30a}$$

Reporting (3.28) in (3.29), substituting $hf/kT$ by $x$ and considering that

$$Ne^{-Nx} = -\frac{\partial\left(e^{-Nx}\right)}{\partial x} \tag{3.30b}$$

then the equation becomes

$$\bar{N} = \frac{1}{e^{hf/kT} - 1} \tag{3.31}$$

Equations (3.25) and (3.31) are combined to provide the average energy per unit length and frequency

$$\overline{u_f} = \rho_f \bar{N} hf = \frac{2}{C'} \frac{hf}{e^{hf/kT} - 1} \tag{3.32}$$

and the total power per frequency unit for both traveling waves is obtained by multiplying the average energy per unit length by the transmission line length, $L$.

The average power $\bar{P}$ generated by each resistance at frequency $f$ in a small bandwidth $\Delta f$, with $\Delta f \ll f$, is given by

$$\bar{P} = \frac{hf\,\Delta f}{e^{hf/kT} - 1} \tag{3.33}$$

where the average power can also be expressed as $v_d^2/4R$, where $v_d$ is the mean-square voltage from each resistor. This mean-square voltage is given by

$$v_d^2 = \frac{4Rhf\,\Delta f}{e^{hf/kT} - 1} \tag{3.34}$$

The variations of $v_d$ are random, and their averages are zero. Therefore, the variance of $v_d$ is zero, and equation (3.34) becomes

$$\overline{v_d^2} = \frac{4Rhf\,\Delta f}{e^{hf/kT} - 1} \tag{3.35a}$$

and since $hf \ll kT$ at radio frequencies, the denominator can be approximated by a power series, with

$$e^{hf/kT} - 1 \cong 1 + hf/kT \tag{3.35b}$$

Equation (3.34) becomes the following, which is the expression presented in (3.16)

$$\overline{v_d^2} = 4RkT\,\Delta f \tag{3.36}$$

Nyquist's derivation elegantly confirms Johnson's experimental findings while also embedding them within a deeper framework that bridges thermodynamics, electromagnetics, and quantum theory. By treating the resistor and transmission line as a one-dimensional analog of black-body radiation, Nyquist was able to rigorously connect random electron motion to the observable mean-square noise voltage. The result, expressed in (3.35), reduces in the radio-frequency limit ($hf << kT$) to the familiar Johnson–Nyquist formula ($4RkT\Delta f$), which underpins virtually all practical noise calculations in RF engineering. This reconciliation of statistical physics with circuit-level measurements not only validates the $4kTR$ expression but also highlights its limitations: at higher frequencies, where $hf$ is no longer negligible compared to $kT$, the full quantum-corrected form must be used. For LNA design, the classical result suffices across microwave and millimetre-wave ranges, yet the quantum derivation provides an essential reminder that even seemingly simple circuit phenomena are grounded in fundamental physical laws. This bridge between microscopic physics and macroscopic circuit behavior sets the stage for analyzing how noise manifests in active devices and how it ultimately shapes amplifier performance.

## UNRESOLVED PARADOXES

Although these paradoxes (Penfield 1966) are intriguing, they do not undermine the validity of the Johnson–Nyquist law itself. Rather, they highlight the subtleties that arise when trying to reconcile idealized circuit theory with the constraints of thermodynamics and statistical physics. In practice, the apparent contradictions are resolved once more detailed models of dissipation, irreversibility, and fluctuation–dissipation balance are considered. For RF engineers, the key takeaway is that thermal noise remains a fundamental limit on performance, but one that can be consistently described within established physical laws. With this foundation in place, we now turn to another intrinsic noise source of equal importance in electronic devices, shot noise, whose origins lie not in thermal agitation but in the discrete nature of charge transport.

## SHOT NOISE

Fluctuations in the rate of pulses generated randomly were first studied by Norman R. Campbell in 1909. The same fluctuations were found by Walter Schottky in vacuum tubes in 1918, who referred to them as "shot noise." Shot noise is due to the quantization of charge and the granular nature of electric current. It can be compared to the sound of raindrops on a tin roof as raindrops hit the roof at unrelated random instants. The power resulting from short noise is given by the variance of the number of carriers (electrons) through the transport space in the time interval $\tau$, where $\tau$ is very small. Several statistically independent events are modeled by a Poisson distribution.

The theory behind shot noise was formulated by W. Schottky and tested on vacuum tubes. This theory provides basic principles for the detection of the minimum detectable signal in the design of receivers and electronic systems (Heffner 1962).

In semiconductors, shot noise is a white noise that occurs when there is a flow of DC due to carriers crossing a potential barrier, such as a p-n junction or a Schottky barrier. Shot noise is not found in linear devices and is expressed as:

$$\overline{i_n^2} = 2qI_{DC}\Delta f \tag{3.37}$$

where $I_{DC}$ is the DC flowing through the device, and $q$ is the charge of the electron ($1.6 \times 10^{-19}$ C). In bipolar transistors, shot noise is generated by the base current and the collector current. In MOSFETs, shot noise is dominant when the device is in the subthreshold region.

Several derivations of the shot noise formula are available, and some derivations are somewhat complicated, for example, derivations in Rice (2016). The Van der Ziel's derivation is presented in this section. It is assumed that the number of carriers crossing the transport space at the rate *n(t)* and in the time interval $\tau$ is $N$. Therefore, $N$ is given by

$$N = \int_0^{\tau} n(t)\,dt \tag{3.38}$$

and because the process is ergodic, time averages are equal to sampling averages. Hence, $\overline{N}$ and $\overline{n}$ are related by

$$N = \overline{n}\tau \tag{3.39}$$

where it is assumed that the time interval $\tau$ is equal to 1 second. Therefore, the DC crossing the transport space is given by the following, where $q$ is the charge of the electron,

$$I_{DC} = q\overline{N} \tag{3.40}$$

and according to elementary probability theory, the variance of a random variable $X$, denoted by $\rho^2$, is defined by

$$\sigma^2 = E\left(\left(X - E(X)\right)^2\right) \tag{3.41}$$

and

$$\rho^2 = E\left(X^2 - 2E(X)X + E^2(X)\right) \tag{3.42}$$

such that the combination of (3.41) and (3.42) yields

$$\rho^2 = E\left(X^2\right) - E^2(X) \tag{3.43}$$

Let us define a new random variable $Y$, with $Y = qN$. The mean of $Y$ is equal to the DC through the transport region, $I_{DC} = \bar{Y} = q\bar{N}$. Applying (3.43) to $Y$, the variance of the random variable $Y$ is given by

$$var\,Y = E\left(\left(qN\right)^2\right) - E^2\left(qN\right) \tag{3.44}$$

and

$$var\,Y = q^2E\left(N^2\right) - q^2E^2\left(N\right) \tag{3.45}$$

and if (3.43) is applied to (3.45), it yields

$$var\,Y = q^2\left(var\,N\right) \tag{3.46}$$

Because $N$ has a Poisson distribution, the variance of $N$ is equal to its mean; therefore, $var\ N = \bar{N}$ and (3.46) becomes

$$var\,Y = q^2\bar{N} = qI_{DC}. \tag{3.47}$$

Hence, the power density in a frequency band $\Delta f$ is equal to $qIDC$. To calculate the total power in $\Delta f$, the power in the symmetric negative band must be considered as well. It follows that the total power in $\Delta f$ is given by the following equation, which is the Schottky formula for shot noise.

$$Noise\ power = \overline{i_n^2} = 2qI_{DC} \tag{3.48}$$

The short noise formula assumes that current fluctuations are independent of each other and have a uniform probability distribution (Rice 2016). Hence, there are no interactions between particles, which assumes low density, high electrical field, and low interactions with atoms in the conduction space. For the vacuum tube diode, as electrons approach the anode, mutual repulsion between electrons creates a space charge effect that reduces shot noise. This reduction can be more than an order of magnitude (Rice 2016). A similar phenomenon is present in metal film resistors, where random fluctuations are reduced because of a charge build-up that repels newly arriving carriers by the coulomb force. In homogeneous conductors, the effects of quantum statistics create correlations between electrons, reducing current fluctuations. However, when electrons cross a potential barrier due to thermal excitation, such as in PN junctions, random fluctuations causing shot noise are not reduced. But a large deviation from the Schottky formula is reported in (Rice 2016) for low currents and at low frequencies in p-n junctions. Shot noise can increase in reverse-biased tunnel diodes due to super-Poissonian statistics. The determination of shot noise due to correlated carriers remains a field of research.

## SHOT NOISE IN PN JUNCTIONS

The conduction in a PN junction is achieved by the injection of minority carriers (electrons are injected in the p-region and holes in the n-region). The injection is followed by the diffusion and recombination of these minority carriers, where electrons recombine with holes in the p-region, and holes recombine with electrons in the n-region. The junction noise can then be evaluated by the sum of the diffusion noise and the generation–recombination noise. The crossing of the junction potential barrier by carriers is random, and Campbell's theorem can be used to evaluate the shot noise due to the junction. However, there is a simple approach that assumes that fluctuations caused by the two components of the current across a PN junction are independent, which provides strangely correct results. This approach is presented in this paragraph. The current $I$ across a PN junction is given by

$$I = I_S\left(e^{\frac{qV}{kT}} - 1\right) \tag{3.49}$$

where $I_S$ is the saturation current and $V$ is the voltage across the junction. Because fluctuations caused by the currents $I_S e^{qV/kT}$ and $-I_S$ are assumed independent, the variance of $I$ is given by

$$\overline{i_n^2} = \left(2qI_S e^{qV/kT} + 2qI_S\right)\Delta f = 2q\left(I + 2I_S\right)\Delta f \tag{3.50}$$

In summary, shot noise arises from the discrete, random nature of charge carrier transport across the PN junction, and its magnitude is fundamentally linked to the junction current. Unlike thermal noise, which depends on temperature and resistance, shot noise scales with the average current flowing through the device. While the derivation above makes simplifying assumptions about the independence of current components, it captures the essential behavior observed in real devices and provides a practical method to quantify the noise contribution of PN junctions in amplifiers and other semiconductor circuits. Understanding shot noise is critical for accurately predicting the overall noise figure of LNAs, particularly when devices operate at low currents or high frequencies, where this noise component can become comparable to, or even dominate over, thermal noise.

## SHOT NOISE IN METAL–SEMICONDUCTOR JUNCTIONS

A junction between a metal and a semiconductor constitutes a Schottky diode, also known as a Schottky barrier diode. Current from the metal to the semiconductor and from the semiconductor to the metal is transmitted by majority carriers, and the variance of its fluctuations is the same as for PN junctions.

## FLICKER NOISE

Flicker noise, or $1/f$ noise or excess noise, is found in all active devices and in some passive components, such as carbon resistors (Vasilescu 2019), and is the dominant noise type at low frequencies. It is caused by traps and crystal defects, and its spectrum density is inversely proportional to the frequency. Flicker noise is expressed by

$$\overline{i^2} = k_1 \Delta f \frac{I_{DC}^a}{f^b} \tag{3.51}$$

where $f$ is the frequency, $k_l$ is a constant for a particular device, $a$ is a constant between 0.5 and 2, and $b$ is a constant close to unity.

## QUANTUM NOISE

Quantum noise is caused by the discrete nature of electromagnetic energy radiation, as demonstrated by quantum mechanics. The energy is exchanged in photons, and the energy of each photon is given by

$$E = hf \tag{3.52}$$

where $h$ is Planck's constant (6.63 $10^{-34}$ Js), and $f$ is the frequency.

When energy is transmitted, the fluctuation of the number of photons generates quantum noise. Quantum noise can be neglected at RF but is important from infrared frequencies ($10^{15}$ Hz). For very high frequencies ($hf \gg kT$), the quantum noise decreases but is limited by Heisenberg's principle. Applying the Heisenberg uncertainty principle to a linear amplifier, it is shown in Heffner (1962) that a noiseless linear amplifier cannot exist. The minimum possible noise temperature $T_n$ of a linear amplifier is given by (Heffner 1962)

$$T_n = \left[ ln \frac{2 - 1/G}{1 - 1/G} \right]^{-1} \frac{hf}{k} \tag{3.53}$$

where $G$ is the gain, $k$ is Boltzmann's constant, and $f$ is the frequency.

## GENERATION–RECOMBINATION NOISE

Random generation, recombination, and trapping of carriers, and the discrete nature of their charge fluctuates randomly, result in the current through a semiconductor. This fluctuation can be viewed as a variation in the semiconductor resistance. The mean square of the fluctuation is given by

$$\overline{i_n^2} = \frac{4\overline{I^2}}{\overline{N}\left(1 + \omega^2 \tau^2\right)} \tag{3.54}$$

where $\overline{I}$ is the DC, $\overline{N}$ is the average number of carriers, $\omega$ is the angular frequency, and $\tau$ is a time constant related to the generation–recombination process. Generation–recombination noise can be neglected when the concentration of carriers is low. This is the case of intrinsic and lightly doped semiconductors.

## BURST (RTS) NOISE

Burst noise is also known as popcorn noise, impulse noise, random telegraph noise (RTN), or random telegraph signal (RTS) noise, and has become prominent in low-frequency noise as semiconductor devices continue to shrink. It consists of pulse-like voltage or current random transitions, which can last several milliseconds. Amplitudes of the transitions vary between discrete values that can reach several microvolts, and the switching appears at audio frequencies, producing a popping sound. Burst noise is generated by ultra-thin gate oxide films and heavily doped emitter junctions and is associated with planar diffusion processes (Vasilescu 2019). Although not well understood, burst noise is generated randomly by carriers that are trapped at film interfaces or in semiconductor defects. These defects are believed to be caused by the manufacturing process. The power spectrum of Burst noise is proportional to $1/f^2$ and, for bipolar transistors, is given by

$$\overline{i_n^2} = \frac{K_1 I_B}{1 + \pi^2 f^2 / 4a^2} \Delta f \tag{3.55}$$

where $a$ is the number of bursts per second, $I_B$ is the base current, and $K_l$ is a constant.

## AVALANCHE NOISE

Avalanche noise is a white noise generated by the ionization of neutral atoms in the depletion area caused by the impact of minority carriers in highly reverse-biased PN junctions. Eventually, the newly released carriers can trigger a chain reaction that multiplies carriers and may result in breakdown. The power spectrum of avalanche noise is the largest for all noise types.

## DIFFUSION NOISE

Charge carriers move from areas of high density to areas of low density when their density is uneven in a semiconductor. The diffusion movement of carriers is random because of collisions with atoms in the semiconductor. It is important to indicate that diffusion noise is dominant when the voltage–current characteristic is not linear, but when this characteristic is linear, such as for a resistor, diffusion noise develops into thermal noise. This is the case of the FET channel in linear mode. In saturation mode, the channel generates diffusion noise.

## KT/C NOISE

This is not a fundamental noise source but rather the noise resulting from the filtering of the thermal noise from a resistor $R$ by a capacitor $C$. The output is the voltage across the capacitor. The total output power is given by

$$\overline{v_n^2} = \frac{1}{2\pi}\int_0^\infty \frac{4kTR}{1+(\omega RC)^2} d\omega = \frac{kT}{C} \tag{3.56}$$

A summary of the main noise sources encountered in basic electronic devices and in typical bias and operation frequency conditions is adapted from Vasilescu (2019) and presented in Table 3.1.

Table 3.1 provides a clear overview of the dominant noise sources in common electronic components under typical bias and operating frequency conditions. It highlights how different types of noise – thermal, shot, flicker (1/*f*), generation–recombination, avalanche, random telegraph signal (RTS), diffusion, and quantum noise – manifest across various devices and frequency ranges. This summary underscores that while thermal noise is ubiquitous, other noise mechanisms become significant depending on the device type, biasing conditions, and frequency of operation. Such an understanding is essential for predicting and

**TABLE 3.1**
**Noise sources in common electronic components**

| Component | Zero Bias | Low Frequency (with Bias) | High Frequency (with Bias) | $f > 1$ THz |
|---|---|---|---|---|
| Resistor | Thermal | 1/*f* | Thermal | Thermal<br>Quantum |
| Diode | Thermal | Shot<br>1/f<br>G-R | Shot<br><br>Avalanche | |
| Bipolar transistor | Thermal | Shot<br>1/f<br>G-R<br>RTS | Shot<br>Thermal | |
| FET | Thermal | Thermal<br>Diffusion<br>Shot<br>G-R | Thermal<br>Diffusion | |
| MOS | Thermal | Thermal<br>Diffusion<br>1/f | | |
| Inductor | Thermal | | | |
| Capacitor | Thermal | | | |
| Laser | | | | Quantum |

managing noise in practical circuits, particularly when designing sensitive analog and RF systems.

## NOISE MEASUREMENTS

Noise measurements can be long-lasting and are prone to errors and uncertainties due to the low power level of noise fluctuations. In general, the noise parameter measured is the NF, which entails power measurements of a noise source and of the intrinsic noise of the device under test (DUT). Two basic types of noise sources are used in measurements: thermal noise and shot noise. The most common noise source consists of a low-capacitance diode that goes into avalanche breakdown when reverse-biased. The diode generates both thermal and shot noise. It generates thermal noise when not biased, and it is equivalent to a thermal noise source at the temperature of the PN junction. This is called the "cold state." Avalanche noise is generated when the diode is in breakdown, corresponding to considerable noise power with noise temperatures between 103 and 105 K (Alimenti *et al.* 2020) and referred to as "hot state." Although shot noise is Poissonian, the statistical distribution of both states is equivalent for measurements.

The mean square value of thermal noise can be evaluated by a wideband oscilloscope, which is the most common way through which engineers may obtain a rapid estimation of noise. The crest factor is defined as the ratio of the peak-to-peak value to the RMS value. For white noise (Gaussian distribution), this factor is equal to 6.6 times the RMS value, 0.10% of the time. The RMS value is obtained by dividing the estimated peak-to-peak value by the crest factor. Several methods can be used to measure the NF. These methods are discussed next.

## Y-FACTOR METHOD

The Y-factor method is the most used method; it requires a noise source and a spectrum analyzer/signal analyzer. The method derives its name from the expression $Y = P_1/P_2$, where $P_1$ and $P_2$ are two noise power levels from the noise source. The noise source is connected to the DUT and operated alternatively in "cold" and "hot" mode. The knowledge of the excess noise ratio (ENR) specification of the noise source, provided by the manufacturer and calibrated, allows us to calculate the NF of the DUT. A low noise amplifier, built-in or external to the spectrum analyzer, is usually required for a very low noise DUT. Measurements in microwave and millimeter wave bands can be performed and, with external mixers, frequencies of up to 110 GHz can be attained.

## THE GAIN OR COLD SOURCE METHOD

This method uses an external "cold" calibrated noise source and a vector network analyzer (VNA) with built-in noise testing features. This method allows automation and rapid measurements of the NF and *S*-parameters.

## NOISE FIGURE METER METHOD

This method is based on a dedicated instrument based on the Y-factor method, in contrast with methods based on spectrum or network analyzers. Noise figure meters are usually equipped with pre-amplifiers and are convenient when several measurements in similar conditions are required.

## VNA-ONLY METHOD

A VNA without a characterized noise source can be used to measure the NF. The VNA supplies a constant wave to the DUT, and filtering allows for extracting the noise due to the DUT only from the total output noise at the output of the DUT. The VNA must be fitted with built-in noise measurement features. In fact, with the rise of millimeter-wave applications, several test equipment manufacturers are offering VNA and spectrum analyzers with built-in noise measuring features.

All NF measurement methods are based on the principle that the noise generated by a DUT and the noise of the source are uncorrelated. Therefore, the power of the noise at the output of the DUT is the sum of the power due to the source and the power of the intrinsic noise of the DUT. This sum is expressed by the gain-noise equation that is solved in the *Y*-factor method. Because the *Y*-factor method is the most used, it is further developed in this section. The gain-noise equation in (3.57) provides the power of the output noise and is represented in

$$N_0 = GkT_0\Delta f + N_A \tag{3.57}$$

where $N_o$ is the noise power at the output of the DUT, $G$ is the power gain of the device under test, $N_A$ is the intrinsic noise of the two-port device at room temperature, and $kT_o$ is the input noise power at room temperature, with $T_o = 290$ K.

The noise source is switched between two levels of noise power, the "cold" noise temperature and the "hot" noise temperature, and the noise factor is given by the expression

$$F = \frac{ENR}{(Y-1)} \tag{3.58}$$

where *ENR* (provided by the manufacturer) is the excess noise ratio of the noise source, and *Y* is the ratio between the hot noise power and the cold noise power at the output of the two-port DUT. It is obvious that, when there is no output signal or when the input signal is attenuated at the output owing to a malfunction of the DUT, the noise factor cannot be measured. Measurements can be performed by dedicated equipment, but remain difficult because the attenuation and noise due to cables and connectors are irreproducible and not well defined. They contribute to the sensitivity and the resolution of measurements.

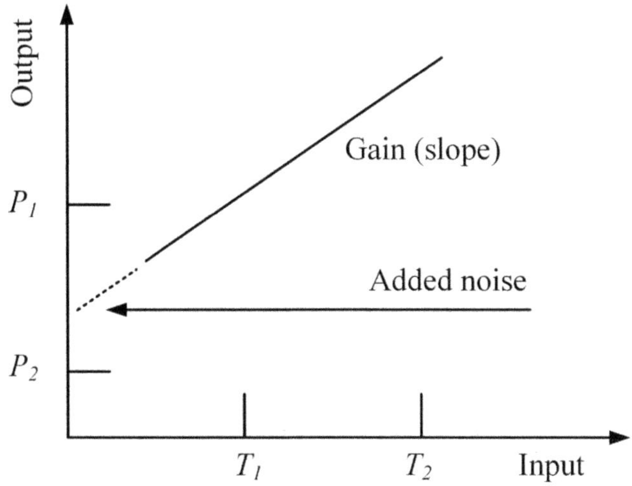


**FIGURE 3.3** Noise-gain equation.

## TWO-PORT NOISE MODEL

An amplifier can be viewed as a black box made of several basic electronic devices and having two ports, one for the input signal and one for the output signal. Noise enters the amplifier from the signal source and is generated internally by its components. The resulting total noise degrades the signal at the output of the amplifier. It is therefore desired to understand how the noise performance (Mumford and Scheibe 1968) of the black box can be characterized and optimized. The two-port noise model defines the noise parameters of a two-port network and indicates how these parameters are related to the NF.

The input signal $S_i$ and noise $N_i$ of a transistor amplifier are subject to the same amplification. The output signal $S_o$ and noise $N_o$ are related to $S_i$ and to $N_i$ by $G$, the power gain of the amplifier, as shown by

$$S_o = GS_i \tag{3.59}$$

and

$$N_o = GN_i \tag{3.60}$$

such that the intrinsic noise of the transistor propagates toward the output and degrades the SNR of the amplifier. The degradation is measured by the noise factor, $F$, which is defined by

$$F = \frac{S_i N_o}{N_i S_o} \tag{3.61}$$

and can be rewritten as

$$F = \frac{N_O}{GN_i} \tag{3.62}$$

The NF of the amplifier is defined by $10log_{10}(F)$ and is expressed in decibels (dB). The main sources of intrinsic noise in bipolar and in field-effect transistors are the thermal noise, the shot noise, and the flicker noise or 1/*f* noise. The flicker noise is a low-frequency noise that can be neglected for RF. The noise model of a noisy two-port transistor amplifier is represented in Figure 3.4.

$F_{min}$ is the minimum achievable noise factor of the noisy amplifier, $R_n$ is the noise resistance, *Y*s is the source admittance ($Y_s = G_s+jB_s$), and $Y_{sopt}$ is the optimum source admittance. $F_{min}$, $R_n$, $B_{sopt}$, and $G_{sopt}$ are the four noise parameters of any two-port noise model, describing completely its behavior in terms of noise performance. The noise resistance $R_n$ determines the sensitivity of *F* to deviations of the source admittance from $Y_{sopt}$.

The two-port noisy amplifier is equivalent to a noiseless device with equivalent current and voltage noise sources reported to the input, as shown in Figure 3.5. Such a representation is convenient because it allows the representation of all possible internal noise sources, which are generally distributed in a circuit, by only two noise sources. In addition, the representation allows one to evaluate the sensitivity of the amplifier and to describe its noise performance by four measurable parameters, as shown by the noise correlation matrix. The measurable parameters are the noise spectral power densities of the current and voltage noise sources at the input and the real and imaginary parts of the correlation of their spectral power densities.

The current and voltage noise sources represent the intrinsic noise of the amplifier and are expressed by the square of the RMS values of the equivalent input noise current ($\overline{i_n^2}$) and input noise voltage ($\overline{v_n^2}$) (Rothe and Dahlke 1954).

If $N(i_n,v_n)$ is the total intrinsic noise of the amplifier referred to the input, then

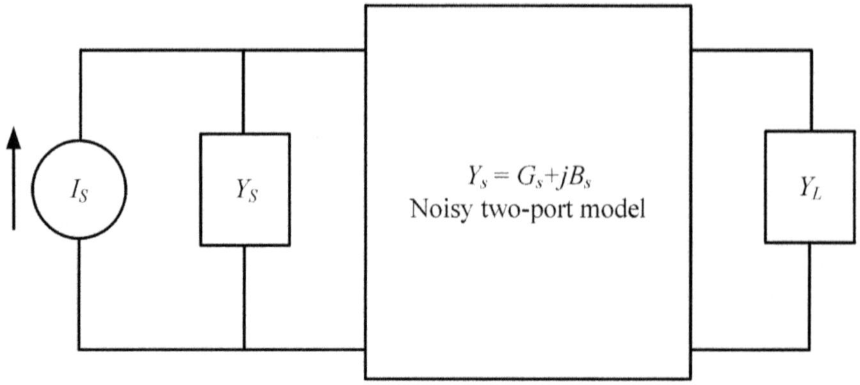


**FIGURE 3.4** Two-port noise model.

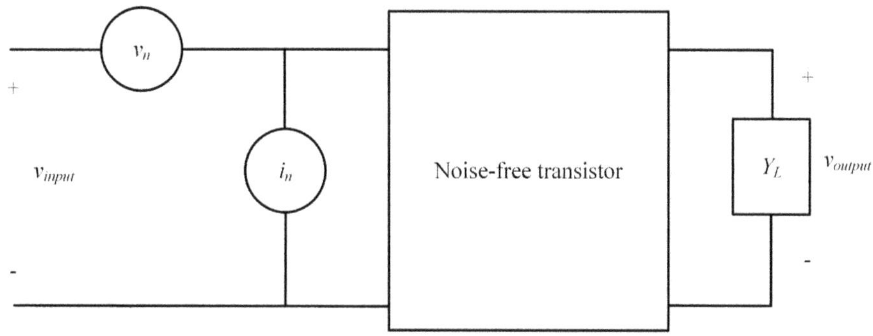


**FIGURE 3.5** Equivalent noise model of a transistor.

$$N_O = GN_i + GN(i_n, v_n) \tag{3.63}$$

and after the substitution of (3.63) into (3.62), the noise factor can be expressed as follows:

$$F = \frac{S_i N_O}{N_i S_O} = \frac{GN_i + GN(i_n, v_n)}{GN_i} = 1 + \frac{N(i_n, v_n)}{N_i} \tag{3.64}$$

The equivalent input referred voltage noise is given by

$$v_{eq} = v_n + Z_S i_n \tag{3.65}$$

where $Z_s$ is the source impedance and is equal to $1/Y_S$.

In general, $v_n$ and $i_n$ are correlated because they are derived from the same physical noise sources, as shown in Figure 3.11. In practice, the correlation is small (Gray *et al.* 2009) and, to quantify it, a technique used in noise analysis involving splitting a noise source into two components can be applied. Using this technique, the input referred current $i_n$ can be split into two components: a component $i_c$ that is correlated to the input referred noise voltage $v_n$ and a component $i_u$ that is uncorrelated to $v_n$ (Lee 2004). This can be illustrated in Figure 3.2 where noise components are not orthogonal: one noise component can then be viewed as the sum of an orthogonal component and a parallel component. These conditions can be expressed by the following expressions:

$$i_n = i_c + i_u \tag{3.66}$$

$$i_n = Y_c v_n \tag{3.67}$$

$$\overline{i_u v_n} = 0 \tag{3.68}$$

where $Y_c$ in (3.66) is the complex correlation admittance.

After combining (3.65), (3.66), (3.67), and (3.68), the equivalent input referred voltage can be written as

$$v_{eq} = v_n\left(1+Z_S Y_c\right)+Z_S i_u \tag{3.69}$$

The variance of the random voltage is given by the following equation. Because $v_{eq}$ is equal to the sum of uncorrelated variables, its variance is equal to the sum of the variances of its uncorrelated components. Hence, the equations

$$\overline{v_{eq}^2} = \overline{v_n^2}\left|1+Z_S Y_C\right|^2 + \left|Z_S\right|^2 \overline{i_u^2} \tag{3.70}$$

and

$$N\left(i_n, v_n\right) = \overline{v_{eq}^2} \tag{3.71a}$$

$$N_i = \left|Z_S\right|^2 \overline{i_S^2} \tag{3.71b}$$

where $\overline{i_S^2}$ is the variance of the thermal noise of the source.

Furthermore, if

$$F = 1+\frac{\overline{v_n^2}\left|1+Z_S Y_C\right|^2 + \left|Z_S\right|^2 \overline{i_u^2}}{\left|Z_S\right|^2 \overline{i_S^2}} \tag{3.72}$$

the noise factor can also be calculated using the equivalent input referred current noise, which is given by

$$i_{eq} = i_n + Y_S v_n \tag{3.73}$$

and after combining (3.73), (3.66), (3.67), and (3.68), the equivalent input referred current can be written as

$$i_{eq} = v_n\left(Y_S + Y_C\right)+i_u \tag{3.74}$$

and the variance of $i_{eq}$ can be expressed as

$$\overline{i_{eq}^2} = \overline{v_n^2}\left|Y_S + Y_C\right|^2 + i_u^2 \tag{3.75}$$

$$N\left(i_n, v_n\right) = \overline{i_{eq}^2}, N_i = \overline{i_s^2} \tag{3.76}$$

$$F = 1 + \frac{\overline{v_n^2}\left|Y_S + Y_C\right|^2 + \overline{i_u^2}}{\overline{i_s^2}} \tag{3.76}$$

which means that (3.72) and (3.77) are equivalent.

The source has a conductance and a susceptance equal to $G_s$ and $B_s$; its admittance is equal to $Y_s = G_s + jB_s$. The correlation admittance $Y_c$ is complex and can be expressed by its real and imaginary components, with $Y_c = G_c + jB_c$. The current power density of the source thermal noise is given by

$$\overline{i_s^2} = 4kTG_s \tag{3.78}$$

where the noise voltage $v_n$ referred to the input can be described by the power density of a thermal noise due to a theoretical resistance $R_n$, therefore

$$\overline{v_n^2} = 4kTR_n \tag{3.79}$$

The power density of the uncorrelated component of the noise current reported to the input can be expressed as a function of a conductance $G_u$

$$\overline{i_u^2} = 4kTG_u \tag{3.80}$$

After combining (3.80), (3.79), and (3.78) with (3.77), the expression of the noise factor becomes

$$F = 1 + \frac{R_n\left[\left(G_s + G_c\right)^2 + \left(B_s + B_c\right)^2\right] + G_u}{G_s} \tag{3.81}$$

and the noise resistance $R_n$, the correlation admittance $Y_c$, and $G_u$ are determined by the noisy network. The noise factor $F$ is a function of the source conductance and is minimal under the condition $\partial F/\partial G_s = 0$, therefore

$$G_s^2 = G_c^2 + \frac{G_u}{R_n} \tag{3.82a}$$

and

$$\frac{\partial F}{\partial B_s} = 0 \tag{3.82b}$$

Resolving (3.82) gives the source admittance $Y_s$ for which the two-port network in Figure 3.10 achieves the minimum noise factor (Lee 2004). The minimum of (3.81) is achieved when equalities (3.83) and (3.84) are satisfied, $B_c$ and $G_c$ being the real and imaginary parts of the noise correlation admittance. It follows that

$$B_s = B_{sopt} = -B_c \tag{3.83}$$

$$G_s = G_{sopt} = \sqrt{\frac{G_u}{R_n} + G_c^2} \tag{3.84}$$

$$Y_{opt} = G_{sopt} + jB_{sopt} \tag{3.85}$$

$$F_{min} = 1 + 2R_n\left(G_{sopt} + G_c\right) \tag{3.86}$$

$$F = F_{min} + \frac{R_n}{G_s}\left[\left(G_s + G_{sopt}\right)^2 + \left(B_s + B_{sopt}\right)^2\right] \tag{3.87}$$

which can be rewritten as

$$F = F_{min} + \frac{R_n}{G_s}\left|Y_s + Y_{sopt}\right|^2 \tag{3.88}$$

In (3.86), $F$ is the noise factor of the two-port amplifier, $F_{min}$ is the minimum achievable noise factor of the noisy amplifier, $R_n$ is the noise resistance, $Y_s$ is the source admittance ($Y_s = G_s + jB_s$), and $Y_{sopt}$ is the optimum source admittance. $F_{min}$, $R_n$, $B_{sopt}$, and $G_{sopt}$ are the four noise parameters of any two-port noise model, describing completely its behavior in terms of noise performance. The noise resistance $R_n$ determines the sensitivity of $F$ to deviations of the source admittance from $Y_{sopt}$. The first term of (3.88) is due to the intrinsic noise of the two-port device, and the second term is due to the mismatch between the admittance of the source $Y_s$ and the optimum noise matching admittance, which is dependent only on the two-port device. The equation shows that the minimum possible noise factor is equal to $F_{min,}$ achieved when the source admittance is equal to $Y_{sopt}$. The admittance of the source and the optimum noise matching admittance being frequency-dependent, the second term of (3.88) is nil at only one frequency. Hence, $R_n$ should be minimized, and $G_s$ maximized to reduce the noise due to source noise mismatch. The dependence of the NF on $Y_s$ is illustrated in Girlando and Palmisano (1999). Equation (3.88) can be expressed in terms of noise temperature by

$$T_e = T_{min} + T_0\frac{R_n}{G_s}\left|Y_s + Y_{sopt}\right|^2 \tag{3.89}$$

where $T_e$ is the equivalent noise temperature, with $F = 1 + Te/To$ and $T_0 = 290$ K (room temperature).

The equation relates the noise resistance and the optimum source conductance:

$$N = R_n G_{sopt} = G_n R_{sopt} \tag{3.90}$$

where $N$ is an invariant that was introduced by Lange and is called the Lange parameter. $N$ is unchanged for any two-port network when lossless transformations are applied. The inequality in (3.91) applies to $N$ for such transformations

$$1 \leq \frac{4NT_0}{T_{min}} < 2 \tag{3.91}$$

and it is shown in Belostotski (2011) that $N \approx (F_{min} - 1)/2$ for HEMTs for which the capacitance between the gate and the source can be neglected and for $\omega T >> \omega$. For several FET devices, $N$ is related to the minimum noise temperature $T_{min}$ by the expression (Pospieszalski 2014): $4NT_0T_{min} \approx 2$. Equation (3.90) indicates that the four parameters of the two-port noise model can be reduced to three parameters only: $T_{min}$ and the real and imaginary components of $Y_{sopt}$. More critical for the designer, (3.90) indicates that achieving the minimum noise factor ($T_{min}$ or $F_{min}$) ensures that the minimum noise resistance $R_n$ is also achieved (Pospieszalski 2014) for FET devices.

## NOISE PARAMETERS AND INPUT-REFERRED SOURCES

The four noise parameters can be expressed as functions of input-referred sources. In the case of a two-port device comprised of a single transistor, the expressions that relate noise parameters to transistor parameters can be derived.

The noise current $i_n$ is related to the noise voltage $v_n$, to $Y_c$, and to $i_u$ by the relation (3.92), from (3.66) and (3.67):

$$i_n = i_u + Y_c v_n \tag{3.92}$$

and by multiplying each side of (3.92) by $v_n^*$ and taking the average, (3.91) yields

$$\overline{i_n v_n^*} = Y_c \overline{v_n^2} \tag{3.93}$$

and we can derive the following:

$$Y_c = \frac{\overline{i_n v_n^*}}{\overline{v_n^2}} \tag{3.94}$$

where the correlation impedance, $Z_c$, is defined by this same equation; therefore, $Z_c$ is not equal to $1/Y_c$. The correlation coefficient is a normalized quantity defined by

$$c = \frac{\overline{i_n v_n^*}}{\sqrt{\overline{v_n^2}\,\overline{i_n^2}}} = Y_c \sqrt{\frac{\overline{v_n^2}}{\overline{i_n^2}}} \tag{3.95}$$

such that the noise resistance $R_n$ is derived from (3.79), which is equivalent to

$$R_n = \frac{\overline{v_n^2}}{4kT} \tag{3.96}$$

$$B_{sopt} = -Im\{Y_c\} = -Im\left\{\frac{\overline{i_n v_n^*}}{\overline{v_n^2}}\right\} \tag{3.97}$$

$$G_{sopt} = \sqrt{\frac{\overline{i_n^2}}{\overline{v_n^2}} + R_e^2\left\{\frac{\overline{i_n v_n^*}}{\overline{v_n^2}}\right\}} = \sqrt{\frac{\overline{i_n^2}}{\overline{v_n^2}} - Im\left\{\frac{\overline{i_n v_n^*}}{\overline{v_n^2}}\right\}} \tag{3.98}$$

and therefore

$$F_{min} = 1 + \frac{\overline{v_n^2}}{2kT}\left(\sqrt{\frac{\overline{i_n^2}}{\overline{v_n^2}} + R_e^2\left\{\frac{\overline{i_n v_n^*}}{\overline{v_n^2}}\right\}} + R_e\left\{\frac{\overline{i_n v_n^*}}{\overline{v_n^2}}\right\}\right) \tag{3.99}$$

or

$$F_{min} = 1 + \frac{\overline{v_n^2}}{2kT}\left(\sqrt{\frac{\overline{i_n^2}}{\overline{v_n^2}} - Im\left\{\frac{\overline{i_n v_n^*}}{\overline{v_n^2}}\right\}} + R_e\left\{\frac{\overline{i_n v_n^*}}{\overline{v_n^2}}\right\}\right) \tag{3.100}$$

In conclusion, expressing noise parameters in terms of input-referred sources provides a powerful framework for analyzing and optimizing the noise performance of a transistor or two-port device. By relating measurable quantities such as noise voltage and current to derived parameters like the correlation impedance, noise resistance, and optimum source admittance, designers can predict the minimum achievable noise figure and identify the conditions under which it occurs. This formalism not only aids in understanding the physical origins of noise but also serves as a practical tool for circuit design, allowing for more informed choices of biasing, topology, and matching to achieve low-noise performance in LNAs and other sensitive electronic systems.

## NOISE CORRELATION MATRIX

A chain representation of the electrical parameters of the two-port network has been used to derive (3.88). $Z$-, $Y$-, or $H$-parameter representations can equally be used for noise analysis. The values of input referred noise sources are related to the $Z$-,

Y-, or H-parameters of the two-port network and can be calculated as shown in this section. Six possible configurations of noise sources relative to a noiseless network can represent any noisy two-port network.

The noiseless parts of the networks in Figure 3.6, Figure 3.7, Figure 3.8, and Figure 3.9 can be represented by the Z-, Y-, and H-parameters.

Using Y-parameters, the voltages and currents in Figure 3.9 are related in the frequency domain as follows:

$$\begin{bmatrix} I_1 + I_{na} \\ I_2 \end{bmatrix} = \begin{bmatrix} Y_{11} & Y_{12} \\ Y_{21} & Y_{22} \end{bmatrix} \begin{bmatrix} V_1 + V_{na} \\ V_2 \end{bmatrix} \tag{3.101}$$

where $I_1$, $V_1$, and $I_2$, $V_2$ are the Fourier transforms of the current and voltage at the input and output, respectively.

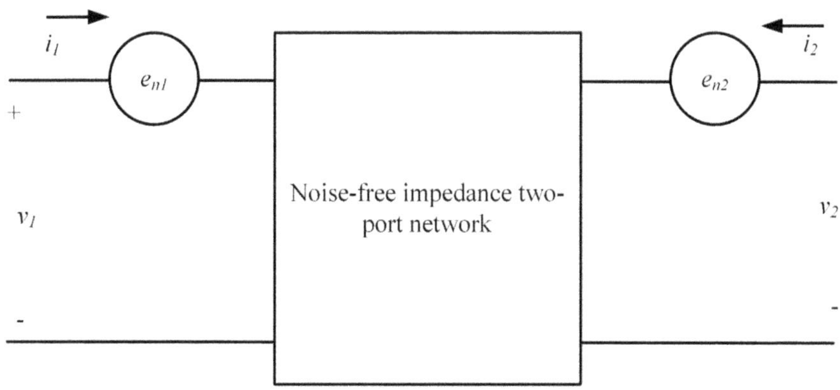


**FIGURE 3.6** Impedance configuration.

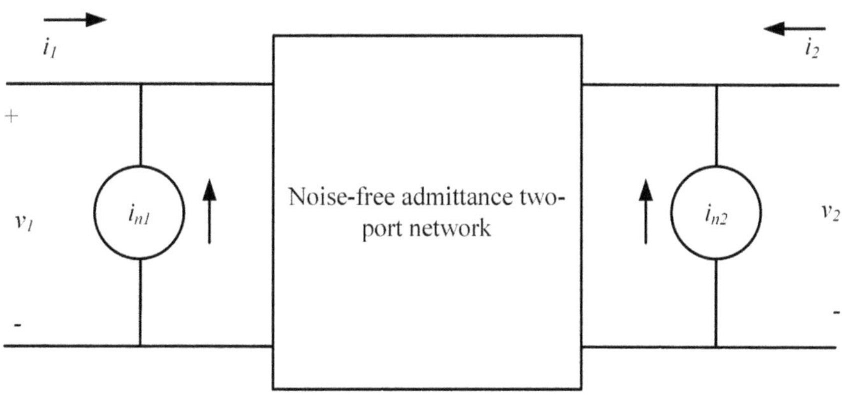


**FIGURE 3.7** Admittance configuration.

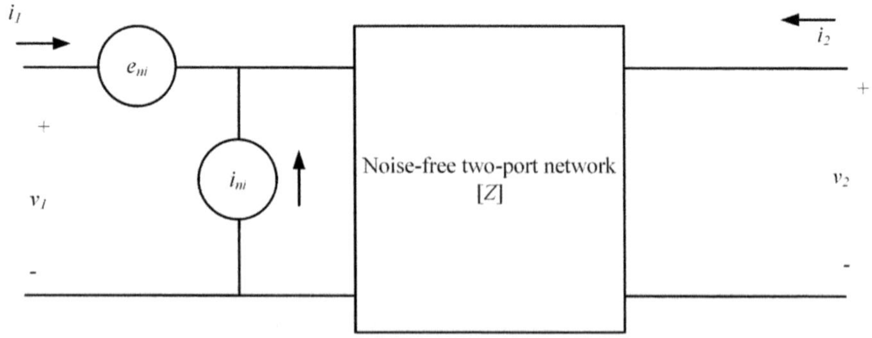


**FIGURE 3.8** Chain representation with *Z*-parameters.

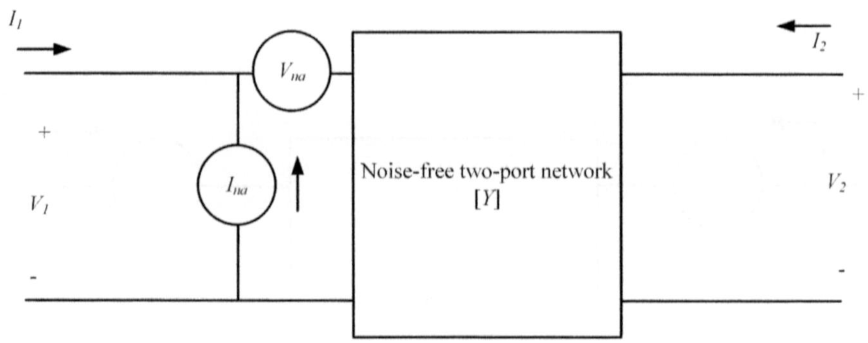


**FIGURE 3.9** Chain representation with *Y*-parameters.

Traditionally, noise sources are replaced by sinusoidal sources of the same power that are used to perform AC analysis for noise power calculations. This method is simple but is restricted to cases where noise sources are uncorrelated or totally correlated. A method that can be used in most cases entails representing each noise source by a linear combination of uncorrelated and correlated sources. However, the method can lead to complicated calculations. Moreover, it does not provide a simple solution for the calculation of the minimum possible noise factor $F_{min}$ and the source impedance required for noise matching.

In the correlation matrix method, a two-port network is derived from basic two-port networks with known characteristics. In Figure 3.6, Figure 3.7, and Figure 3.8, the two-port noise networks are fully characterized by their correlation matrix *C*, whose elements are the self- and cross-power spectral densities of their noise sources in the bandwidth $\Delta f$ centered at *f*, the operating frequency. The correlation matrix of two noise sources, $v_n$ and $i_n$, is a 2 × 2 matrix whose elements are the self- and cross-average powers of the noise sources.

The self- and cross-power densities of the noise sources are the Fourier transforms of the self- and cross-correlation functions of the noise sources. The power spectral density of a signal with $F(j\omega)$ as the Fourier transform is given by the expression

$S_\omega = (1/2\pi) \times |F(j\omega)|^2$. The cross-power spectral density of two signals $s_1(t)$ and $s_2(t)$ of Fourier transforms $S_1$ and $S_2$ is defined as the Fourier transform of their cross-correlation function

$$i_n(s_1, s_2) = F\left[R_{s_1 s_2}(\tau)\right] \tag{3.102}$$

and the cross-correlation function of $s_1(t)$ and $s_2(t)$ is defined as follows:

$$R_{s_1 s_2}(\tau) = \lim_{T\to\infty} \frac{1}{2T} \int_{-T}^{T} s_1(t) s_2(t+\tau)\,dt \tag{3.103}$$

For the noise sources $v_n$ and $i_n$ in Figure 3.11, the correlation matrix is

$$\xi_A = \begin{bmatrix} \xi_{11} & \xi_{12} \\ \xi_{21} & \xi_{22} \end{bmatrix} = \frac{1}{2\Delta f} \begin{bmatrix} \overline{v_n v_n^*} & \overline{v_n i_n^*} \\ \overline{v_n^* i_n} & \overline{i_n i_n^*} \end{bmatrix} \tag{3.104}$$

where

$$\xi_A = 2kTC_A = 2kT \begin{bmatrix} C_{A11} & C_{A12} \\ C_{A21} & C_{A22} \end{bmatrix} \tag{3.105}$$

therefore

$$\begin{bmatrix} C_{A11} & C_{A12} \\ C_{A21} & C_{A22} \end{bmatrix} = \frac{1}{4kT\Delta f} \begin{bmatrix} \overline{v_n v_n^*} & \overline{v_n i_n^*} \\ \overline{v_n^* i_n} & \overline{i_n i_n^*} \end{bmatrix} \tag{3.106}$$

Where the subscript $A$ in $C_A$ refers to the chain representation, the asterisk refers to the complex conjugate, and the matrix terms on the right-hand side of (3.106) are the self- and cross-correlation terms between the noise sources referred to the input and are functions of the frequency $f$.

The correlation matrix has several well-defined mathematical properties. Its diagonal elements are always real, meaning that the imaginary parts of $C_{11}$ and $C_{22}$ vanish, a condition that holds for impedance, admittance, and chain representations alike. The off-diagonal elements are related by complex conjugation, so that $C_{21} = C_{12}^*$. Both diagonal entries are nonnegative, with $C_{11} >= 0$ and $C_{22} >= 0$, and the determinant of the matrix is also constrained to be nonnegative, specifically $det\{C\} = C_{11}C_{22} - |C12|^2 > = 0$. When all noise sources are modeled as current sources, the elements of the correlation matrix can be constructed directly from circuit topology. In this case, $C_{11}$ is given by the total contribution of noise currents connected to node 1 (the input), while $C_{22}$ reflects the sum of contributions connected to node 2 (the output). The cross terms, $C_{12}$ and $C_{21}$, take on negative values corresponding to the noise sources that bridge node 1 and node 2. A noise source tied to ground contributes only to

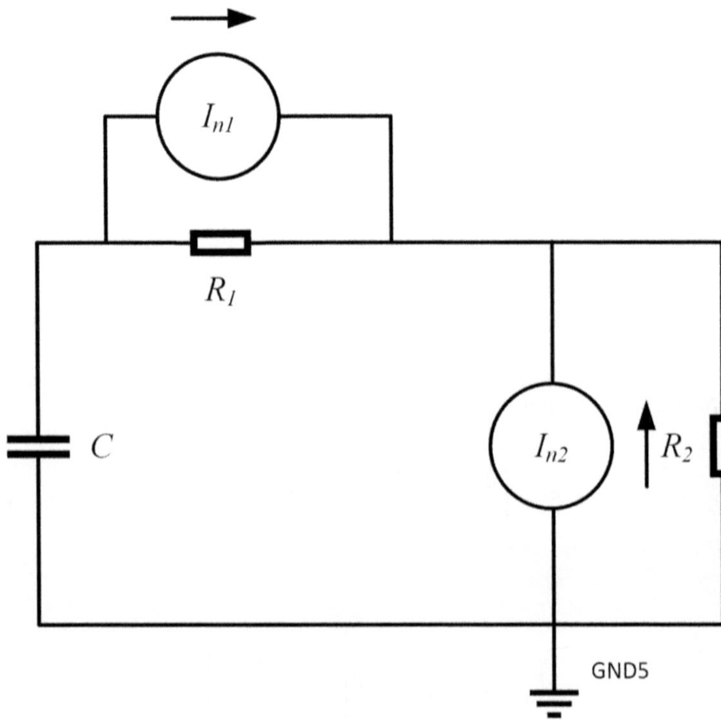


**FIGURE 3.10** Passive network with current noise sources.

one diagonal term, whereas an ungrounded source influences all matrix elements. An illustration of this procedure is shown in Figure 3.10, in the case of a two-port network consisting of only passive components.

The correlation matrix of Figure 3.10 is shown in

$$C_Y = \frac{1}{\Delta f}\begin{bmatrix} \overline{i_{n1}^2} & \overline{-i_{n1}^2} \\ \overline{-i_{n1}^2} & \overline{i_{n1}^2} + \overline{i_{n2}^2} \end{bmatrix} \tag{3.107}$$

where the elements of this correlation matrix can also be obtained by using the equations of the correlation matrix for a two-port network with only passive elements:

$$C_Z = 2kTRe\{Z\} \tag{3.108}$$

and

$$C_Y = 2kTRe\{Y\} \tag{3.109}$$

where $k$ is Boltzmann's constant, and $T$ is the absolute temperature.

Basic two-port networks can be interconnected in parallel, series, or cascade to form more elaborate two-port networks. This technique is well-suited for computer-based noise analysis. Noise sources from separate two-port networks are uncorrelated. Therefore, the correlation matrix $C_Y$ of two parallel two-port networks $C_{Y1}$ and $C_{Y2}$, each network in admittance configuration, can be expressed by the equation

$$C_Y = C_{Y1} + C_{Y2} \tag{3.110}$$

where for two-port networks in series $C_{Z1}$ and $C_{Z2}$, each using the series configuration, the resulting correlation matrix is expressed by the equation

$$C_Z = C_{Z1} + C_{Z2} \tag{3.111}$$

In the case of two two-port networks in cascade, of correlation matrices $C_{A1}$ and $C_{A2}$, each in a chain representation, the correlation matrix of the resulting cascade network is represented by

$$C_A = A_1 A_{A2} A_1^* + C_{A1} \tag{3.112}$$

where $A_1$ is the chain or *ABCD* matrix of the first two-port network. The noise factor can be calculated from $C_A$ using the expression

$$F = 1 + \frac{Z^+ C_A Z}{2kTRe\{Z_S\}} \tag{3.113}$$

where $Z_s$ is the impedance of the source, $Z = \begin{bmatrix} 1 \\ Z_S^* \end{bmatrix}$ and $Z^+$ is the Hermitian conjugate of $Z$. The Hermitian conjugate $Z^+$ is obtained by transposing $Z$ and converting its elements to their complex conjugates.

In cases where the $Y$ representation is used, the noise factor can also be calculated using the expression

$$F = 1 + \frac{Y^+ C_A Y}{2kTRe\{Y_S\}} \tag{3.114}$$

where $Y^+$ is the Hermitian conjugate of $Y$ and $Y_s$ is the admittance of the source.

## TRANSFORMATION BETWEEN REPRESENTATIONS

Conjugate decomposition simplifies the analysis of arbitrary two-port networks into basic two-port networks. After the decomposition, transformations between series, parallel, and chain representations are often desired. The transformations allow one to convert a correlation matrix $C$ in each representation into a new correlation matrix

**TABLE 3.2**
**Impedance, admittance, hybrid, and chain matrix transformations**

| | $C_Y$ | $C_Z$ | $C_A$ | $C_H$ |
|---|---|---|---|---|
| $C'_Y$ | $\begin{bmatrix}1 & 0\\0 & 1\end{bmatrix}$ | $\begin{bmatrix}Y_{11} & Y_{12}\\Y_{21} & Y_{22}\end{bmatrix}$ | $\begin{bmatrix}-Y_{11} & 1\\-Y_{21} & 0\end{bmatrix}$ | $\begin{bmatrix}-Y_{11} & 0\\-Y_{21} & 1\end{bmatrix}$ |
| $C'_Z$ | $\begin{bmatrix}Z_{11} & Z_{12}\\Z_{21} & Z_{22}\end{bmatrix}$ | $\begin{bmatrix}1 & 0\\0 & 1\end{bmatrix}$ | $\begin{bmatrix}1 & -Z_{11}\\0 & -Z_{21}\end{bmatrix}$ | $\begin{bmatrix}1 & -Z_{12}\\0 & -Z_{22}\end{bmatrix}$ |
| $C'_A$ | $\begin{bmatrix}0 & A_{12}\\1 & A_{22}\end{bmatrix}$ | $\begin{bmatrix}1 & -A_{11}\\0 & -A_{21}\end{bmatrix}$ | $\begin{bmatrix}1 & 0\\0 & 1\end{bmatrix}$ | $\begin{bmatrix}1 & A_{12}\\0 & A_{22}\end{bmatrix}$ |
| $C'_H$ | $\begin{bmatrix}-H_{11} & 0\\-H_{21} & 1\end{bmatrix}$ | $\begin{bmatrix}1 & -H_{12}\\0 & -H_{22}\end{bmatrix}$ | $\begin{bmatrix}1 & -H_{11}\\0 & -H_{21}\end{bmatrix}$ | $\begin{bmatrix}1 & 0\\0 & 1\end{bmatrix}$ |

$C'$ in a new representation. The new correlation matrix is obtained below, referred to as the congruence transformation

$$[C'] = [T][C][T^+] \tag{3.115}$$

where $T$ is the transformation matrix and $T^+$ is its Hermitian transform. The table of transformation matrices between impedance, admittance, hybrid, and chain representations is provided in Table 3.2. Additional information on transformation matrices in noise calculations is provided in Hillbrand and Russer (1976).

## NOISE PARAMETERS AND CORRELATION MATRIX

The relations between noise parameters and the elements of the noise correlation matrix in chain representation are provided in this section. The noise resistance $R_n$ has been defined by the expression $R_n = \overline{v_n^2} / 4kT$, which is equivalent to $R_n = v_n(f)v_n^*(f)/2kT$ and corresponds to the term $CA_{11}$ of the correlation matrix. The term $CA_{12}$ of the correlation matrix is equal to $\overline{i_n i_n^*}$, which is the noise conductance, $g_n = \overline{i_n i_n^*}/4kT$ and is calculated below.

The equivalent current noise at the input is given by

$$i_{neq} = \frac{v_n}{Z_S(\omega)} + i_S + i_n \tag{3.116}$$

and the power density of the noise is given by

$$\overline{i_{neq}^* i_{neq}} = \frac{\overline{v_n^* v_n}}{Z_S^2(\omega)} + \frac{\overline{\left(i_n^* v_n\right)}^*}{Z_S(\omega)} + \frac{\overline{i_n^* v_n}}{Z_S(\omega)} + \overline{i_n^* i_n} + \overline{i_S^* i_S} \tag{3.117}$$

or

$$\overline{i_{neq}^{*} i_{neq}} = \overline{v_n^{*} v_n} Y_S^2(\omega) + 2Re\left\{\overline{i_n^{*} v_n} Y_S(\omega)\right\} + \frac{\overline{i_n^{*} v_n}}{Z_S(\omega)} + \overline{i_n^{*} i_n} + \overline{i_S^{*} i_S} \tag{3.118}$$

where

$$\overline{i_S^{*} i_S} = 4kTRe\left\{Y_S(\omega)\right\} \tag{3.119}$$

such that

$$F = 1 + \frac{\overline{v_n^{*} v_n} Y_S^2(\omega) + 2Re\left\{\overline{i_n^{*} v_n} Y_S(\omega)\right\} + \overline{i_n^{*} i_n}}{4kTRe\left\{Y_S(\omega)\right\}} \tag{3.120}$$

and

$$F = 1 + \frac{R_n\left(G_S^2 + B_S^2\right)}{G_S} + 2\frac{Re\left\{\overline{i_n^{*} v_n}\right\}}{4kT} - 2\frac{Im\left\{\overline{u_n i_n^{*}}\right\}}{4kTG_S} B_S + \frac{g_n}{G_S} \tag{3.121}$$

From (3.119), when $Y_s$ tends to zero, the term

$$\frac{\overline{i_n^{*} i_n}}{4kTRe\left\{Y_S(\omega)\right\}}$$

tends to infinity, and when $Y_s$ tends to infinity, the term

$$\frac{\overline{v_n^{*} v_n} Y_S^2(\omega)}{4kTRe\left\{Y_S(\omega)\right\}}$$

also tends to infinity. As $F$ tends to infinity when $Y_s$ tends to zero and when $Y_s$ tends to infinity, there is a value of $Y_{sopt}$ for which $F$ reaches a minimum.

The admittance of the input is $Y_s = G_s + jB_s$. The optimum source admittance $Y_{sopt}$ is calculated by differentiating $F$ as follows:

$$\frac{\partial F}{\partial G_S} = 0 \text{ and } \frac{\partial F}{\partial B_S} = 0 \tag{3.122}$$

and the differentiation of $F$ versus $B_s$ provides the second condition for minimum $F$

$$B_{sopt} = \frac{Im\left\{\overline{v_n i_n^{*}}\right\}}{4kTR_n} \tag{3.123}$$

The differentiation of $F$ versus $G_s$ provides the first minimum condition

$$G_{sopt}^2 = \frac{\overline{i_n i_n^*} + \overline{v_n v_n^*} B_S^2}{\overline{v_n v_n^*}} \tag{3.124}$$

such that

$$G_{sopt} = \sqrt{\frac{g_n + R_n B_S^2 - \frac{im\left\{\overline{v_n i_n^*}\right\}}{kT} B_S}{R_n}} \tag{3.125}$$

where

$$g_n = R_n \left(G_S^2 + B_S^2\right) \tag{3.126}$$

and

$$g_n = R_n \left|Y_{sopt}\right|^2 \tag{3.127}$$

so the term $C_{A22}$ of the correlation matrix is then

$$C_{A22} = 4kTg_n = 4kTR_n \left|Y_{sopt}\right|^2 \tag{3.128}$$

and the term $\overline{v_n i_n^*}$ is determined as follows:

For $Y_s = Y_{sopt}$, equation (3.121) can be rewritten as below, after introducing the expression (3.125),

$$F_{min} = 1 + \frac{2g_n}{G_{sopt}} + \frac{G_{sopt} Re\left\{\overline{v_n i_n^*}\right\}}{2kTG_{sopt}} - \frac{Im\left\{\overline{v_n i_n^*}\right\}}{2kTG_{sopt}} B_{sopt} \tag{3.129}$$

Using the relation

$$B_{sopt} = \frac{Im\left\{\overline{v_n i_n^*}\right\}}{4kTR_n}$$

then (3.129) yields

$$F_{min} = 1 + \frac{2kTg_n}{2kTG_{sopt}} + \frac{G_{sopt} Re\left\{\overline{v_n i_n^*}\right\}}{2kTG_{sopt}} - \frac{4kTR_n B_{sopt}^2}{2kTG_{sopt}} \tag{3.130}$$

and

$$Re\left\{\overline{v_n i_n^*}\right\} = 2kT\left[\frac{F_{min}-1}{2} - R_n G_{sopt}\right] \tag{3.131}$$

and again using the expression for $B_{sopt}$, (3.131) yields

$$Re\left\{\overline{v_n i_n^*}\right\} + Im\left\{\overline{v_n i_n^*}\right\} = 2kT\left[\frac{F_{min}-1}{2} - R_n\left(G_{sopt} - jB_{sopt}\right)\right] \tag{3.132}$$

therefore

$$\overline{v_n i_n^*} = 2kT\left[\frac{F_{min}-1}{2} - R_n\left(Y_{sopt}^*\right)\right] \tag{3.133}$$

and since $\left(\overline{v_n i_n^*}\right)^* = \overline{i_n v_n^*}$ it follows that

$$\overline{v_n i_n^*} = 2kT\left[\frac{F_{min}-1}{2} - R_n Y_{sopt}\right] \tag{3.134}$$

The chain correlation matrix is defined in (3.135) by the two-port noise parameters, as

$$C_A = 2kT\begin{bmatrix} R_n & \left[\frac{F_{min}-1}{2} - R_n\left(Y_{sopt}^*\right)\right] \\ \left[\frac{F_{min}-1}{2} - R_n Y_{sopt}\right] & R_n\left|Y_{sopt}\right|^2 \end{bmatrix} \tag{3.135}$$

Inversely, the two-port noise model parameters can be obtained from the chain correlation matrix by the relations

$$R_n = \frac{C_{A11}}{2kT} \tag{3.136}$$

$$Y_{sopt} = \sqrt{\frac{C_{A22}}{C_{A11}} - \left[Im\left(\frac{C_{A12}}{C_{A11}}\right)\right]^2} + jIm\left(\frac{C_{A12}}{C_{A11}}\right) \tag{3.137}$$

$$F_{min} = 1 + \frac{C_{A12} + C_{A11} Y_{sopt}^*}{kT} \tag{3.138}$$

and (3.138) can also be written as

$$F_{min} = 1 + 2\left(Re\left\{C_{A12}\right\} + C_{A11}G_{sopt}\right) \tag{3.139}$$

with $C_{A11} = \dfrac{\overline{v_n v_n^*}}{4kT\Delta f}$ and $C_{A12} = \dfrac{\overline{v_n i_n^*}}{4kT\Delta f} = \dfrac{\overline{i_n v_n^*}}{4kT\Delta f}$.

## TRANSFER OF NOISE SOURCES TO THE INPUT IN ABCD REPRESENTATION

The transfer of the noise source to the input is facilitated using a chain (ABCD) representation. A two-port ABCD representation is provided by

$$v_1 = Av_2 + Bi_2 \tag{3.140}$$

and

$$i = Cv_2 + Di_2 \tag{3.141}$$

A noisy ABCD two-port network, with all noise sources placed at the output, is represented in Figure 3.11.

Equations (3.140) and (3.141) are rewritten as follows, when noise sources are considered:

$$v_1 = Av_2 + Bi_2 - Av_n - Bi_n \tag{3.142}$$

and

$$i_1 = Cv_2 + Di_2 - Cv_n - Di_n \tag{3.143}$$

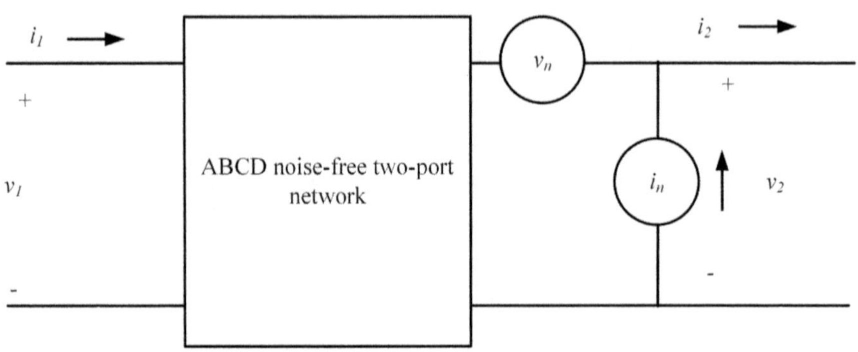


**FIGURE 3.11** ABCD noise-free network with noise sources at the output.

where equation (3.142) is equivalent to two voltage noise sources in series $A_{vn}$ and $B_{in}$ at the input, and two current noise sources $C_{vn}$ and $D_{in}$ at the input. The equivalent voltage and current noise source at the input are correlated. Equations (3.142) and (3.143) are used in referring to the input noise (Tartarin *et al.* 2019) sources of a common-emitter bipolar transistor (Voinigescu *et al.* 1997).

## TRANSFER OF NOISE SOURCES TO THE INPUT IN *Y* REPRESENTATION

In the case of the *Y* representation, noise sources can be represented by current sources at the input and output of the two-port representation, as shown in Figure 3.12. This is the case of the shot noise sources for the BJT.

The admittance equations in Figure 3.12 are written as

$$i_1 = Y_{11}v_1 + Y_{12}v_2 - i_{ni} \tag{3.144}$$

and

$$i_2 = Y_{21}v_1 + Y_{22}v_2 - i_{n2} \tag{3.145}$$

When the noise sources are reported to the input, the two-port noise model is shown in Figure 3.13, where $i_{nl}$ and $v_{nl}$ are the current and voltage equivalent noise sources.

The admittance equations in Figure 3.13 are written as

$$i_1 = Y_{11}\left(v_1 + v_{ni}\right) + Y_{12}v_2 - i_{ni} \tag{3.146}$$

and

$$i_2 = Y_{21}\left(v_1 + v_{ni}\right) + Y_{22}v_2 \tag{3.147}$$

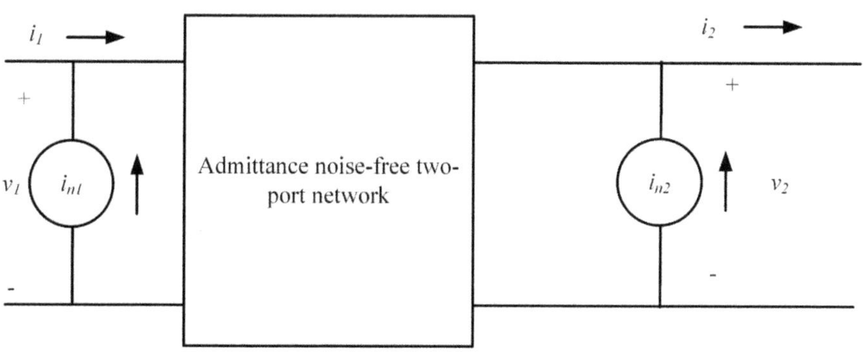


**FIGURE 3.12** An admittance noise-free network with noise sources at the input and output.

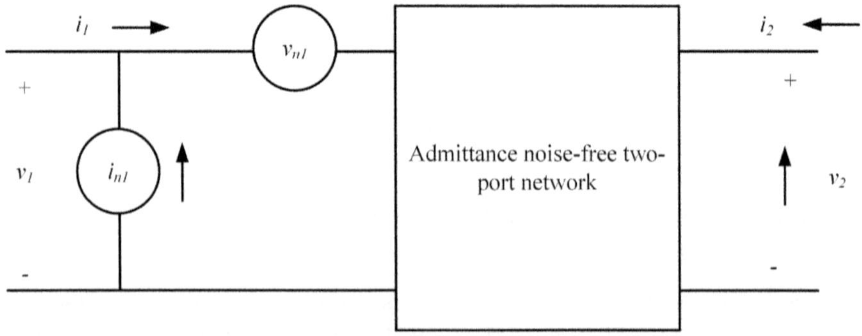


**FIGURE 3.13** An admittance noise-free network with noise sources at the input.

After combining equations (3.144), (3.145), (3.146), and (3.147), the equivalent input voltage and current sources referred to the input are given by

$$v_{ni} = -\frac{i_{n2}}{Y_{21}} \tag{3.148}$$

and

$$i_{ni} = i_{n1} - \frac{Y_{11}}{Y_{21}} i_{n2} \tag{3.149}$$

In summary, transferring noise sources to the input in the *Y* representation allows all noise contributions of a two-port network to be represented as equivalent input-referred voltage and current sources. This approach simplifies noise analysis and design, as it enables the evaluation of overall noise performance from the perspective of a single reference point at the input. By using the derived relations for $v_{ni}$ and $i_{ni}$, designers can systematically account for both internal device noise and noise coupling through the network, facilitating accurate prediction of the noise figure and supporting optimal low-noise circuit design.

## ELECTRONIC NOISE APPLICATIONS

To close this chapter, the reader should be aware that electronic noise can be desired in situations other than measurements and that noise is not always a nuisance. Because noise is caused by the random motion of carriers, noise can be used in the study of microscopic processes, such as the determination of the lifetime of carriers in semiconductors (Gupta 1975) and other constants. This can be demonstrated in the calculation of Boltzmann's constant and the electron charge. Noise sources can be used to simulate random quantities in radar, seismic, and mechanical applications. Electronic noise is used as a countermeasure in radar applications to generate false echoes or to mask the expected echo. In digital applications, noise sources are used for the generation of random numbers instead of pseudo-random numbers generated by related algorithms. Random numbers of applications include cryptography and computer simulations (e.g., Monte Carlo simulations). In the biological and medical domain, noise is used in the study of neurons and physiological systems. Noise can be used to induce sleep and anesthesia.

In electronic systems, noise sources are used for the measurement of noise, bandwidth, linearity, and intermodulation. Cosmic noise is used for the measurement of antenna characteristics (focal point, radiation pattern, efficiency, etc.). White noise is used to determine the insertion loss of a two-port network. The thermal noise of a resistor can be used as a reference signal for measuring microwave RF power. White noise is also used to measure the impulse response of linear two-port networks.

Linear systems can be characterized by their response to sinusoidal signals or their impulse response. The characterization of nonlinear systems is more complex, and their complete description based on Wiener kernels is difficult. However, the response of nonlinear systems to Gaussian random signals can be used to provide their incomplete descriptions, as Gaussian noise can be viewed as a sum of all possible signals (Barrett and Coales 1956; Pervozvanskiĭ 1965). Performance parameters of electronic systems, such as NF, gain, distortion, and bandwidth, can be checked during fabrication and when they are in service, owing to the use of noise sources. On-chip noise sources such as integrated avalanche diodes can be implemented to benefit from the numerous advantages of integration and to facilitate the testing (Alimenti *et al.* 2020).

The low-frequency noise measurements of passive components can be used to characterize material imperfections and to determine the stability and reliability of electronic components. They are used to characterize micro- and nanotechnology fabrication methods (Stadler *et al.* 2015). They can be used to develop carrier transport models and to contribute to fundamental science. Other applications where electronic noise is desired are as follows:

- Noise helps to trigger oscillations according to the Barkhausen criterion.
- Noise is useful in the training phase of neural networks as it improves their generalization capabilities.
- Noise can improve the performance of nonlinear systems through stochastic resonance (Ando *et al.* 2001).

## CONCLUSION

Focusing on LNAs and the transistors that constitute them, this chapter has explored noise in electronic systems in detail. It began by introducing the mathematical tools needed to analyze stochastic processes in circuits, starting from the basic definitions and statistical characterization of noise. Key concepts such as RMS values, autocorrelation, and power spectral density were discussed, providing a foundation for understanding how noise behaves in practical systems. The physical origins, probability distributions, and impact on circuit performance of the primary intrinsic noise sources – thermal (Johnson) noise, shot noise, flicker noise, and generation–recombination noise – were examined. Historical perspectives, including Planck's law and Nyquist's derivation, were also highlighted to connect fundamental physics with contemporary electronic noise modeling.

The chapter then turned to noise measurement and characterization in two-port devices. Input-referred sources, noise parameters, and their relationships with correlation, optimal matching, and device admittance were carefully detailed. Techniques for referring noise sources to the input were presented as practical tools for simplifying the analysis of complex amplifier circuits. Throughout, attention was paid to the interplay between theory, statistical modeling, and real-world circuit considerations, emphasizing how each type of noise affects the overall noise figure and bandwidth of LNAs.

In summary, this chapter lays a solid groundwork for understanding, modeling, and mitigating noise in electronic circuits. The principles and methods discussed here form the basis for later exploration of noise optimization, noise matching, and advanced LNA design, equipping designers with the tools to predict and control noise performance in practical systems.

## REFERENCES

Abbott, D., Davis, B.R., Phillips, N.J., and Eshraghian, K., 1996. Simple Derivation of the Thermal Noise Formula Using Window-Limited Fourier Transforms and Other Conundrums. *IEEE Transactions on Education*, 39 (1), 1–13.

Alimenti, F., Simoncini, G., Brozzetti, G., Maistro, D.D., and Tiebout, M., 2020. Millimeter-Wave Avalanche Noise Sources Based on p-i-n Diodes in 130 nm SiGe BiCMOS Technology: Device Characterization and CAD Modeling. *IEEE Access*, 8, 178976–178990.

Ando, B., Graziani, S., and Pitrone, N., 2001. Hysteresis Shaping in Stochastic Driven Systems. *IEEE Transactions on Instrumentation and Measurement*, 50 (5), 1264–1269.

Barrett, J.F., and Coales, J.F., 1956. An Introduction to the Analysis of Non-linear Control Systems with Random Inputs. *Proceedings of the IEE Part C: Monographs*, 103 (3), 190.

Belostotski, L., 2011. On the Number of Noise Parameters for Analyses of Circuits with MOSFETs. *IEEE Transactions on Microwave Theory and Techniques*, 59 (4), 877–881.

Girlando, G., and Palmisano, G., 1999. Noise Figure and Impedance Matching in RF Cascode Amplifiers. *IEEE Transactions on Circuits and Systems II: Analog and Digital Signal Processing*, 46 (11), 1388–1396.

Gray, P.R., Hurst, P.J., and Meyer, R.G., 2009. *Analysis and Design of Analog Integrated Circuits*. 5th ed. Wiley.

Gupta, M.-S., 1975. Applications of Electrical Noise. *Proceedings of the IEEE*, 63 (7), 996–1010.

Heffner, H., 1962. The Fundamental Noise Limit of Linear Amplifiers. *Proceedings of the IRE*, 50 (7), 1604–1608.

Hillbrand, H., and Russer, P., 1976. An Efficient Method for Computer Aided Noise Analysis of Linear Amplifier Networks. *IEEE Transactions on Circuits and Systems*, 23 (4), 235–238.

Lee, T.H., 2004. *The Design of CMOS Radio-Frequency Integrated Circuits*. Cambridge University Press.

Mumford, W., and Scheibe, E., 1968. *Noise Performance Factors in Communication Systems*. 1st ed. Horizon House.

Penfield, P., 1966. Unresolved Paradox in Circuit Theory. *Proceedings of the IEEE*, 54 (9), 1200–1201.

Pervozvanskiĭ, A.A.., 1965. *Random Processes in Nonlinear Control Systems*. Elsevier Science.

Pospieszalski, M.W., 2014. On Noise Properties of Transistors and Amplifiers a Critical Review. In: *2014 20th International Conference on Microwaves, Radar and Wireless Communications (MIKON)*. IEEE, 1–5.

Rice, F., 2016. A Frequency-Domain Derivation of Shot-Noise. *American Journal of Physics*, 84 (1), 44–51.

Rothe, H., and Dahlke, W., 1954. The Theory of Noisy Four-Poles. *Transactions of the IRE Professional Group on Electron Devices*, ED-1 (4), 258–259.

Stadler, A.W., Kolek, A., Mleczko, K., Zawiślak, Z., Dziedzic, A., and Nowak, D., 2015. Noise Properties of Thick-Film Conducting Lines for Integrated Inductors. *Metrology and Measurement Systems*, 22 (2), 229–240.

Tartarin, J.-G., Graffeuil, J., and Laurent, E., 2019. Low Frequency Noise Deviation from Schottky Theory in p-n Junctions. In: *25th International Conference on Noise and Fluctuations (ICNF 2019)*. ICNF.

Vasilescu, G., 2019. *Electronic Noise and Interfering Signals: Principles and Applications*. Scholars Portal.

Voinigescu, S.P., Maliepaard, M.C., Showell, J.L., Babcock, G.E., Marchesan, D., Schroter, M., Schvan, P., and Harame, D.L., 1997. A Scalable High-Frequency Noise Model for Bipolar Transistors with Application to Optimal Transistor Sizing for Low-Noise Amplifier Design. *IEEE Journal of Solid-State Circuits*, 32 (9), 1430–1439.

# 4 Integrated Passive Devices and Their Role in Low-Noise Amplifier Design

## PASSIVE COMPONENTS

Passive components form the structural backbone of every RF circuit. They appear in nearly every functional block: impedance-matching networks that govern power transfer, filters that implement selectivity, attenuators that stabilize gain, decoupling paths that isolate bias points, and resonant inductor-capacitor (LC) tanks that define oscillator frequency and phase noise (Soltanian and Kinget 2005, Lambrechts and Sinha 2017). These elements can often be modeled as idealized lumped values at audio or low-frequency operation, with parasitics essentially ignored.

The situation changes radically in the RF and mm-wave regime. Here, the physical implementation of a capacitor or inductor, its geometry, its relation to the substrate, and the conductivity of surrounding materials cannot be decoupled from its electrical behavior. A 10-nH inductor on paper may provide only a fraction of that inductance in practice once parasitic capacitances, skin effect, and substrate losses are included. What is more, these imperfections often dominate the system-level performance, imposing hard limits on achievable noise figure (NF), linearity, and efficiency. This is why in modern monolithic RF integrated circuits (RFICs), the passives rather than the transistors frequently define the performance ceiling.

A useful starting point in classifying passive components is the distinction between lumped and distributed implementations. Lumped elements are those that can be approximated as concentrated components of inductance, capacitance, or resistance, with dimensions much smaller than the operating wavelength. This approximation holds reasonably well at lower RF frequencies, but it begins to degrade as frequency increases and the physical size of interconnects, vias, and device geometries becomes a significant fraction of the wavelength. In contrast, distributed elements such as transmission lines, stubs, and coupled lines inherently rely on their geometric length relative to wavelength. Their impedance and resonance properties cannot be captured by a simple lumped *L*, *C*, or *R* but must be described by transmission line theory.

Historically, distributed circuit techniques became standard in microwave engineering during the 1940s and 1950s, when radar and satellite systems operated at frequencies where lumped approximations were no longer valid. In modern monolithic integration, the transition from lumped to distributed is increasingly encountered as CMOS and SiGe circuits extend into the 60–300 GHz range. Here, even short


DOI: 10.1201/9781003459385-4

interconnect segments can behave as resonant structures, and the "$\lambda/20$ rule" (where physical dimensions greater than one-twentieth of a wavelength should be treated as distributed) is often exceeded. Designers are thus forced to balance the compactness and tunability of lumped passives with the more predictable high-frequency performance of distributed implementations.

Despite decades of research, achieving high-$Q$ passives in standard silicon processes remains difficult. For example, integrated inductors in BiCMOS processes typically offer $Q \approx 10$ (Tilmans *et al.* 2003). Specialized process steps such as removing silicon beneath the inductor or adding insulating layers can boost $Q$ to approximately 20 or 30, but even this falls short of the higher than 30 $Q$ desired for modern RF frontends (Amakawa 2010).

As a result, RFIC designers often face trade-offs among three inductor implementations:

- Off-chip inductors can achieve high $Q$ but suffer from parasitic bond-wire inductances (> 0.5 nH, poorly controlled) that become prohibitive at mm-wave frequencies.
- Bond wires themselves as inductors, which offer surprisingly high $Q$ but lack precise control.
- On-chip spiral inductors, which are compact and integrable but intrinsically lossy.

Short transmission lines are commonly used as inductors at mm-wave frequencies (30–300 GHz), particularly in III-V technologies. For instance, 1.25 mm is needed for a 60-GHz quarter-wave ($\lambda/4$) impedance transformer (Gordon 2007, Božanić and Sinha 2018), which is manageable in compound semiconductors but excessive in CMOS at lower frequencies.

The following sections examine the two essential categories of lumped passives: resistors and capacitors, followed by inductors, transformers, and diodes. We emphasize their noise mechanisms, equivalent models, and RF design implications.

## INTEGRATED RESISTORS

Resistors are among the simplest and most familiar circuit elements, yet their role in integrated RF design is more subtle than often appreciated. In practice, they serve a broad range of functions: bias stabilization, current limiting, voltage division, feedback, and impedance control. In oscillators and amplifiers, they frequently provide the quiet biasing backbone on which the performance of the active devices depends. Despite this conceptual simplicity, integrated resistors are far from ideal. Their implementation in silicon processes introduces non-idealities that can significantly impact circuit linearity, noise, and stability. In many RFICs, the humble resistor can make or break a design just as much as the more celebrated transistor.

As circuits have grown, resistors have become increasingly important. Resistors are more important for precise biasing and noise shaping in advanced CMOS nodes, where transistor gain is lower, and voltage headroom is reduced. Furthermore, resistor design frequently becomes a limiting factor in achieving reproducible performance across process corners and operating conditions because passive accuracy does not improve with lithographic scaling.

## IMPLEMENTATION AND PROCESS OPTIONS

A range of structures is used to realize resistors in integrated circuits, each with trade-offs in area efficiency, accuracy, linearity, and parasitics:

- Diffused resistors: These resistors are formed by doping a small region of the silicon substrate using a process similar to that of integrated transistors. These are relatively easy and inexpensive to implement, using existing diffusion processes, but suffer from poor tolerance, strong substrate coupling, and pronounced voltage dependence – especially at high bias voltages. Junction leakage currents can become significant at high temperature or reverse bias and affect the performance of low-current circuits.
- Polysilicon resistors: In CMOS, these are created by doping polycrystalline silicon to achieve a desired, and well-controlled, resistance. Polysilicon resistors exhibit better isolation from the substrate compared to diffusion resistors; however, they introduce additional parasitic capacitances to the substrate and have a noticeable voltage (and temperature) dependence.
- Metal thin-film resistors: These resistors have a thin metal layer that is the resistive element, and they also have a non-conducting (insulating) body. They provide good linearity, lower parasitic coupling, and they can be trimmed, post-fabrication (e.g., laser trimming) for high-precision (low tolerance) applications. These steps introduce additional complexity and cost, seen as a trade-off for their high precision.

Absolute tolerance is often poor (on the order of 20%), but careful layout techniques allow matched resistors to achieve <0.1% variation across a die. This matching capability is far more valuable in RF design than absolute accuracy, since biasing and gain-setting networks depend on ratios rather than absolute resistance values.

Beyond intrinsic device choice, resistor precision is increasingly determined by layout. Common-centroid, interdigitated, and dummy-enclosed arrangements are routinely used to suppress gradients and edge effects. The trade-off is silicon area: achieving <0.1% matching can consume significant die space, leading designers to weigh layout economy against statistical performance.

## NONIDEALITIES AND NONLINEAR EFFECTS

At first glance, a resistor is expected to exhibit a simple, linear I-V characteristic. In practice, integrated resistors deviate from this ideal under high fields and elevated temperatures. Mechanisms such as carrier velocity saturation, depletion-layer pinching, and self-heating can introduce nonlinearity. In RF front ends, where linearity is often the bottleneck specification, such effects can upconvert into distortion products that degrade spurious-free dynamic range. Thus, a poorly chosen or modeled resistor can directly compromise system performance.

These nonlinearities are not merely second-order details: they set hard limits on linearity metrics such as IP3, often at levels comparable to or worse than the active devices themselves. This places resistors firmly in the signal path of RF performance rather than in the background.

Another often-overlooked effect is substrate coupling. Integrated resistors are not isolated components; they sit near other devices and interconnects. Capacitances to the substrate and leakage through junctions create parasitic pathways. These parasitics may appear small (tens of femtofarads), yet at gigahertz frequencies they provide low-impedance leakage routes that can inject noise into sensitive nodes.

## NOISE MECHANISMS

Resistors are also an unavoidable source of noise. Two fundamental mechanisms dominate:

1. Thermal noise: A white, frequency-independent noise proportional to *4kTR*, where *k* is Boltzmann's constant, *T* is the absolute temperature, and *R* is the resistance (Abbott *et al.* 1996).
2. Flicker ($^1/_f$) noise or excess noise: Arising from imperfections and inhomogeneities in the material. It dominates at low frequencies and scales with the square of the DC voltage across the resistor.

The experimentally established expression for the power spectral density of excess noise in a resistor *R* is:

$$\overline{v_{nex}^2} = \frac{CV_{DC}^2}{f} \tag{4.1}$$

where *C* depends on process and material quality, $V_{dc}$ is the DC voltage across the resistor, and *f* is the frequency. It is worth noting that excess noise, unlike thermal noise, does not scale down with device dimensions. In fact, as resistors are implemented with smaller geometries to save area, surface and grain-boundary effects can increase, making material and process selection even more critical in scaled nodes. This demonstrates how material uniformity and excess noise are closely related: metal-film and thin-film resistors perform significantly better because of their uniform microstructure, whereas carbon resistors have high excess noise. For a bandwidth of $\Delta f = f_2 - f_1$, the total noise from the resistor is equal to

$$\overline{v_n^2} = \int_{f_1}^{f_2} \frac{CV_{DC}^2}{f} df + 4kTR\left(f_2 - f_1\right) \tag{4.2}$$

which simplifies to

$$\overline{v_n^2} = \frac{CV_{DC}^2}{f} \ln\frac{f_2}{f_1} + 4kTR\left(f_2 - f_1\right) \tag{4.3}$$

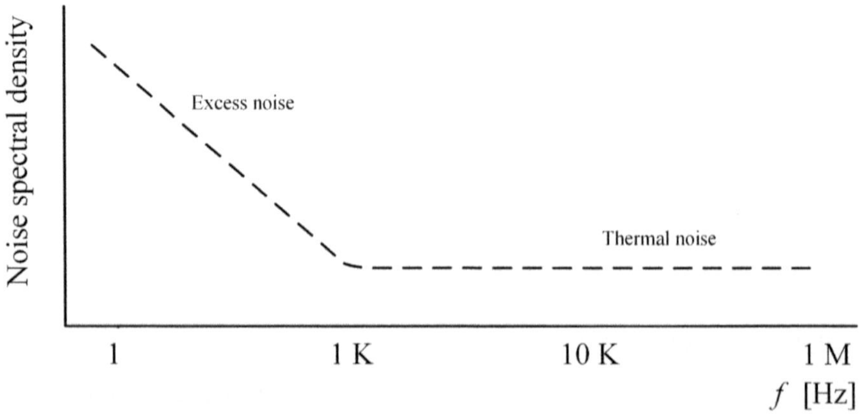


**FIGURE 4.1** Noise power spectral density of a resistor.

This decomposition shows how flicker noise dominates at low frequencies, while thermal noise sets a flat floor at higher frequencies. The interplay between the two determines the usable noise performance of the resistor in each band. The noise power spectral density of the resistor is plotted in Figure 4.1.

Alternatively, the excess noise in a resistor can be evaluated from the noise index (NI). The noise index is defined as the ratio of $v_{nex}$, the RMS value of the flicker noise in μV (Sepke *et al.* 2009), to the voltage DC drop across the resistor, $V_{DC}$, in a frequency band of one decade, such that

$$\overline{v_{nex}^2} = \frac{V_{DC}^2}{f} \frac{10^{[NI]_{dB}/10}}{\ln 10} \left[\mu V^2\right] \tag{4.4}$$

where NI is expressed in dB, $f$ is a frequency between $f_1$ and $f_2$, and $f_2 = 10 \times f_1$.

At RF, resistors cannot be treated as a pure lumped element. Instead, equivalent circuit models are required to capture the distributed parasitic capacitances and leakage paths that become significant at high frequencies. For a polysilicon resistor, a commonly used and simplified representation is the π-model shown in Figure 4.2.

Two tiny capacitances, $C_1$ and $C_2$, which link the resistor terminals to the substrate, are highlighted in this model. Although these capacitances are usually on the order of femtofarads in practice, their significance should not be understated. Because they provide a coupling path into the substrate, they can divert a portion of the signal energy away from the intended circuit nodes. Importantly, however, they do not substantially attenuate the thermal noise across the resistor itself. This might seem counterintuitive if one recalls that the total noise of an RC circuit is often expressed as *kT/C*. That result represents an upper bound, valid when the capacitor dominates across the entire frequency range. In the case of the π-model, capacitances $C_1$ and $C_2$ are small enough that they primarily serve as leakage paths rather than as dominant storage elements. The net effect is that the resistor's noise contribution is still largely

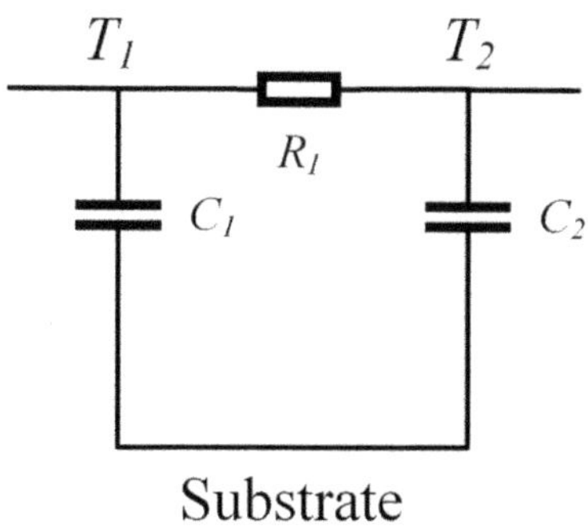


**FIGURE 4.2** Simplified equivalent circuit of a polysilicon resistor.

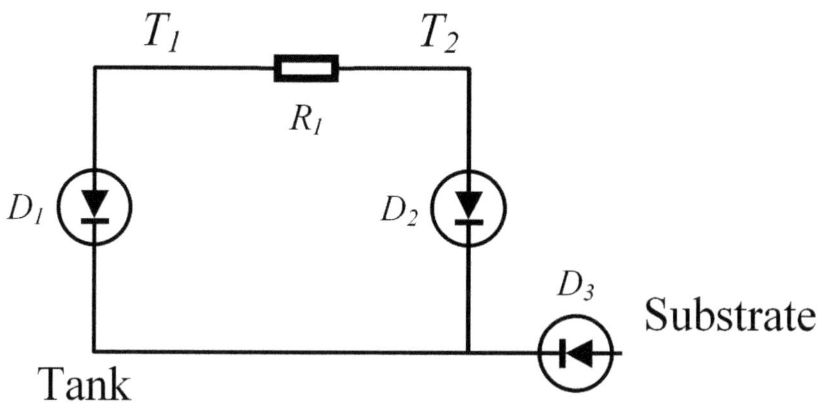


**FIGURE 4.3** Simplified equivalent circuit of a diffused resistor.

preserved at its terminals, while the parasitic capacitors act to inject some of this noise into the substrate.

A different situation arises when resistors are implemented as diffused structures. In this case, the resistor body is formed within a semiconductor region, and the resulting junctions to the substrate cannot be ignored. The simplified equivalent circuit for a diffused resistor, shown in Figure 4.3, therefore includes not only parasitic capacitances but also junction diodes $D_1$-$D_3$.

These parasitic diodes arise naturally from the p-n junctions between the diffused region and the surrounding substrate. They still contribute to leakage currents and junction capacitances that couple signals and noise into the substrate, even though they are typically reverse-biased in normal operation. The presence of these nonlinear elements also means the resistor may exhibit voltage-dependent behavior, particularly under large swings, further complicating RF modeling. The important observation is that, as with polysilicon resistors, the parasitic network surrounding a diffused resistor does not suppress the noise across $R_1$. Instead, the parasitics form secondary conduction paths, shunting energy and noise away from the circuit of interest and into the substrate.

Thus, in both cases, whether polysilicon or diffused, the designer is confronted with a similar reality: parasitic elements do not reduce the intrinsic noise of the resistor but instead determine where that noise flows. In practical RFIC design, this insight is critical, since substrate-injected noise can easily couple into sensitive nodes elsewhere in the chip, degrading overall system performance.

To account for these effects in a tractable way, the noise power density of monolithic and thin-film resistors is often expressed in the compact form

$$\overline{v_n^2} = \frac{CV_{DC}^2}{f}\Delta f + 4kTR\Delta f \tag{4.5}$$

where $\Delta f$ is the frequency band of interest. Flicker noise and thermal noise contributions are conveniently separated in this formulation, which also implicitly recognizes that parasitic components largely influence the distribution of noise rather than its magnitude. Even though (4.5) is seemingly straightforward, it has been very helpful in practice, giving designers a solid framework for anticipating resistor noise in the RF regime without using complete electromagnetic simulation.

## DESIGN IMPLICATIONS

The shortcomings of integrated resistors serve as a reminder that, despite being frequently regarded as incidental parts, they have a significant impact on circuit performance. The achievable linearity of an amplifier, the amount of phase noise in an oscillator, or the degradation of isolation through substrate coupling can all be caused by a poorly modeled resistor. On the other hand, considerable performance improvements can be obtained by carefully choosing the resistor type, geometry, and layout. For example, elongated geometries improve matching at the expense of area, while guard rings and well ties decrease substrate coupling. Similarly, when low excess noise is crucial, thin-film or high-resistivity polysilicon resistors might be worth the money. Finally, even though resistors are rarely the main feature of an RFIC design, they are silently incorporated into almost all bias networks and signal paths. Ignoring their nuances could compromise the functionality of the much more sophisticated active devices they support. Understanding the capabilities and limitations of resistors is still crucial for creating reliable and high-performing integrated circuits, just like it is for inductors and capacitors.

As RFICs migrate into mm-wave and 5G/6G bands (Sinha *et al.* 2025), resistor parasitics once considered negligible now sit inside the signal bandwidth. This trend is pushing designers toward either specialized thin-film options or careful electromagnetic extraction of resistor layouts. Far from a commodity device, the integrated resistor continues to demand careful modeling and creative layout to keep pace with modern RF performance requirements.

## INTEGRATED CAPACITORS

Integrated capacitors represent one of the most fundamental passive devices available to the circuit designer. Their role is multifaceted: they store energy in the form of an electrostatic field, act as coupling or bypass elements for AC signals, participate in filter networks, and determine the oscillation frequency of resonant circuits. As with resistors, their importance is often underestimated, yet in many high-frequency designs it is the capacitor (not the transistor) that ultimately dictates performance limits.

Despite their central role, integrated capacitors remain limited in achievable values because their physical realization on silicon consumes a relatively large chip area. The maximum capacitance that can be implemented on-chip seldom exceeds a few hundred picofarads, with values in the microfarad range almost always delegated to external discrete components. This area-disadvantage urges designers to use integrated capacitors sparingly, mainly in large-scale RFICs and SoCs, where semiconductor real estate is valuable.

## STRUCTURAL IMPLEMENTATIONS

The parallel-plate structure, which has two conductive electrodes spaced apart by a dielectric, is the most widely used implementation of an integrated capacitor. The process technology determines which dielectric is used; the most common options are composite oxide-nitride stacks, silicon dioxide ($SiO_2$), and silicon nitride ($Si_3N_4$).

A different implementation uses the depletion region of a reverse-biased p–n junction, yielding the so-called junction capacitor. Junction capacitors can achieve higher capacitance density but suffer from strong nonlinearity since the depletion width varies with the applied bias.

Circuits with substantial signal swing are distorted by this voltage dependence. Therefore, these capacitors work best in applications like RF tuning or ESD protection structures where linearity is not a crucial factor. Metal-insulator-metal (MIM) capacitors, which offer better linearity, more precision, and a comparatively higher density than previous oxide-polysilicon alternatives, are also made possible by modern CMOS processes. MIM capacitors frequently provide the best trade-off between electrical performance and area efficiency because they use stacked metal layers as plates, separated by a thin, premium dielectric.

## PRECISION AND MATCHING

Because of process variances and voltage-dependent effects, integrated capacitors typically have tolerances of ±20%, which is a modest level of absolute precision. Although this might seem less effective than discrete capacitors at first, the ratio of capacitors matters more in many analog and radio frequency circuits than their absolute value. Voltage dividers, switched-capacitor filters, and charge redistribution ADCs, for instance, depend on matched ratios rather than precise accuracy.

Layout strategies are also critical to accommodate proper matching. Techniques such as common-centroid placement, interdigitated structures, and dummy devices at array edges help suppress gradient-induced mismatch. In practice, well-designed capacitor arrays can achieve matching in the 0.1–0.2% range, sufficient for demanding analog applications.

## PARASITICS AND EQUIVALENT CIRCUITS

Every physical capacitor is accompanied by parasitic elements that influence its behavior. These parasitics are captured in the so-called noise equivalent circuit, shown in Figure 4.4.

In Figure 4.4, $R_s$ models the series resistance due to contacts and leads, while $R_p$ represents the dielectric leakage resistance. In larger-value film capacitors, parasitic inductance associated with leads or interconnects may also become significant, creating unwanted resonances. These parasitics are typically smaller but not negligible for integrated capacitors, particularly as frequencies approach the mm-wave regime.

If $R_s$ is neglected, the thermal noise contribution can be attributed to $R_p$. Since the product of $C$·and $R_p$ is usually very large, noise contributions only become visible at extremely low frequencies. For this reason, capacitors are often considered "noiseless" in the RF domain.

For integrated structures, where careful processing suppresses $R_s$ and $R_p$, the simplified equivalent circuit is shown in Figure 4.5.

The coupling of the bottom plate to the silicon substrate, denoted by $C_2$, is the predominant parasitic in this instance. This coupling can create signal leakage paths at GHz frequencies, even though it is frequently insignificant at low and

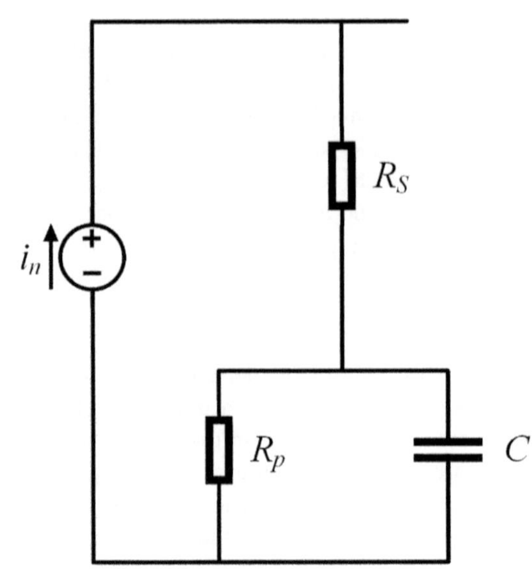


**FIGURE 4.4** . Noise equivalent circuit of a capacitor.

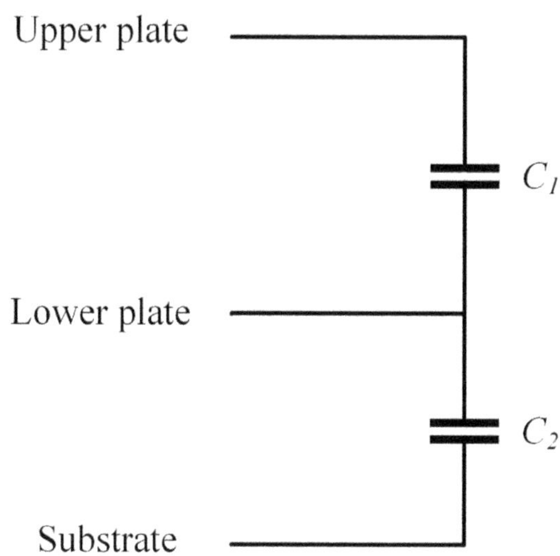


**FIGURE 4.5** Simplified equivalent circuit of an integrated capacitor.

mid- frequencies. Substrate coupling may introduce undesired noise or permit spurious oscillations in delicate circuits like VCOs, mixers, or LNAs. This problem can be lessened by employing strategies like guard rings, deep N-well isolation, or top-metal MIM capacitors.

## NONLINEARITY

Unlike ideal capacitors, physical integrated capacitors have voltage dependence. Junction capacitors are strongly nonlinear, but even MIM and thin-film capacitors display small, measurable voltage coefficients. This nonlinearity manifests as harmonic distortion, which can degrade performance in high-linearity RF front ends. Designers often reserve the best capacitors (MIM) (Wenger *et al.* 2006; Mu *et al.* 2018) for linear signal paths and relegate nonlinear types to biasing or decoupling roles where linearity is less critical.

## INTEGRATED CAPACITOR AREA IMPLICATIONS

To appreciate the design trade-offs inherent in integrated capacitors, consider the case of implementing a 1 pF capacitor in a modern CMOS process using a MIM structure. Assume a dielectric constant ($\varepsilon_r$) of 7 for the insulator, a thickness of 10 nm, and effective plate dimensions determined by the available top-metal layers.

The capacitance per unit area can be estimated as:

$$\frac{C}{A} = \frac{\varepsilon_r \varepsilon_0}{d} \tag{4.5}$$

which equates to roughly 6.2 fF/μm². Therefore, to achieve a capacitance of 1 pF, an active plate area of approximately

$$A = \frac{C}{C/A} \approx 161\,\mu m^2 \tag{4.6}$$

is required. Although this value might seem relatively attainable, if the capacitance is scaled to larger values, the area implications quickly become impractical. For example, a 100 pF capacitor would require about 16,000 μm², a significant section of chip area in any IC, especially at RF and mm-wave. This simple calculation illustrates why on-chip capacitors are rarely used for bulk energy storage and why careful budgeting of capacitance is crucial in analog and RF designs. Instead, designers reserve integrated capacitors for functions where precision, matching, and high-frequency operation matter more than absolute value.

## ROLE IN RF AND MIXED-SIGNAL CIRCUITS

Capacitors are indispensable in almost every RF block. In VCOs, they regulate the oscillation frequency and tuning range. In filters they are used to set the corner frequencies as well as the stop-band attenuation. In LNAs and mixers, they serve as AC coupling and bypass elements, controlling stability and gain. Even in digital systems, decoupling capacitors stabilize supply rails, preventing logic switching noise from corrupting sensitive analog nodes.

## INTEGRATED INDUCTORS

Inductors occupy a unique place in RF integrated circuit design. Unlike resistors and capacitors, which are relatively easy to implement with acceptable predictability, inductors are inherently large, lossy, and technologically constrained. Yet, despite these difficulties, they are indispensable in many of the most noise- and performance-critical blocks of a receiver. LNAs, oscillators, and filters all rely heavily on inductors to provide impedance matching, resonance, and bias stabilization. In LNAs, inductors are often the single most influential passive component: the input matching network that governs NF is almost always inductor-dominated, while output loads and degeneration structures also require carefully chosen inductance values. A low-quality inductor directly translates into degraded noise performance, reduced gain, or inferior bandwidth.

The importance of inductors in LNAs was recognized early in the transition from discrete RF circuits to monolithic RFICs. In discrete designs, designers could draw on a vast catalog of air-core and ferrite-core inductors with $Q$-factors exceeding 100, allowing near-ideal matching and resonance. But as RF design shifted into silicon, especially with the rise of CMOS RFICs in the late 1980s and early 1990s, it became apparent that integrated inductors would be the limiting element. While transistors steadily improved with lithographic scaling, offering higher cutoff frequencies and better noise performance, the integrated inductor remained stubbornly constrained by fundamental electromagnetic considerations. This mismatch between the pace of transistor scaling and the stagnation of passive performance has been described by many as the "Achilles' heel" of RF integration.

Historically, the feasibility of monolithically integrated inductors was first convincingly demonstrated in 1990, when Nguyen and Meyer reported spiral inductors fabricated in a 0.8-μm BiCMOS process (Niknejad and Meyer 1998). Their work showed that such inductors could indeed resonate with on-chip capacitors to form useful LC tanks, paving the way for RF CMOS as a practical technology. The use of square spirals became standard: easy to lay out, relatively compact, and compatible with standard back-end-of-line (BEOL) metallization. However, even at this early stage, the limitations were clear. Spiral inductors required a large die area and provided only modest inductance values, typically no more than a few tens of nanohenries. More critically, their $Q$ were severely limited, rarely exceeding 10 in standard CMOS processes. This was primarily due to conductor losses in the thin aluminum interconnects, capacitive coupling to the silicon substrate, and eddy currents induced in the substrate itself.

For LNAs, these nonidealities are not just inconvenient; they are fundamental performance bottlenecks. The input-referred NF of a narrowband LNA, for example, is highly sensitive to the $Q$ of the input matching inductor. A low-$Q$ device broadens the resonance, increases resistive loss, and introduces additional noise directly into the signal path. Similarly, inductive degeneration, often used to achieve simultaneous noise and input matching, becomes less effective when the inductor itself introduces significant series resistance. Thus, inductor performance often defines the ceiling of achievable LNA NF in each process.

Off-chip inductors offer a tempting alternative, and indeed they were widely used in early RFICs for input matching and load networks. Off-chip air-core coils or ceramic inductors can easily achieve $Q$-factors of 50 to 100, far surpassing their integrated counterparts. Bond wires themselves have also been exploited as inductors, with inductances in the range of 1–2 nH/mm and surprisingly high $Q$ (up to 60 at 2 GHz for a typical 1 mil wire). These "free" inductors, available simply by careful package design, played an important role in RFIC practice through the 1990s and remain attractive in some applications today. However, reliance on off-chip passives or uncontrolled package parasitics fundamentally undermines the promise of monolithic integration. As frequencies pushed into the tens of gigahertz, the distributed nature of bond wires, pads, and interconnects became too significant to ignore, and the variability from assembly processes made such solutions difficult to scale reliably.

At microwave and mm-wave frequencies, the challenges are magnified. As inductance values shrink into the nanohenry range, even small parasitics dominate performance. A bond wire may provide sufficient inductance for matching at 30 GHz, but its variability and poor modeling accuracy severely limit reproducibility. On-chip spiral inductors, meanwhile, suffer increasing substrate loss due to the reduced skin depth and higher dielectric loss at these frequencies. Designers have therefore often abandoned lumped inductors entirely in favor of distributed alternatives, short transmission lines, stubs, or coupled-line sections, which behave more predictably in the mm-wave regime. Still, even in this range, compact integrated inductors retain an important role in bias chokes, tuned resonators, and hybrid LC-transmission-line topologies.

The enduring challenge of integrated inductors is that their performance is bound by geometry and materials rather than lithographic scaling. While transistors benefit enormously from shrinking feature sizes, inductors do not: reducing the linewidth of a spiral conductor simply increases resistance and reduces $Q$, while shrinking its diameter reduces inductance. Due to this scaling discrepancy, inductors do not "scale for free" like active devices do. Rather, designers use system-level trade-offs (where feasible, replacing transmission lines or active inductors), process innovations (low-resistivity substrates, micro-electromechanical suspensions, or even magnetic integration), and meticulous layout optimization (wide metals, thick top layers, patterned ground shields).

This has various consequences for LNAs. Initially, designers need to carefully balance die area and performance: larger spirals use valuable silicon real estate but produce higher inductance and better $Q$. Second, achievable NFs are directly impacted by the limited $Q$ of on-chip inductors, which frequently necessitates trade-offs between bandwidth, linearity, and gain. Finally, because inductors are difficult to model accurately, especially when substrate interactions are strong, their presence increases the burden of electromagnetic simulation in the design flow.

In short, integrated inductors embody the tension at the heart of RFIC design: they are essential, but they resist integration. Their significance in LNAs cannot be overstated; whether as input matching elements, degeneration inductors, or resonant loads, they often determine the lower bound on noise performance. At the same time, their stubborn limitations force designers to innovate at both the circuit and process levels, making them one of the most intensely studied and creatively engineered passive components in RFIC history.

Conventional on-chip inductors are planar spiral devices implemented on a single metal layer. The equivalent circuit of a square spiral inductor is shown in Figure 4.6 (Ellinger 2008). A simplified equivalent circuit is given in (Yue *et al.* 1996).

In this model, the inductance value is $L_s$, the resistance of the spiral is $R_s$, and $C_s$ represents the equivalent capacitance between the spiral lines and the center-tap underpass. The spiral lines are coupled to the substrate through the oxide capacitance $C_{ox}$, while $R_{sub}$ and $C_{sub}$ represent the substrate resistance and capacitance, respectively.

In theory, substrate losses through eddy currents vanish when the substrate resistance is either zero or infinite. In these cases, either the voltage or the current is zero, leading to zero loss. Losses due to induced eddy currents in the substrate are not included in Figure 4.6, but for frequencies below 10 GHz and substrate resistivities near 10 Ω-cm, they are negligible (Niknejad and Meyer 2001).

## BASIC STRUCTURES

The most common implementation is the planar spiral inductor, realized using the top metal layers of a CMOS process. Square, octagonal, and circular layouts are

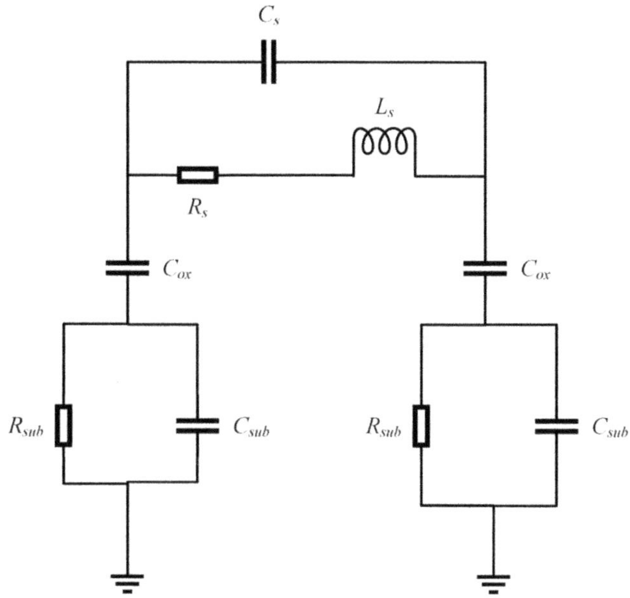


**FIGURE 4.6** Spiral inductor equivalent circuit.

variations that were selected to balance inductance density, $Q$-factor, and routing simplicity. The geometry, number of turns, spacing, conductor width, and overall area of such a structure are the main determinants of its inductance. Wheeler's formula provides a commonly used empirical approximation for a square spiral's inductance:

$$L_s \approx \frac{9\mu_0 n^2 d_{mean}^2}{11d_{out} - 7d_{mean}} \tag{4.7}$$

where $n$ is the number of turns, $d_{out}$ and $d_{in}$ are the outer and inner diameters of the spiral, respectively, and $d_{mean} = \sqrt{(d_{out} \times d_{in})}$. While approximate, this equation is sufficient for early-stage design estimates and gives insight into how layout dimensions influence inductance. The resistance is approximated by

$$R_s \approx \frac{1}{w\sigma\delta\left(1 - e^{-t/\delta}\right)} \tag{4.8}$$

where $w$ is the width of the spiral conductor, $\sigma$ is its conductivity, $\delta$ is its skin depth, and $t$ is the thickness of the spiral. $R_s$ is a function of the operating frequency, although the dependence of the skin depth on frequency is due to eddy current. The skin depth is given by $\delta=\sqrt{2/\omega\mu\sigma}$, where $\mu$ is the magnetic permeability of the conductor.

Although these equations can be used for first-order approximations, they can have errors of up to 25%. More accurate values can be attained using EM simulators, measurements, or in CAD tools such as ADS, SpectreRF, MW Office, or Genesys.

When the skin effect is neglected, and one terminal of the inductor is grounded, the quality factor can be approximated (Ellinger 2008) by

$$Q = \frac{\omega L_S}{R_S} k_{SL} k_{SR} \tag{4.9}$$

where $k_{SL}$ is the substrate loss factor, $k_{SL} \leq 1$, and $k_{SR}$ is the self-resonance loss factor with $k_{SR} \leq 1$.

Integrated inductors have a self-resonance frequency, $f_{SR}$, resulting from parasitic capacitance between the metal lines and the substrate. For consistent performance, the operating frequency must be kept sufficiently below $f_{SR}$. In group III–V semiconductors such as GaAs and InP, the substrate loss factor $k_{SL}$ is close to unity and significantly higher than in Si-based substrates. At the self-resonance frequency $f_{SR}$, the self-resonance factor reduces to zero, and practical design guidelines limit the maximum operating frequency to approximately 80% of $f_{SR}$.

At low frequencies, losses that degrade the quality factor $Q$ are dominated by the series resistance $R_s$. At higher frequencies, substrate coupling through $C_{ox}$ and $R_{sub}$ becomes the primary source of loss. In this regime, $Q$ can be improved by increasing the distance $d_{ms}$ between the spiral conductor and the substrate. Conversely, at lower frequencies, increasing the width and thickness of the spiral conductor reduces losses.

A fundamental trade-off exists between size and performance. Currents in the inner spiral flow in opposite directions, which reduces the net inductance. Increasing the size of the inner spiral alleviates this effect but results in a physically larger inductor for the same inductance value. Such inductors are termed hollow spiral inductors. Substrate losses can also be mitigated by introducing a shielding plate between the spiral and the substrate. While this reduces magnetic field penetration into the substrate, it induces counter-flowing eddy currents in the shield, which in turn reduce inductance. This problem is addressed using a patterned ground shield, illustrated in Figure 4.7 (Yue *et al.* 1996).

Copper metallization, MEMS-based structures, low-dielectric-constant interlayer dielectrics, and Si-on-insulator (SOI) substrates are further methods to improve inductor performance. To fabricate passive devices on a low-loss substrate without compromising the performance of active devices, a thin insulating layer is placed between the active components and the bulk substrate in SOI processes. Selective etching beneath the spiral in MEMS implementations widens the air gap, which lowers substrate coupling. Larger separation and air's high dielectric constant combine to produce high $Q$-factors that are on par with those obtained on III–V substrates. However, the high cost and mechanical instability of MEMS-based methods limit their use. Similar advantages are obtained by low-dielectric metal stack processes, which place low-$k$ dielectrics underneath the spiral conductor. Although copper metallization significantly lowers sheet resistance, problems like leakage or shorts through isolation layers arise because of its interaction with silicon.

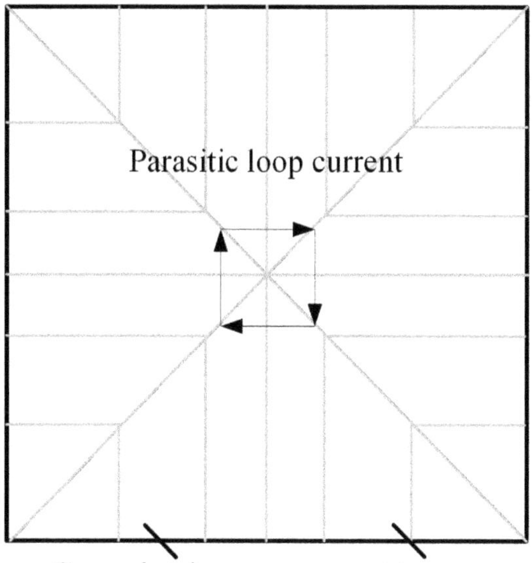


**FIGURE 4.7** Patterned ground shield.

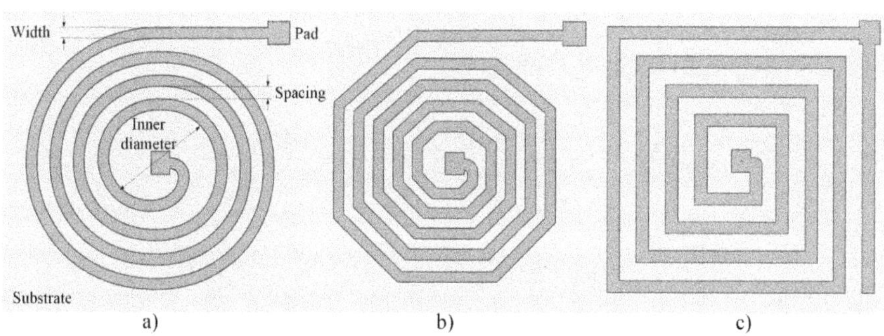


**FIGURE 4.8** Inductor geometries: (a) circular, (b) octagonal, and (c) stacked. (Adapted from Razavi 1997.)

By using geometries other than the traditional square spiral, inductor performance can also be customized. The most common layouts are shown in Figure 4.8 (Razavi 1997). Among these, circular geometries typically achieve the highest $Q$, followed by octagonal and then square spirals.

The intrinsic noise of an integrated inductor arises primarily from its series resistance $R_s$ and other parasitic losses. Additionally, inductors are susceptible to electromagnetic coupling, which introduces noise via external magnetic fields. A simplified equivalent circuit is shown in Figure 4.9, where $C_p$ represents parasitic capacitances and $R_p$ captures resistive and substrate losses.

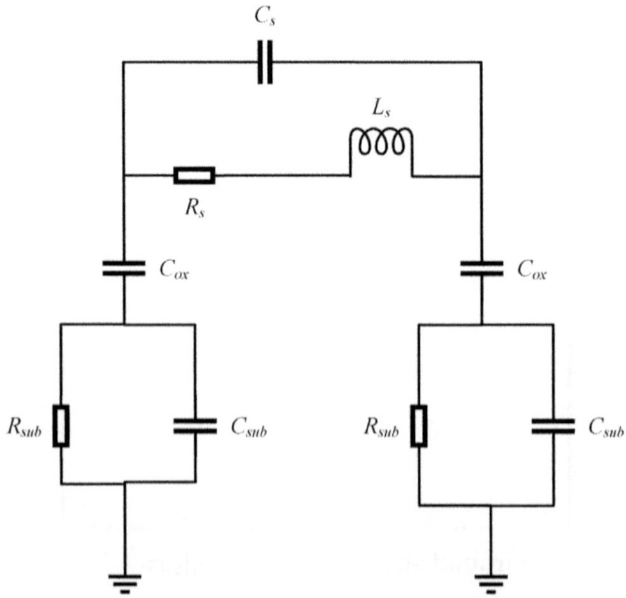


**FIGURE 4.9** Equivalent circuit of an integrated inductor.

The impedance of the circuit in Figure 4.10 is given by (4.10), where the numerator can be simplified when $R_p$ is small.

$$Z_{in}(s) = \frac{sL + R_p}{s^2 LC_p + sR_pC_p + 1} \cong \frac{sL}{s^2 LC_p + sR_pC_p + 1} \tag{4.10}$$

The $Q$-factor of Figure 4.9 is given by the equation (among other equivalent definitions):

$$Q = \frac{\omega L}{R_p} \tag{4.11}$$

where $\omega$ is the angular frequency with $\omega = 1\sqrt{LC_p}$.

To gain further insight, consider the alternative circuit of Figure 4.11, where the inductor and resistor are modeled in parallel rather than in series. The inductance value remains unchanged, and the objective is to determine the value of $R$ such that Figures 4.10 and 4.11 exhibit equivalent impedances.

The impedance of the circuit in Figure 4.10 is given by

$$Z'_{in}(s) = \frac{sL}{s^2 LC_p + sL / R + 1} \tag{4.12}$$

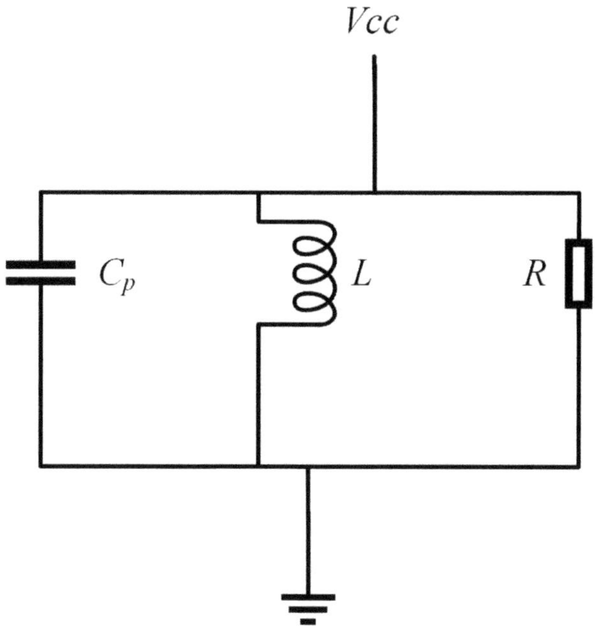


FIGURE 4.10 Equivalent circuit of an integrated inductor.

Both circuits have equal impedances when $R=L/C_pR_p$. After replacing, in the expression of $R$, $L$ by $QR_p/\omega$, then $1/\omega C_p$ by $L\omega$, and finally $L\omega$ by $QR_p$, the expression of $R$ is given by

$$R = Q^2 R_p \tag{4.13}$$

The circuit in Figure 4.10 is equivalent to a tank circuit, commonly used at the output of RF narrow-band amplifiers, for loading and biasing reasons. The thermal noise generated by this circuit is given by

$$\overline{v^2} = 4kTRe\left\{Z'_{in}(s)\right\}\Delta f \tag{4.14}$$

where $Re\{Z'in(s)\}$ is the real part of $Z'in(s)$. The real part of $Z'in(s)$ is equal to $R$ at resonance. Therefore, at resonance, (4.14) becomes

$$\overline{v^2} = 4kTQ^2 R_p \Delta f \tag{4.15}$$

so the equivalent current noise of the tank circuit at resonance is given by the expression (Leung 2011) $\overline{i^2} = 4kT\Delta f/Q^2R_p$, where $Q$ is the $Q$-factor of the inductor, and $R_p$ is the parasitic series resistance of the inductor. This expression shows that the current noise decreases as $Q$ increases (Leung 2011).

In many RF blocks, such as LNAs or oscillators, this noise contribution is not negligible. In fact, an inductor with poor $Q$ may degrade the overall NF as much as, or more than, the active device itself. Thus, inductors must be evaluated not only for their inductance value but also for their noise behavior.

## INTEGRATED TRANSFORMERS

Integrated transformers occupy a unique place among passive devices in RFICs. While transistors often receive most of the design attention, transformers can determine whether a circuit ultimately meets its performance goals, especially in highly integrated radios. Fundamentally, an integrated transformer is realized by arranging two or more inductors in close proximity so that a time-varying magnetic field in one winding induces a voltage in the other, consistent with Faraday's law of induction and Lenz's law. The most common implementation in CMOS is the interwound planar spiral transformer, but variations include vertically stacked windings (leveraging multiple metal layers), partially overlapping spirals, and, in specialized cases, bond-wire transformers that exploit the inductance of packaging interconnects. Historically, these structures emerged in the late 1990s and early 2000s as RF CMOS matured, when researchers realized that transformers could replace bulky off-chip components, enabling full transceiver integration.

The utility of on-chip transformers spans a remarkably wide range of RFIC building blocks. They serve as baluns for differential-to-single-ended conversion, as impedance transformers in matching networks, as coupling elements between circuit stages, and as feedback devices in oscillators. In mixers, they facilitate balanced drive and port-to-port isolation; in power amplifiers, they allow compact load transformation. Figure 4.11 shows a canonical planar transformer layout with interleaved, symmetric windings. The symmetry is not purely aesthetic; it maintains the common-mode rejection ratio (CMRR), suppresses even-order distortion, and guarantees balanced signal paths, properties that are crucial in differential circuits. Without such

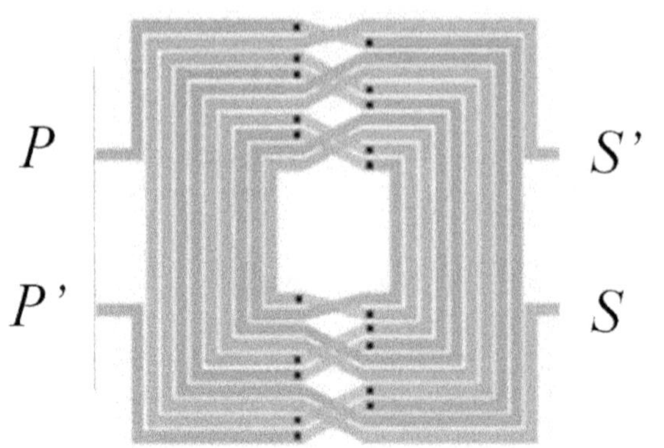


**FIGURE 4.11** Integrated transformer layout.

symmetry, mismatches degrade noise performance and allow unwanted coupling paths to dominate.

The geometry of the transformer dictates nearly all its RF behavior. In planar interleaved layouts, multiple vias connect the top metals, providing symmetry at the cost of added process steps and via resistance. Symmetry is especially prized in baluns and differential circuits, since any imbalance increases common-mode leakage and erodes linearity. In practice, achieving high symmetry requires compromises: routing complexity rises, resistance increases, and die area expands. Nonetheless, designers almost always prioritize symmetry because imbalance-related penalties far outweigh the cost in layout effort.

Planar transformers inevitably occupy significant die area, which is a critical penalty in cost-sensitive SoCs. To address this, researchers explored three-dimensional (3D) topologies, in which one winding is placed above the other across stacked metal layers. This approach increases magnetic flux linkage and coupling coefficient $k$, while reducing lateral footprint. However, the same vertical proximity that strengthens magnetic coupling also boosts parasitic capacitance, lowering the self-resonance frequency ($f_{SR}$). As a result, many 3D transformers exhibit higher coupling but narrower bandwidth, an unfavourable trade-off in wideband or mm-wave systems.

Implementing 3D transformers also strains process capabilities. Accurate control of via dimensions, inter-layer spacing, and dielectric properties is mandatory, and many mainstream CMOS nodes are not optimized for such features. Industrial practice, therefore, tends to favor planar interleaved structures: although less compact, they offer predictable performance, lower process risk, and better reproducibility across wafers. In commercial designs, manufacturability often outweighs advantages and improvements in theoretical performance.

When integrated on silicon, transformers inherit many of the same efficiency challenges seen in on-chip inductors, with a few additional complications unique to their two-port nature. Loss mechanisms are well understood and usually fall into four categories. First, there is the unavoidable series resistance of the metal tracks. Even though modern CMOS processes provide thick top-metal layers, the finite conductivity of aluminum or copper, and the skin effect at RF, ensures that Joule heating remains a nonnegligible contributor to loss. Second, energy is dissipated into the silicon substrate itself. Magnetic fields penetrate the substrate, where they induce eddy currents in what is essentially a lossy medium. Special measures, such as using high-resistivity wafers, silicon-on-insulator (SOI), or deep trench isolation, are often necessary to suppress this mechanism. Third, parasitic capacitances come into play. Coupling between adjacent turns, between different windings, and between the windings and the substrate provides leakage paths for high-frequency signals. These effects both lower the quality factor and bring down the self-resonance frequency of the device. Finally, there is the matter of eddy currents in ground shields. While patterned ground shields are frequently used to reduce substrate coupling, they can inadvertently form closed loops that oppose the transformer's magnetic field. The result is a reduction in effective inductance.

Unlike simple inductors, transformers must be treated as genuine two-port devices. Their description requires more than a single quality factor; instead, a richer set of parameters is used to capture performance. The familiar quality factor still applies, but now each winding has its own $Q$, and mutual coupling between windings complicates the picture further. The coupling factor $k$, which in standard CMOS processes tends to lie between 0.5 and 0.85, is a central measure of how efficiently energy is transferred. The self-resonance frequency $f_{SR}$ sets the upper limit of usable operation: beyond this point, parasitic capacitances dominate, and the device ceases to behave inductively. In practice, designs are kept comfortably below this limit, typically no more than about 70–80% of $f_{SR}$.

Another important quantity is the mutual inductance $M$. This tells us how strongly the magnetic flux from the primary winding links to the secondary one. Increasing overlap and reducing spacing between windings generally improve $M$, but at the expense of higher capacitance, which pulls down the resonance frequency. Therefore, balancing these trade-offs is a constant in transformer design. Furthermore, the power gain $G$ becomes important, particularly in applications where insertion loss is crucial, such as baluns or impedance-matching networks. The device's effective impedance level is determined by the transformer characteristic resistance (TCR), which is frequently used to inform circuit-level design choices. Lastly, the efficiency of transformers in terms of silicon area is also examined because they are one of the largest passive structures in RFICs. Metrics such as inductance per unit area or coupling factor per unit area are often reported to benchmark designs.

Fundamentally, designing an integrated transformer involves considering various trade-offs. Each improvement typically comes at a cost elsewhere in the system, and the target frequency band, process technology, and intended circuit application all influence these trade-offs. The link between coupling and self-resonance serves as an illustration. Windings must be placed closer together to achieve a stronger magnetic coupling, but this invariably lowers the self-resonance frequency and increases parasitic capacitances. In a similar vein, pursuing a higher quality factor typically results in wider traces, which lowers resistance but increases inter-winding capacitance and die area consumption.

Another area where compromise is inevitable is layout symmetry. Balance is improved by symmetry but frequently requires longer interconnects and higher series resistance. Even choosing between 3D and planar structures requires carefully accommodating advantages and disadvantages. Stronger coupling and smaller footprints are promised by 3D integration, but there are additional capacitive limitations and fabrication complexity.

These decisions are not arbitrary. The application determines the transformer topology in actual circuit design. When implemented in bulk CMOS, SOI, or InP, a device designed for mm-wave 5G systems will look very different from one operating at a few gigahertz, like in WLAN. The final architecture is driven by requirements like wideband matching, low noise, or a small footprint. Transformers are essential components of RF front ends; they are not just optional extras. They look like baluns, and their inherent 180° phase shift allows for differential-to-single-ended conversion without the additional noise penalty of active circuitry. To maximize power transfer, they are employed in impedance transformation, where the turns ratio determines the effective source and load impedances. Transformers play a key role in mixers because they provide balanced drive and isolation, two things that are essential for linearity. Transformer-based feedback networks allow for differential outputs and low-phase-noise LC designs in oscillators. The galvanic isolation they offer is crucial for stopping the propagation of DC offsets between stages, even in basic coupling networks.

At higher frequencies, the useful benefits of transformers become particularly evident. Transmission-line baluns take up far too much chip space, and lumped LC baluns become too lossy in the mm-wave range, say 60 GHz and above. Transformer-based baluns predominate in these situations. Transformers are also essential in power amplifiers. They allow the integration of the entire RF transceiver without the need for external matching networks by enabling load-pull impedance matching directly on-chip.

To minimize substrate losses, research groups are experimenting with MEMS-assisted structures, which use suspended layers or micromachining. To increase inductance density, others are investigating thin-film magnetic materials; however, issues with magnetic loss and compatibility with CMOS processes are still unresolved. To relieve pressure on the silicon area, a parallel line of development is moving transformers partially out of the die itself and into sophisticated packaging substrates like system-in-package (SiP) or fan-out wafer-level packaging. With machine learning algorithms optimizing transformer layouts for a desired balance of $Q$, coupling factor, and resonance frequency, even AI-based tools have started to emerge in more recent times.

After discussing transformers' basic behavior and the trade-offs that come with their design, it is beneficial to take a closer look at the secondary parameters that are useful in actual circuits. In addition to helping characterize performance, metrics like the $Q$-factor, self-resonance frequency, coupling coefficient, TCR, power gain, and NF are also significant figures of merit in optimization. In addition to these, simplified equivalent-circuit models are still frequently used in the literature because they offer a simple means of connecting circuit-level results with physical design decisions.

## THE INDUCTOR AND TRANSFORMER $Q$-FACTOR

The quality factor, or $Q$-factor, is a cornerstone parameter in the characterization of integrated passive components. While its definition originates in resonant LC circuits, it plays an equally critical role in describing both inductors and transformers, where it quantifies the balance between stored energy and dissipated energy per oscillation cycle. In integrated transformers, each winding, primary and secondary, exhibits its own $Q$-factor, but the coupled nature of the device means that these individual quantities cannot be treated in isolation. Instead, the effective transformer $Q$ must be interpreted in terms of coupled circuits, where losses in either winding or in the substrate manifest jointly in the overall performance.

Formally, using network parameters, the $Q$-factor of a single winding can be expressed as:

$$Q = \frac{Im\left(\frac{1}{Y_{11}}\right)}{Re\left(\frac{1}{Y_{11}}\right)} = -\frac{Im(Y_{11})}{Re(Y_{11})} \tag{4.16}$$

where $Y_{11}$ is the input admittance of the winding under consideration. This expression captures the ratio of reactive to dissipative energy, thus, how much energy is (temporarily) stored in the magnetic field relative to how much is dissipated per oscillation. A high $Q$-factor reflects an efficient storage of energy with minimal losses, whereas a low $Q$-factor indicates that ohmic resistance, substrate effects, or parasitic radiation govern component performance.

In integrated transformers, several distinct mechanisms conspire to limit the achievable $Q$-factor. The first is ohmic resistance in the metal traces, which introduces unavoidable Joule heating. At radio and mm-wave frequencies, this resistance is exacerbated by the skin effect, which confines current to a thin surface layer of the conductor, effectively increasing the resistance with the square root of frequency. Using wider or thicker conductors can reduce ohmic losses, but such strategies also increase parasitic capacitance, thereby lowering the self-resonant frequency. A second and equally important source of loss is substrate coupling. Time-varying magnetic fields can penetrate the silicon substrate, where they induce eddy currents that dissipate energy. Standard CMOS substrates are particularly problematic because of their conductivity, but the use of high-resistivity silicon, silicon-on-insulator, or GaAs platforms can suppress such losses significantly. Micromachining approaches, where substrate material is etched away beneath the transformer, are also frequently reported as a way of reducing eddy current formation.

Another major factor is capacitive coupling, both between windings and between each winding and the substrate. These parasitic capacitances form unintended leakage paths, especially at high frequencies, and effectively short-circuit the reactive energy that should otherwise be stored in the magnetic field. The resulting reduction in both $Q$-factor and self-resonant frequency is a recurring limitation in integrated transformer design. To counteract capacitive coupling to the substrate, patterned ground shields are often introduced. While they can be effective, they must be carefully optimized, since continuous or poorly segmented shields may themselves support eddy currents, thereby introducing additional magnetic loss channels and further lowering $Q$.

Transformers present further complications compared to single inductors. Because they consist of at least two coupled windings, their $Q$-factors are interdependent. A poor $Q$ in one winding inevitably drags down the performance of the other, and the effective $Q$ of the transformer reflects this combined behavior. Asymmetry between windings, such as differences in width, length, or layout, can lead to unequal $Q$-factors, which in turn degrade balance and increase common-mode error in transformer applications such as baluns. Direct parasitic capacitance between the primary and secondary windings does more than just shift resonance; it also opens cross-coupled loss paths that are unique to transformers. These additional channels blur the picture of what the quality factor really represents, making its interpretation less straightforward than in the case of a simple inductor.

Because of this, extracting a reliable $Q$-value is not trivial. One common workaround is to measure a single winding in isolation by grounding the opposite terminal, then deriving the parameters from the input admittance. A more complete view comes from treating the device as a two-port network and defining $Q$ in terms of $S$-parameters or $Y$-parameters, which naturally fold in the effects of coupling and loading. In all cases, the $Q$-factor is strongly frequency-dependent. It generally rises with frequency, peaks just below the self-resonant point, and then collapses rapidly as parasitic effects dominate. For this reason, reported $Q$ values must always be contextualized by the operating frequency.

In terms of performance levels, transformers typically exhibit lower $Q$-factors than stand-alone inductors because of the compounded parasitic pathways. In CMOS processes at frequencies between 2 and 5 GHz, single-winding $Q$-factors in the range of 8–15 are common, whereas effective transformer $Q$ values are often between 4 and 10. At mm-wave frequencies above 30 GHz, $Q$-factors tend to fall below 5 as both skin effect and capacitive coupling intensify. More advanced platforms, such as SOI, GaAs, or InP, consistently outperform bulk CMOS and in some cases achieve single-winding $Q$-factors exceeding 20.

A variety of strategies are employed to push these values higher. Foundries now provide ultra-thick top-metal layers to reduce ohmic losses, while micromachining and substrate trenching reduce effective permittivity and conductivity beneath the windings. Optimized patterned ground shields have proven useful in suppressing substrate coupling when carefully designed to prevent closed current loops. Finally, geometric optimization, such as adjusting the spacing between windings, can lower capacitive coupling, although this comes at the expense of reduced coupling factor and must be balanced against system-level requirements.

In summary, the $Q$-factor of integrated transformers should not be viewed as a fixed property of the device but rather as the outcome of a complex set of trade-offs among conductor quality, substrate characteristics, parasitic couplings, and operating frequency. Unlike inductors, where $Q$ is often defined unambiguously and improved by incremental layout adjustments, transformers involve inherently coupled loss mechanisms that make their $Q$ particularly sensitive to design and process decisions. As such, achieving a high $Q$-factor requires not only careful layout but also consideration of materials, process technology, and electromagnetic environment. Continued advances in process features such as thick metals, SOI substrates, and micromachining remain some of the most effective levers for enhancing transformer $Q$ while maintaining compact integration.

## THE SELF-RESONANCE FREQUENCY ($f_{SR}$)

Another central parameter in the characterization of integrated transformers is the self-resonance frequency ($f_{SR}$) (Lafferty 1982). Unlike the idealized case of an inductor that stores energy purely in its magnetic field, every practical winding is accompanied by a network of parasitic capacitances. These capacitances arise between adjacent turns of the same coil, across overlapping conductors in multi-layer structures, and between the windings and the underlying substrate. The result is that each winding behaves not as a simple inductive element, but rather as an LC resonator with a finite resonance frequency. At frequencies approaching or exceeding this self-resonance point, the inductive reactance no longer dominates; instead, the device transitions to capacitive behavior. This transition renders the winding unsuitable for most RF applications, since impedance transformation, energy transfer, and magnetic coupling all rely on the assumption of inductive operation.

When working with integrated transformers, keeping the self-resonance frequency ($f_{SR}$) high enough is not a luxury; it's a necessity. Both windings need to remain comfortably in the inductive region across the band of interest; otherwise, the device starts to misbehave. If resonance creeps too close to the operating frequency, the apparent inductance falls away, the coupling weakens, and insertion loss climbs far more quickly than one would like. For this reason, designers usually aim to push $f_{SR}$ at least two or three times higher than the frequency of operation. A transformer intended for 2.4 GHz, for example, should ideally resonate above 6 or 7 GHz to leave room for parasitics and maintain a reliable margin.

The factors that limit $f_{SR}$ are rooted in layout and process. The resonance frequency is lowered by parasitic capacitance between windings, which is increased by tight conductor spacing, wide overlap regions, or turns that are too closely spaced. By introducing more capacitive paths and performance-degrading loss channels, the silicon substrate itself may exacerbate the situation. This is why layout choices matter so much. Wider spacing between turns, reduced overlap, or even cutting back the number of turns altogether can all push $f_{SR}$ higher, although these same changes inevitably reduce coupling and inductance density. In practice, it becomes an exercise in balancing priorities at the circuit level rather than chasing a single metric.

Process technology opens further levers. A thicker oxide between metal layers reduces the capacitance of overlapping conductors, which helps lift $f_{SR}$ without heavily penalizing inductance. Using high-resistivity silicon or bypassing it entirely with insulating substrates like SOI or glass can also prevent the bulk from dragging resonance down. More advanced approaches, such as trench isolation or micromachining under the winding area, physically strip out substrate volume and with it the parasitic capacitance and eddy currents that hurt performance. With the rise of 3D integration, it is now possible to place windings on different vertical planes, reducing overlap while maintaining a compact footprint.

It is worth emphasizing that $f_{SR}$ is not just an academic number buried in the datasheet. In many ways, it sets the practical ceiling on how useful a transformer can be. Once the operation pushes too close to resonance, everything else, $Q$, coupling, and insertion loss, starts to roll off in a way that can give a misleading picture if not interpreted carefully. Reported $Q$-factors or coupling coefficients measured at or near $f_{SR}$ may look attractive on paper, but are unlikely to hold up in circuit conditions. A solid transformer design, therefore, is one that not only maximizes $Q$ in the band of interest but also leaves resonance comfortably out of the way, ensuring predictable and repeatable behavior across real-world conditions.

In summary, the self-resonance frequency captures the inevitable consequence of parasitic capacitances in integrated windings and defines the upper limit of useful inductive behavior. Its value is determined jointly by layout geometry, conductor spacing, substrate properties, and process features such as oxide thickness and metal stacking. Maintaining a sufficiently high $f_{SR}$ requires a careful balance between minimizing capacitance and preserving adequate coupling, inductance, and density. As technology nodes advance and operating frequencies move deeper into the mm-wave regime, strategies such as thicker dielectrics, high-resistivity substrates, micromachining, and 3D coil geometries remain essential tools for extending the practical frequency range of integrated transformers.

## TRANSFORMER COUPLING COEFFICIENT

The degree to which magnetic flux generated in the primary winding is effectively connected to the secondary winding is measured by the transformer coupling coefficient, denoted as $k$. This parameter (Deng *et al.* 2025) is directly related to the mutual inductance $M$ according to the relation

$$k = \frac{M}{\sqrt{L_p L_s}} \tag{4.17}$$

where $L_p$ and $L_s$ are the self-inductances of the primary and secondary windings, respectively. A value of $k = 1$ indicates perfect coupling, corresponding to an ideal transformer in which all magnetic flux generated by the primary links to the secondary. In practice, however, $k$ is typically much lower, ranging between 0.5 and 0.8 for well-optimized on-chip transformers.

Because it directly controls bandwidth, insertion loss, and impedance transformation, the coupling coefficient is one of the characteristics that make an integrated transformer unique. Less energy moves from the primary to the secondary when there is weak coupling, which manifests as worse matching and higher insertion loss. Stronger coupling does the opposite: it improves transfer efficiency, boosts gain in tuned circuits, and extends the usable bandwidth.

Reaching high coupling, however, is not without its costs. The usual methods, such as tight spacing between coils, interleaving turns, or overlapping windings across metal layers, strengthen the magnetic link, but at the same time, they increase parasitic capacitance between the windings. The extra capacitance pulls the self-resonance frequency downward, which can undercut high-frequency operation. As a result, transformer design almost always comes down to a trade-off: pushing the geometry to maximize magnetic coupling while holding the capacitive side effects in check.

Designers use a variety of tricks to navigate this balance. Differential layouts, patterned ground shields, and careful distribution of turns can all help to ease the capacitive penalty without giving up too much coupling. Process technology matters as well. A thicker dielectric between metal levels, or the use of high-resistivity substrates, can reduce capacitive loading enough to allow tighter coupling while keeping self-resonance at a safe distance. In this sense, the coupling coefficient acts as both a measure of transformer efficiency and a practical guidepost for layout, since it reflects the interplay between the magnetic and electric fields that coexist in the integrated environment.

## TRANSFORMER CHARACTERISTIC RESISTANCE

Traditional metrics such as insertion loss or maximum available gain are useful but assume that both source and load are perfectly matched, which is rarely the practical case. Tuned circuits, oscillators, LNAs, and narrowband filters often operate under impedance conditions that depart significantly from this assumption (Carrara *et al.* 2006). In these situations, the TCR becomes a more meaningful figure of merit. TCR ties directly to how the transformer contributes to circuit gain and overall performance. In a tuned configuration, the primary winding is usually placed in parallel with a capacitor so that its reactive part is canceled, making the resistive behavior, as captured by the TCR, the dominant factor.

In a tuned circuit configuration, the primary winding is typically connected in parallel with a capacitor, canceling its reactive component. The resulting tank circuit has an effective reactance given by

$$X_{opt} = -\omega L_1 \left(1 + \frac{1}{Q_1^2} + k^2 \frac{Q_2}{Q_1}\right) \tag{4.18}$$

where $L_1$ is the inductance of the primary turns, $Q_1$ and $Q_2$ are the quality factors of the primary and secondary turns, respectively.

The TCR arises naturally when analyzing a transformer embedded within a resonant tank. Typically, the primary winding is placed in parallel with a capacitor, canceling its reactive component and forming a parallel LC resonance. The equivalent impedance of this tank circuit can be decomposed into a reactive term and a resistive term. The resistive part, directly linked to energy transfer efficiency, is quantified through the TCR, which is defined as

$$TCR = \omega Q_{EQ} L_{EQ} \tag{4.19}$$

where $Q_{EQ}$ is the equivalent quality factor and $L_{EQ}$ is the equivalent inductance, given by

$$Q_{EQ} = Q_1 \frac{k^2 Q_1 Q_2}{1 + k^2 Q_1 Q_2} \tag{4.20}$$

$$L_{EQ} = L_1\left(1+\frac{1}{Q_1^2}+k^2\frac{Q_2}{Q_1}\right) \tag{4.21}$$

Maximizing the TCR directly enhances the gain of the tuned transformer. The power loss factor $A$, defined as the ratio of input to output power, approaches a limiting value of approximately two under optimized conditions:

$$A = 2\frac{k^2Q_1Q_2+2}{k^2Q_1Q_2} \approx 2 \tag{4.22}$$

From a design standpoint, pushing the TCR higher is effectively the same as pushing for more gain in a tuned configuration. As the effective quality factor improves and the inductance is tuned properly, the resistive part of the tank's impedance grows, which lifts voltage gain and trims down power loss.

This link between TCR and efficiency can also be described through the power loss factor, $A$, which, under ideal conditions, approaches a limiting value of about two. That limit serves as a reminder: losses can be reduced, but never entirely removed, because the balance among $Q$, coupling coefficient, and parasitics always imposes constraints.

The real value of TCR shows up in narrowband RF circuits, where gain and selectivity matter more than wideband efficiency. In oscillators, a higher TCR eases startup and helps stabilize the oscillation amplitude. In bandpass filters, it improves insertion gain while sharpening selectivity. Unlike broadband metrics that spread attention across the whole spectrum, TCR is tightly focused on the operating band, making it the more practical figure of merit for tuned resonant systems.

In the end, TCR provides the missing link between device-level parameters, like $Q$ and coupling, and circuit-level goals such as gain, selectivity, and stability. It packages the underlying physics of inductance, quality factor, and coupling into a single number that maps directly onto system performance. As such, TCR remains an indispensable design consideration whenever integrated transformers are deployed in resonant topologies.

## MAXIMUM AVAILABLE POWER GAIN ($G_{max}$)

Because transformers are lossy devices, they can be characterized by the power gain $G$, which is equal to the ratio of the power at the output to the power at the input. Because this ratio is dependent on termination impedances, the maximum available gain $G_{max}$ is more favorable for measuring the efficiency of a monolithic transformer. $G_{max}$ measures the gain when the source and the load impedances are conjugately matched to $Z_{11}$ and $Z_{22}$, the input and output impedances of the transformer, respectively. The maximum available power gain is given by

$$G_{max} = 1+2\left(x-\sqrt{x^2-x}\right) \tag{4.23}$$

where $x$ is given by (4.24), which is based on the $Z$-parameters of the transformer viewed as a two-port network.

$$x = \frac{Re(Z_{11})Re(Z_{22}) - |Re(Z_{12})|^2}{|Im(Z_{12})|^2 + |Re(Z_{12})|^2} \tag{4.24a}$$

$G_{max}$ can also be expressed in terms of $S$-parameters, as follows [20]:

$$G_{max} = \left|\frac{S_{21}}{S_{12}}\right|\left(k - \sqrt{k^2 - 1}\right) \tag{4.24b}$$

$$k = \frac{1 - |S_{11}|^2 - |S_{22}|^2 + |\Delta|^2}{2|S_{12}S_{21}|} \tag{4.24c}$$

$$\Delta = S_{11}S_{22} - S_{12}S_{21} \tag{4.24d}$$

These formulations highlight the dependence of gain on both intrinsic transformer properties (such as coupling coefficient and $Q$-factor) and extrinsic network interactions (such as impedance matching). In practice, designers must evaluate $G_{max}$ alongside parameters such as $f_{SR}$ and $TCR$ to ensure the transformer delivers acceptable performance across the intended frequency range and application.

## THE SPIRAL TRANSFORMER MODEL

Let us consider the layout of the spiral transformer in Figure 4.12. Each pair of parallel coupled microstrip line segments can be modelled by a circuit like Figure 4.6, and the pairs are connected in series to complete the transformer model (Yue *et al.* 1996), as shown in Figure 4.6.

Several simplifications of this model are done in Yo-Sheng Lin *et al.* (2005) and presented in Figure 4.13. The values in Figure 4.13 are given below and derived in Yo-Sheng Lin *et al.* (2005).

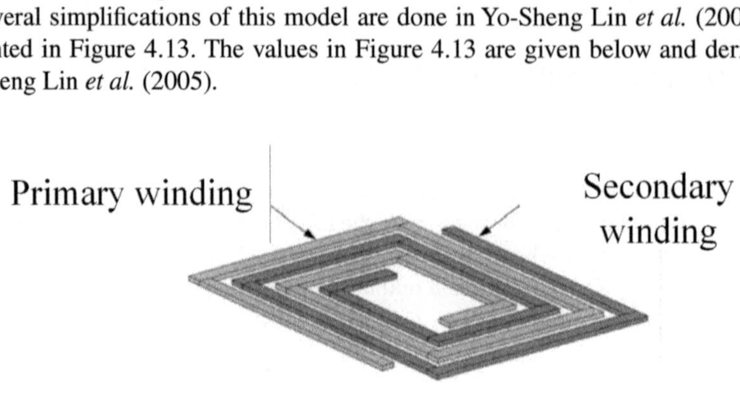


**FIGURE 4.12** Spiral transformer layout.

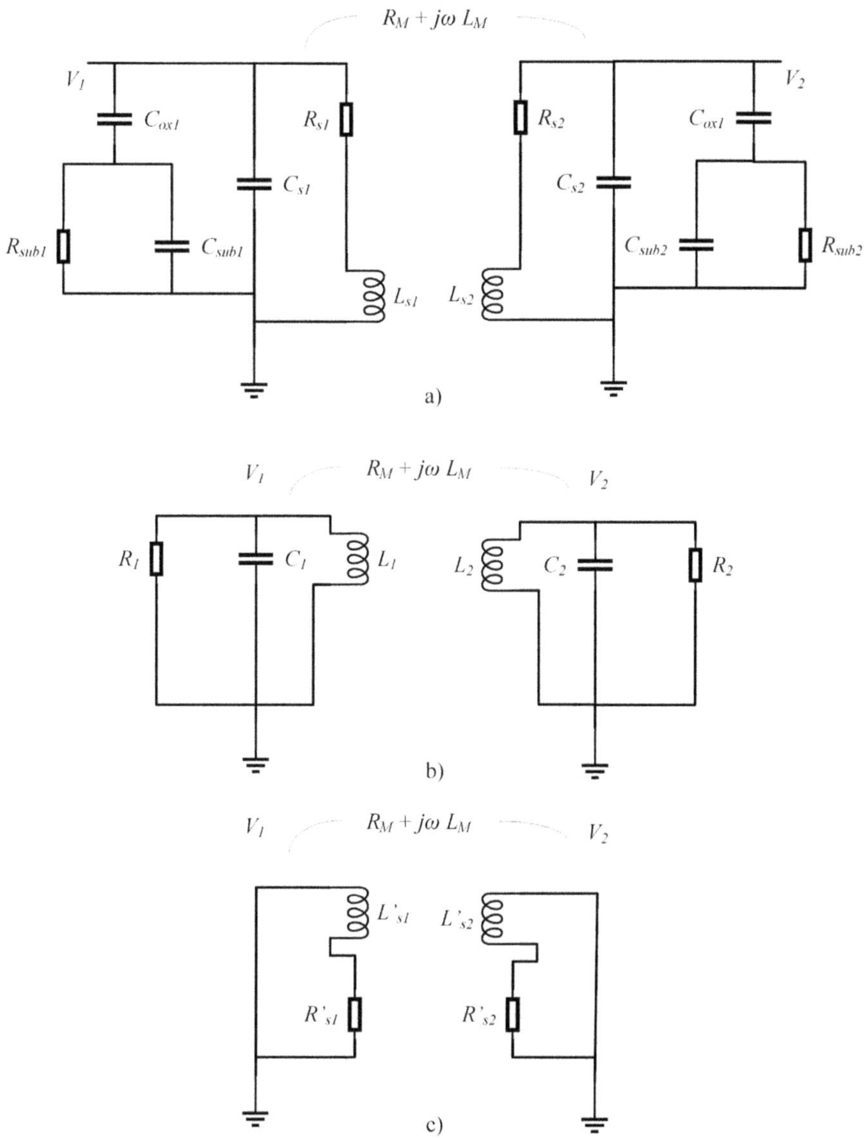


**FIGURE 4.13** Simplified spiral transformer models.

$$R'_{s1} = \frac{\dfrac{(\omega L_1)^2}{R_1}}{(1-\omega^2 L_1 C_1)^2 + \dfrac{(\omega L_1)^2}{R_1}} \tag{4.25}$$

$$L'_{s1} = \frac{L_1\left(1-\omega^2 L_1 C_1\right)}{\left(1-\omega^2 L_1 C_1\right)^2 + \dfrac{\left(\omega L_1\right)^2}{R_1}} \tag{4.26}$$

$$R_1 = \left(\frac{1}{R'_{sub1}} + \frac{R_{s1}}{\left(\omega L_{s1}\right)^2 + R_{s1}^2}\right)^{-1} \tag{4.27}$$

$$C_1 = C_{s1} + C'_{sub1} \tag{4.28}$$

$$L_1 = \frac{R_{s1}^2 + \omega^2 L_{s1}^2}{\omega^2 L_{s1}} \tag{4.29}$$

$$R'_{sub1} = \frac{1}{\omega^2 C_{ox1}^2 R_{sub1}} + \frac{R_{sub1}\left(C_{ox1} + C_{sub1}\right)^2}{C_{ox1}^2} \tag{4.30}$$

$$C'_{sub1} = C_{ox1}\frac{1+\omega^2\left(C_{ox1}+C_{sub1}\right)C_{sub1}R_{sub1}^2}{1+\omega^2\left(C_{ox1}+C_{sub1}\right)^2 R_{sub1}^2} \tag{4.31}$$

## TRANSFORMER NOISE FIGURE

While transformer theory is extensively treated in the literature, there is comparatively little discussion on the noise performance of integrated transformers. Understanding transformer noise is critical in low-noise amplifiers and other sensitive RF circuits, where every resistive and coupling loss contributes to the overall NF. The NF of a transformer can be expressed as (Yo-Sheng Lin *et al.* 2005)

$$NF = -10\log G \tag{4.32}$$

where $G$ is the gain above. The $Z$-parameters of Figure 4.15(c) in the gain expression are given by

$$Z_{11} = R'_{s1} + j\omega L'_{s1} \tag{4.33}$$

$$Z_{22} = R'_{s2} + j\omega L'_{s2} \tag{4.34}$$

$$Z_{12} = R_M + j\omega L_M \tag{4.35}$$

Also,

$$x = \frac{1 - k_{Re}^2}{k_{Im}^2 Q_1 Q_2 + k_{Re}^2} \tag{4.36}$$

where

$$k_{Im} = \sqrt{\frac{\left[Im(Z_{12})\right]^2}{Im(Z_{11})Im(Z_{22})}} = \frac{L_M}{\sqrt{L'_{s1}L'_{s2}}} \tag{4.37}$$

$$k_{Re} = \sqrt{\frac{\left[Re(Z_{12})\right]^2}{Re(Z_{11})Re(Z_{22})}} = \frac{R_M}{\sqrt{R'_{s1}R'_{s2}}} \tag{4.38}$$

where $k_{Re}$ and $k_{Im}$ are the mutual resistive and reactive coupling factors, respectively, capturing how both energy loss and stored energy in the primary turns affect the secondary turns. This framework provides a practical basis for evaluating transformer noise contributions in integrated RF circuits and highlights the importance of both resistive and reactive coupling in determining overall performance.

## INTEGRATED DIODES

Integrated diodes are fundamental devices in semiconductor technology and play a wide range of roles in analog, digital, and RF circuits. Their operation is based on the behavior of PN junctions or metal–semiconductor junctions, and their implementation in integrated processes has been optimized for various functions, including current steering, voltage reference generation, high-frequency detection, and protection. Within RFIC design, diodes are not only passive rectifying elements but also active circuit components, shaping the performance of mixers, detectors, limiters, and biasing networks.

In modern bipolar processes, three main diode types are commonly encountered: the diode-connected transistor, the base–emitter Zener diode, and the Schottky diode. Although a BJT contains two PN junctions, they are generally unsuitable for direct diode implementation due to parasitic effects. Specifically, when the base–collector junction is exploited, a parasitic PNP transistor is formed with the substrate; when the base–emitter junction is used, a parasitic NPN transistor appears due to the enclosing tank. Both parasitic devices result in significant carrier injection into the substrate or tank, leading to unwanted leakage and loss. For this reason, carefully designed structures are used to obtain practical diode characteristics.

The diode-connected transistor, illustrated in Figure 4.14, represents a simple but effective way of realizing a diode in integrated form. In this configuration, the base is shorted to the collector, and current flows between the emitter and collector terminals. The current–voltage behavior mimics that of a diode, with the added benefit that

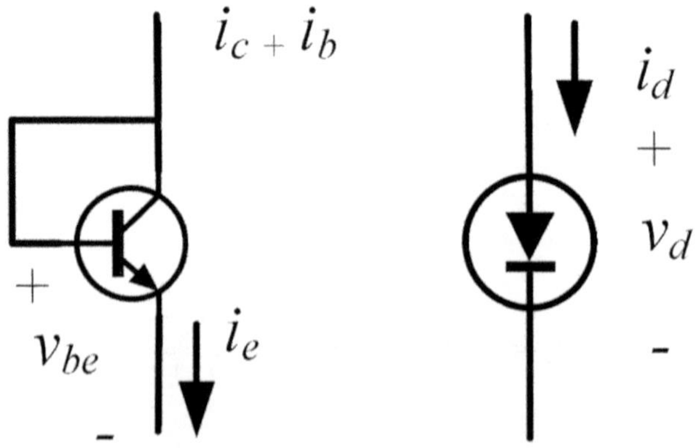


**FIGURE 4.14** Diode-connected transistor.

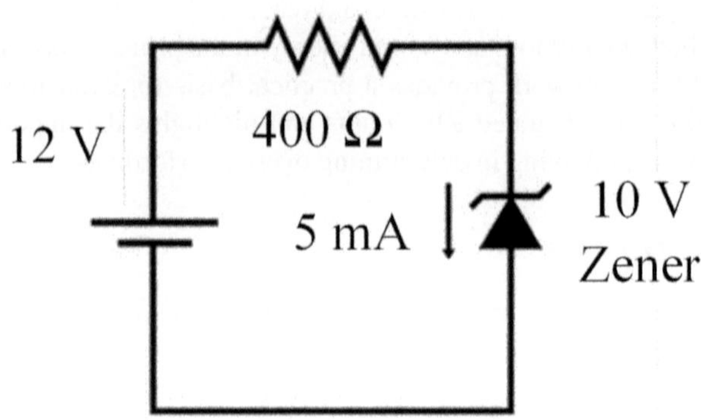


**FIGURE 4.15** Base-emitter Zener diode.

the base current is negligible due to the transistor effect. These devices are widely used in biasing networks, bandgap reference circuits, and temperature-dependent voltage generation. At 25 °C, a diode-connected transistor provides a forward drop of approximately 0.65 V, making it useful as a predictable voltage reference. However, its breakdown voltage is relatively low, typically between 6 and 8 V, which limits its use in high-voltage applications.

Another important device is the base–emitter Zener diode, which operates under reverse bias when the applied voltage exceeds a critical breakdown value. The mechanism of conduction depends on the breakdown voltage: for values below approximately 6 V, current is dominated by quantum tunneling, and the device is called a true Zener diode. For breakdown voltages above 6 V, avalanche multiplication of carriers becomes dominant, leading to what is technically an avalanche diode. In practice, the term *Zener diode* is often used loosely to refer to both types. A schematic of such a device is shown in Figure 4.15. These diodes are used extensively for voltage regulation, overvoltage protection, and as precise reference elements in integrated circuits.

The Schottky diode, by contrast, relies on a metal–semiconductor junction rather than a PN junction. At the heart of a Schottky diode is the barrier that forms between the metal and semiconductor, which allows carriers to cross with a much lower forward voltage drop than in a conventional PN junction. Instead of the usual 0.7 V seen in silicon diodes, Schottky diodes typically operate in the 0.2–0.4 V range. This seemingly small difference has a big impact: it makes them highly attractive in high-speed digital logic, where they serve as Baker clamps in BJTs, in RF mixers and detectors where fast switching is vital, and in power circuits where every fraction of a volt saved translates to better efficiency.

Another advantage is their lack of minority-carrier storage. Because only majority carriers are involved in conduction, there is almost no reverse recovery delay, which gives Schottky diodes their reputation for speed at high frequencies. This property is often the deciding factor in RF and switching applications.

## DIODE CURRENT–VOLTAGE CHARACTERISTICS

The conduction process in PN diodes is more involved, since it draws on a combination of carrier generation, recombination, diffusion, and drift. Under forward bias, majority carriers are pushed across the depletion region, contributing to the forward current and shaping the familiar exponential I–V relationship. The total current $I$ in a forward-biased PN junction is generally expressed as the sum of two components:

- $I_F$, the forward diffusion current due to majority carriers (electrons moving from the N region to the P region, and holes moving in the opposite direction).
- $I_S$, the reverse saturation current caused by the diffusion of minority carriers across the junction.

Thus, the diode current can be written as $I = I_F - I_S$. An exponential relationship is given by the Shockley diode equation, also known as the diode law:

$$I = I_S\left(e^{qV/nkT} - 1\right) \tag{4.38}$$

where $I_S$ is the reverse saturation current, $q$ is the electron charge, $V$ is the applied voltage, $k$ is the Boltzmann constant, $T$ is the absolute temperature, and $n$ is the ideality factor (typically between 1 and 2 depending on the device structure and fabrication process). For an ideal diode, $n = 1$. This exponential relationship captures the strongly nonlinear behavior of the diode and serves as the basis for modeling in circuit simulators. In real diodes, additional effects such as series resistance, recombination in the depletion region, and high-level injection modify this ideal behavior.

## NOISE PROPERTIES OF DIODES

In addition to their I–V characteristics, diodes introduce noise into circuits, which is a critical concern in RF design. The two principal types of noise in diodes are shot noise and thermal noise, although flicker noise and generation–recombination noise also appear under certain conditions.

Shot noise originates from the discrete and random nature of carrier transport across the potential barrier. For a diode current $I$, the mean-square shot noise current spectral density is

$$\overline{i_n^2} = 2q\left(I + 2I_s\right)\Delta f \tag{4.39}$$

which reduces to

$$\overline{i_n^2} = 2qI\Delta f \tag{4.40}$$

when the forward current dominates over the reverse saturation current. The small-signal conductance $Y$ of a PN diode can be derived as the derivative of the current with respect to voltage

$$Y = \frac{\partial I}{\partial V} = \frac{q}{kT}\left(I + I_s\right) \cong \frac{q}{kT} I \tag{4.41}$$

which ties directly to the diode's noise model. Based on these expressions, the diode can be represented as a parallel combination of a conductance $Y$ and a noise current source, as shown in Figure 4.16. This simplified model neglects the series resistances due to contacts and bulk regions, but it provides a first-order approximation useful for circuit analysis.

Interestingly, the diode's noise power can also be related to thermal noise considerations, leading to the alternative expression:

$$\overline{i_n^2} = 2kTY\Delta f \tag{4.42}$$

This shows that the diode noise has a thermal interpretation, although the origin is in shot noise due to carrier discreteness.

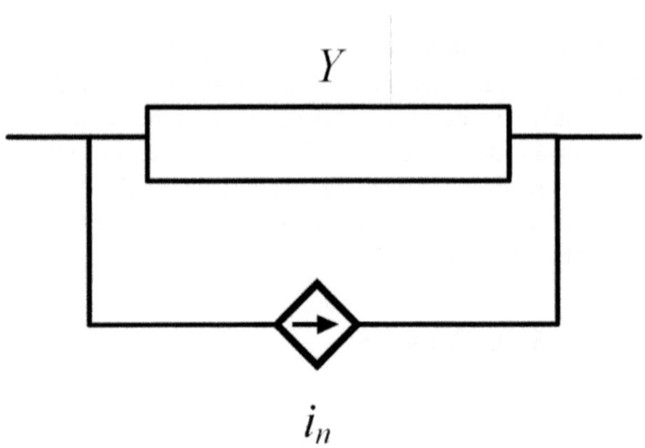


**FIGURE 4.16** Noise model of a PN diode.

At higher frequencies or under reverse bias near breakdown, additional noise contributions appear. For instance, Zener and avalanche diodes exhibit excess noise due to tunneling and avalanche multiplication processes, respectively. Flicker noise, with a $^1/_f$ spectral density, results from traps and defects at the semiconductor surface and bulk, while thermal noise is generated by ohmic resistances and contact regions.

When the diode is reverse-biased, and the bias voltage is less than the breakdown voltage, the power of the noise generated by the diode is equal to $2qI_S$. At breakdown, the diode noise is due to the Zener effect and to avalanche multiplication, which are not discussed in this chapter.

The PN diode and the Schottky diode are nonlinear devices. White noise in diodes can be viewed as resulting from thermal noise in nonlinear resistive devices, except for the tunnel diode. It is shown in (Dasgupta *et al.* 2015) that the noise spectral density of these devices can be derived from Gupta's theorem. This theorem demonstrates that the spectral power densities of the thermal noise voltage $v_n$ and current $i_n$ of a nonlinear resistor. Shot noise in tunnel and Schottky diodes is given by.

$$\overline{v_n^2} = 4kT\left(\frac{dV}{dI} + \frac{1}{2}I\frac{d^2V}{dI^2}\right) \tag{4.43a}$$

$$\overline{i_n^2} = 4kT\left(\frac{dI}{dV} + \frac{1}{2}I\frac{d^2I}{dV^2}/\frac{dI}{dV}\right) \tag{4.43b}$$

where $V$ and $I$ are the voltage and the current across the diode, respectively.

Although only shot noise has been discussed, other types of noise are present in PN junctions. This is due to the simplified model in Figure 4.16, which does not include thermal and flicker noise. Thermal noise is produced by ohmic contacts and by bulk P and N regions. The impedance $Z_s$ due to contacts and bulk regions is not linear and shows a reactive component at high frequencies. Therefore, the voltage noise density due to $Z_s$ is given below. Flicker noise is caused by defects in the bulk and on the surface, resulting in carriers falling into traps before being released (Vasilescu 2019).

$$\overline{v_s^2} = 4kTRe\left\{Z_s\right\}\Delta f \tag{4.44}$$

Shot noise is caused by the random nature of the crossing of carriers of the potential barrier in the depletion zone, as discussed in Chapter 2. In the depletion region, carriers are subject to thermal diffusion and generation–recombination. Generation–recombination noise is related to the instantaneous disappearance or appearance of electrons. At RF, the main noise types are thermal noise and shot noise, and the diode can be decomposed into an ideal diode and a series resistor. If reactive components are neglected, the noise model of the diode at RFs is shown in Figure 4.16, where $r_{ser}$ is the series resistance, $r_d$ is the junction space-charge differential resistance, *e(t)* the thermal noise voltage source associated with $r_{ser,}$ and *i(t)* the shot noise current source. Thermal diffusion noise is about two-thirds of the shot noise, and generation-recombination noise is about two-thirds of the shot noise.

The voltage noise power spectral density at the diode terminals in Figure 4.20 is given by (4.45), where $I$ is the current. It can be noted that there is no thermal noise associated with $r_d$. Equation (4.445) is particularly applicable to tunnel diodes [23]

$$\overline{v_n^2} = 4kTr_{ser} + 2qIr_d^2 \tag{4.45}$$

The current noise power spectral density is given by

$$\overline{i_n^2} = \frac{4kTr_{ser}}{\left(r_{ser} + r_d\right)^2} + \frac{2qIr_d^2}{\left(r_{ser} + r_d\right)^2} \tag{4.46}$$

For small currents, $r_d \gg r_{ser}$ and the diode noise reduces to the shot noise approximation (39). At higher currents, the contribution of thermal noise from $r_{ser}$ becomes more significant.

## DESIGN CONSIDERATIONS AND RF APPLICATIONS

Although the underlying physics of PN junctions applies to all diodes, circuit designers rarely treat them as interchangeable parts. In practice, the choice between PN, Schottky, and Zener diodes comes down to their electrical behavior; forward drop, speed, noise, breakdown, and parasitics all weigh into the decision.

PN diodes are the workhorses of standard bipolar processes. They are straightforward to fabricate and show a forward drop of about 0.65–0.7 V at room temperature. Their weaknesses are well known: minority-carrier storage makes them relatively slow, and their junction capacitance is higher than that of Schottky's. For this reason, they find their niche in biasing networks and low-frequency circuits, but are less attractive when speed or RF performance is required.

Schottky diodes behave differently. Because the current flows across a metal–semiconductor barrier, the forward drop is much lower, typically around 0.3–0.45 V. This saves headroom in low-voltage designs and reduces conduction loss. Just as important, conduction involves majority carriers only, so there is no charge storage to slow switching. That combination makes Schottky's the obvious choice for clamp circuits in BJTs, RF rectifiers, and other fast-switching roles. The compromises are also clear: their reverse leakage is higher, and breakdown voltages are lower, so designers must weigh speed and efficiency against leakage and robustness, especially in hot or high-voltage environments.

Another dimension is added by Zener diodes. The Zener tunneling mechanism, which provides sharp clamping with little temperature drift, predominates at low breakdown voltages (below roughly 6 V). Avalanche breakdown takes over at higher voltages, where carriers accumulate sufficient energy to cause impact ionization. ICs frequently use these effects to implement biasing circuits, ESD clamps, and voltage references.

Capacitance is another recurring theme in diode design. Every diode has a depletion capacitance that varies with applied reverse bias. At RF, this parasitic capacitance becomes a bottleneck, reducing bandwidth and introducing unwanted phase shift. Mixer diodes, detectors, and rectifiers are all limited in this way. Designers can respond by reducing device area, choosing lateral or vertical layouts, or relying on Schottkys, which naturally offer lower capacitance. In advanced nodes, dedicated low-capacitance diodes are even offered for multi-GHz applications.

Noise is also unavoidable. Under forward bias, shot noise dominates; series resistances introduce thermal noise; and at low frequencies, generation–recombination and flicker noise appear. In RFICs, these noise sources can lower receiver sensitivity, reduce mixer linearity, and narrow the dynamic range of analog front ends. Equivalent-circuit models are therefore widely used to capture noise contributions and guide optimization.

Practical layouts often rely not on single diodes but on networks. Current mirrors, voltage dividers, and biasing ladders all use diode drops to establish predictable operating points. Matching is crucial here: in analog ICs, mismatched PN or Schottky diodes lead to offsets and nonlinearity. Common-centroid layouts, interdigitated structures, and thermal balancing are standard tricks to reduce mismatch from process variation or gradients.

Process advances continue to broaden the designer's toolkit. SOI or high-resistivity substrates reduce parasitics, while compound semiconductors like GaAs or InP enable Schottky's with very high speed and very low noise, suitable for mm-wave circuits. Even MEMS-based diodes have been explored for ultra-low capacitance in niche radar and communications systems.

Protection and sensing are equally important roles. On-chip ESD networks depend on diodes to safely shunt transients, but must be sized carefully to avoid loading the signal path. Temperature sensing is another practical use: the forward drop of a diode changes predictably with temperature, which makes it a simple but reliable thermal monitor inside ICs.

In short, diodes in integrated circuits are far more than rectifiers. They enable biasing, referencing, switching, RF detection, noise shaping, and protection. Extracting the best performance requires understanding not only the textbook physics but also the parasitics, noise, and layout effects that appear in silicon. When chosen and implemented carefully, PN, Schottky, and Zener diodes remain indispensable to analog, mixed-signal, and RF design.

## INTEGRATED VARACTORS

Varactor diodes, often referred to as varicaps or tuning diodes, are a class of semiconductor devices that make use of the way a PN junction's capacitance changes when reverse-biased. Unlike ordinary diodes, which are valued mainly for rectification or switching, varactors are intentionally designed to operate as voltage-controlled capacitors. They do not provide gain or amplification and are therefore considered passive components, but they store charge electrostatically with a capacitance that can be adjusted through the applied reverse voltage.

Even though they are passive, varactors are central to a wide range of active RF and microwave circuits. Their ability to offer electronic tunability makes them indispensable in voltage-controlled oscillators (VCOs), frequency multipliers, tunable filters, and phase-locked loops (PLLs). In all these applications, they provide reconfigurability without the need for mechanical tuning elements, which are bulky and impractical in integrated systems.

The tunability of a varactor stems from the depletion region formed at the PN junction. When a reverse bias is applied, the majority carriers are pushed away from the junction, creating a charge-free depletion region. As the reverse bias increases, the depletion width widens, which in turn reduces the capacitance. This direct link between bias voltage and capacitance is what allows the varactor to serve as a controllable reactive element in RF design. This capacitance is often modeled by the junction capacitance expression

$$C_j(V) = \frac{C_0}{\left(1 + \frac{V}{\Phi}\right)^m} \tag{4.46}$$

where $C_0$ is the junction capacitance at zero bias, $V$ is the applied reverse-bias voltage, $\Phi$ is the built-in potential of the junction, and $m$ is the grading coefficient that depends on the doping profile (e.g., $m=1/2$ for an abrupt junction and $m=1/3$ for a linearly graded junction). This nonlinear dependence is the fundamental mechanism that makes varactors useful in frequency-tunable circuits.

## VARACTOR APPLICATIONS IN RFIC DESIGN

In RF integrated circuits, varactors are widely used to provide voltage-controlled tuning. A classic example is the VCO, where the varactor is connected either in parallel or in series with the resonant tank. Adjusting the control voltage shifts the effective resonant frequency, giving the circuit frequency agility without the need for mechanical tuning. This electronic approach is fast, low-power, and compact, making it indispensable in modern communication systems such as LTE, 5G, and emerging mm-wave platforms. Beyond VCOs, varactors also find application in frequency multipliers, where the nonlinear capacitance–voltage relationship generates harmonics of an input signal, allowing high-frequency outputs to be derived from lower-frequency sources. They are equally important in tunable filters, where varactors embedded in LC resonators enable dynamic adjustment of center frequency and bandwidth, and in phase shifters or mixers, particularly in adaptive beamforming systems, where they help achieve reconfigurable phase responses. These diverse roles highlight how varactors support both reconfigurability and miniaturization in integrated RF systems, where space and power budgets are always under pressure.

The performance of a varactor in these applications is determined by several interrelated parameters. The most indicative is the capacitance range and tuning ratio, defined as $C_{max}/C_{min}$. A wide tuning ratio (typically between 2:1 and 5:1) directly translates into broader frequency tunability in oscillators and filters. Achieving such ratios depends strongly on junction design, the doping profile, and the maximum reverse voltage the device can sustain before breakdown. Another critical parameter is the quality factor, or $Q$. As with inductors and transformers, series resistance reduces $Q$, and in integrated varactors, this is often exacerbated by substrate losses, especially on bulk CMOS. Low $Q$ degrades phase noise in oscillators and increases insertion loss in filters, so processes such as silicon-on-insulator (SOI) or GaAs are sometimes preferred,. since they can support much higher $Q$ values. Linearity is also a major consideration: because the capacitance–voltage characteristic is inherently nonlinear, large RF drive signals can generate distortion and intermodulation products that compromise system performance. Differential configurations or harmonic suppression techniques are often used to mitigate these effects. Finally, practical varactors are never free of parasitics. Additional capacitances and resistances from the device geometry, contacts, and layout can shift the tuning range, depress the self-resonance frequency, and limit high-frequency operation. Careful device scaling, shielding, and layout optimization are therefore crucial to preserving performance in real RFIC implementations.

## ON-CHIP VARACTOR IMPLEMENTATION AND IMPACT ON SYSTEM PERFORMANCE

In integrated circuits, varactors are most realized as either junction varactors or MOS varactors, each with distinct characteristics. Junction varactors are formed from reverse-biased PN junctions and are valued for their relatively high capacitance density and straightforward integration in BiCMOS or SiGe processes. On standard silicon substrates, however, their performance is often limited by junction leakage and moderate $Q$ factors, which can constrain high-frequency operation. MOS varactors, by contrast, exploit the voltage-dependent capacitance between the gate and channel of a MOS transistor. By biasing the MOS capacitor in either the accumulation or depletion regime, designers can achieve a tunable capacitance. MOS varactors typically offer higher tuning ratios than junction varactors, although this comes at the cost of increased non-linearity, which must be carefully managed in sensitive RF applications. Emerging approaches, such as MEMS-based tunable capacitors, promise very high $Q$ factors and wide tuning ranges, but their adoption in mainstream CMOS RFICs remains limited due to the additional fabrication complexity they introduce.

The choice and design of the varactor have a direct impact on system-level performance. In VCOs, for instance, the varactor's quality factor is a key determinant of phase noise, as described by Leeson's equation. Low-$Q$ varactors contribute to increased phase noise and reduced spectral purity, potentially degrading the performance of the entire communication link. Similarly, in tunable filters or adaptive phase-shifting circuits, varactor $Q$, tuning range, and linearity collectively influence insertion loss, bandwidth, and distortion, emphasizing the need for careful device selection and layout. Properly optimized varactors, therefore, play a crucial role not only as passive tuning elements but also as enablers of stable, high-performance RF systems.

## CONCLUSION

This chapter has examined the physical, electrical, and system-level behavior of integrated passive and semi-passive devices that fundamentally shape the performance limits of RF CMOS circuits. While often treated as secondary to active devices, elements such as inductors, transformers, diodes, and varactors emerge here as decisive contributors to noise, linearity, bandwidth, and efficiency. Their behavior cannot be abstracted away without consequence; instead, it must be understood through a combination of electromagnetic insight, circuit modeling, and technology-aware design practice.

A central theme throughout the chapter is that parasitics are not incidental imperfections but defining characteristics of integrated RF components. Series resistance, substrate coupling, parasitic capacitance, and magnetic loss collectively determine quality factor, self-resonance frequency, and ultimately usable bandwidth. The treatment of inductors illustrated how noise, loss, and resonance are intimately linked through equivalent-circuit representations, and how even seemingly simple quantities such as Q encode multiple physical mechanisms. These same ideas extend naturally to transformers, whose two-port nature introduces additional degrees of freedom, coupling, symmetry, mutual inductance, and balance that must be negotiated carefully. The discussion of transformer geometries, coupling coefficients, characteristic resistance, and maximum available power gain highlights how electromagnetic layout decisions translate directly into circuit-level performance metrics.

Across these passive structures, the recurring trade-off between confinement and coupling appears in many guises. Increasing magnetic coupling or inductance density tends to raise parasitic capacitance and lower self-resonance frequency; improving symmetry and balance often increases resistance and area; reducing substrate loss through shielding or isolation can introduce new eddy-current paths or fabrication complexity. These competing effects explain why no universally optimal structure exists and why practical RFIC design relies on carefully tuned compromises rather than idealized models. The emphasis on self-resonance frequency, transformer coupling, and characteristic resistance reflects the reality that modern RF systems operate increasingly close to the physical limits of on-chip passives.

The discussion of transformer noise and maximum available gain further reinforces the inseparability of noise and impedance transformation. In contrast to active devices, where noise can often be traded against bias or transconductance, passive noise originates directly from loss mechanisms embedded in geometry and materials. As a result, transformer and inductor noise must be managed structurally rather than electrically. This perspective aligns naturally with system-level requirements in low-noise amplifiers, mixers, and oscillators, where passive losses frequently dominate the achievable noise figure once transistor performance reaches maturity.

The chapter then extends these ideas to nonlinear passive devices, namely integrated diodes and varactors. Their current–voltage behavior, junction capacitance, and noise properties introduce controlled nonlinearity that can be exploited for tuning, frequency conversion, and reconfigurability. At the same time, these devices impose limits on phase noise, linearity, and dynamic range. The treatment of varactors emphasizes their dual role as both enabling elements for tunable systems and sources of sensitivity to voltage noise, process variation, and parasitic coupling. In highly integrated RF systems, their influence propagates upward, affecting oscillator stability, filter selectivity, and overall system robustness.

Taken together, the material in this chapter highlights a unifying principle: in RF CMOS, device physics, layout geometry, and system architecture are inseparable. Passive components cannot be treated as ideal abstractions appended to otherwise complete circuits; they are structural elements that shape performance at every level. Their limitations motivate architectural choices, constrain achievable bandwidth and noise, and often dictate whether a given specification is feasible in a particular technology. Understanding their behavior, quantitatively and physically, is therefore essential not only for circuit optimization, but also for informed architectural decision-making.

This chapter thus provides the foundation needed to interpret and evaluate RF front-end designs in later sections. Grounding system-level behavior in the realities of integrated passive devices, it establishes a coherent bridge between semiconductor physics, electromagnetic modeling, and practical RF circuit design.

## REFERENCES

Abbott, D., Davis, B.R., Phillips, N.J., and Eshraghian, K., 1996. Simple Derivation of the Thermal Noise Formula Using Window-Limited Fourier Transforms and Other Conundrums. *IEEE Transactions on Education*, 39 (1), 1–13.

Amakawa, S., 2010. A Thru-Only De-Embedding Method Foron-Wafer Characterization of Multiport Networks. In: *Advanced Microwave Circuits and Systems,* edited by Vitaliy Zhurbenk. IntechOpen (specifically under the InTech imprint at the time).

Božanić, M., and Sinha, S., 2018. *Millimeter-Wave Low Noise Amplifiers*. Springer.

Carrara, F., Italia, A., Ragonese, E., and Palmisano, G., 2006. Design Methodology for the Optimization of Transformer-Loaded RF Circuits. *IEEE Transactions on Circuits and Systems I: Regular Papers*, 53 (4), 761–768.

Dasgupta, A., Khandelwal, S., and Chauhan, Y. S., 2015. Surface Potential Based Modeling of Thermal Noise for HEMT Circuit Simulation. *IEEE Microwave and Wireless Components Letters*, 25 (6), 376–378. doi: 10.1109/LMWC.2015.2422472

Deng, Y., Wang, F., Yin, X., Yang, Y., Yu, N., and Li, Y., 2025. A High Coupling Coefficient and Symmetric Transformer Based on TSV. *IEEE Transactions on Components, Packaging and Manufacturing Technology*, 15 (1), 165–172.

Ellinger, F., 2008. *Radio Frequency Integrated Circuits and Technologies*. Springer.

Gordon, M., 2007. *Design Methodology for Millimeter-Wave Low-Noise Amplifiers*. Library and Archives Canada = Bibliothèque et Archives Canada.

Lafferty, R., 1982. Measuring the Self-Resonant Frequency of Capacitors. *IEEE Transactions on Components, Hybrids, and Manufacturing Technology*, 5 (4), 528–530.

Lambrechts, W., and Sinha, S., 2017. *SiGe-Based Re-engineering of Electronic Warfare Subsystems*. Springer International Publishing Imprint: Springer.

Leung, B., 2011. *VLSI for Wireless Communication*. Springer US.

Lin, Y.-S., Liang, H.-B., Wang, T., and Lu, S.-S., 2005. Temperature-Dependence of Noise Figure of Monolithic RF Transformers on a Thin (20 /spl mu/m) Silicon Substrate. *IEEE Electron Device Letters*, 26 (3), 208–211.

Mu, J., Chou, X., Ma, Z., He, J., and Xiong, J., 2018. High-Performance MIM Capacitors for a Secondary Power Supply Application. *Micromachines*, 9 (2), 69.

Niknejad, A.M., and Meyer, R.G., 1998. Analysis, Design, and Optimization of Spiral Inductors and Transformers for Si RF ICs. *IEEE Journal of Solid-State Circuits*, 33 (10), 1470–1481.

Niknejad, A.M., and Meyer, R.G., 2001. Analysis of Eddy-Current Losses over Conductive Substrates with Applications to Monolithic Inductors and Transformers. *IEEE Transactions on Microwave Theory and Techniques*, 49 (1), 166–176.

Razavi, B., 1997. *RF Microelectronics*. Prentice Hall PTR.

Sepke, T., Holloway, P., Sodini, C.G., and Lee, H.-S., 2009. Noise Analysis for Comparator-Based Circuits. *IEEE Transactions on Circuits and Systems I: Regular Papers*, 56 (3), 541–553.

Sinha, S., Lambrechts, J.W., Bimana, A., and Ashipala, A., 2025. Augmented Artificial Intelligence in 5G, 6G, and Beyond: A Quantum Leap. *Computer*, 58 (1), 24–32.

Soltanian, B., and Kinget, P., 2005. A Tail Current-Shaping Technique to Reduce Phase Noise in LC VCOs. In: *Proceedings of the IEEE 2005 Custom Integrated Circuits Conference, 2005*. IEEE, 574–577.

Tilmans, H.A.C., Raedt, W. De, and Beyne, E., 2003. MEMS for Wireless Communications: From RF-MEMS Components to RF-MEMS-SiP. *Journal of Micromechanics and Microengineering*, 13 (4), S139–S163.

Wenger, C., Oberbeck, L., Schroeder, U., Mussigmussig, H.-J., Lippert, G., Sorge, R., Schroeder, T., Mane, A.U., Lupina, G., Dabrowski, J., Zaumseil, P., and Fan, X., 2006. High-Quality MIM Capacitors for RF Applications. *IEEE Transactions on Electron Devices*, 53 (8), 1937–1939.

Yue, C.P., Ryu, C., Lau, J., Lee, T.H., and Wong, S.S., 1996. A Physical Model for Planar Spiral Inductors on Silicon. In: *International Electron Devices Meeting. Technical Digest*. IEEE, 155–158.

# 5 Integrated Active Devices and Their Role in Low-Noise Amplifier Design

## HISTORICAL BACKGROUND

The first practical electronic active device was the vacuum tube, originating from Sir John A. Fleming's 1900 invention of the vacuum tube diode, and later extended by Lee De Forest (Geddes and Roeder 2003) in 1907 with the addition of a control grid to create the triode. This innovation allowed the regulation of electron flow from a heated cathode to the anode, with the grid voltage controlling the current between them. The triode's ability to amplify weak signals and generate oscillations made it a cornerstone of early radio communications.

Despite these advantages, vacuum tubes had significant drawbacks. They exhibited high shot noise, which limited their effectiveness in amplifying very weak signals. Additionally, they consumed large amounts of power and required substantial physical space. For example, the ENIAC, the first large-scale digital electronic computer, used 18,000 vacuum tubes distributed across 49-foot-tall cabinets, consuming 174 kilowatts of power and necessitating a massive ventilation system.

Interestingly, many design techniques that are standard today in modern transistor circuits, such as inductive degeneration, cascode topologies, and inductive peaking, originated from innovations developed for vacuum tubes in the 1950s. Even now, vacuum tubes find use in specific niches, including high-end audio systems, RF transmitters, and military equipment, prized for their linearity and robustness against electromagnetic interference, such as from explosions.

In the context of noise analysis, vacuum tubes illustrate early lessons on how intrinsic device physics, here, shot noise and thermal noise, directly limit circuit sensitivity. In its simplest form, thermal noise can be understood by considering a resistive element connected to an otherwise ideal electrical environment. Any resistor at a finite temperature exhibits random microscopic motion of charge carriers, driven by thermal agitation. This motion gives rise to voltage and current fluctuations even in the absence of an applied signal. When such a resistor is connected to a matched load, these fluctuations manifest as a measurable noise power that is transferred from the resistor to the load. The phenomenon is fundamentally unavoidable and arises directly from thermodynamic principles rather than from imperfections in fabrication or design. As a result, thermal noise represents a lower bound on the noise performance of any electronic system.

DOI: 10.1201/9781003459385-5

From a circuit perspective, a resistor can therefore be modeled not only as a passive dissipative element, but also as an equivalent noise source. This noise is commonly represented either as a voltage source in series with the resistor or as a current source in parallel, both descriptions being mathematically equivalent. The mean-square noise voltage generated across a resistance over a given bandwidth is proportional to the absolute temperature and the resistance value, a relationship that directly links microscopic physical processes to macroscopic electrical behavior. When the resistor is impedance-matched to its load, half of the generated noise power is delivered to the load, establishing a well-defined reference condition that is widely used in noise analysis.

This simple resistive example plays a foundational role in the development of noise theory because it isolates thermal noise from other mechanisms such as shot noise or flicker noise. It also provides a convenient reference for defining noise temperature and noise figure, which are later extended to more complex active devices. Even though practical circuits rarely consist of a lone resistor connected to an ideal load, the conceptual model captures the essence of how unavoidable thermal agitation limits sensitivity. In this sense, it serves as the conceptual ancestor of more elaborate noise models used for transistors, amplifiers, and complete receiver front ends.

Importantly, the thermal noise generated by a resistor is spectrally white over a wide frequency range, meaning that its power spectral density is approximately constant with frequency up to very high limits. This property makes it especially useful as a reference when comparing devices and systems operating across different bandwidths. It also explains why bandwidth plays such a central role in determining total noise power: increasing bandwidth directly increases the amount of thermal noise collected, even if the underlying device remains unchanged.

By highlighting thermal noise at this early stage, the discussion establishes a conceptual bridge between classical circuit theory and the later treatment of semiconductor devices. The same fundamental noise mechanisms reappear in vacuum tubes, bipolar transistors, metal–oxide–semiconductor field-effect transistors (MOSFETs), and passive RF structures, albeit expressed through different physical processes. In all cases, noise originates from random carrier motion and energy dissipation, reinforcing the idea that noise is not merely a secondary imperfection but a fundamental consequence of operating at finite temperature. This insight becomes especially important when transitioning from historical devices such as vacuum tubes to modern solid-state technologies, where shrinking dimensions and higher operating frequencies make noise considerations increasingly central to system-level performance.

The advent of the transistor marked a transformative moment in electronics. In December 1947, John Bardeen, Walter Brattain, and William Shockley at Bell Laboratories made a groundbreaking discovery: a small germanium point-contact device could function as an amplifier. Interestingly, their initial research was motivated by efforts to build a MOSFET, but the result was a new class of active device altogether. The point-contact transistor was remarkably compact, roughly 1/50th the size of a conventional vacuum tube, offering the promise of dramatically smaller, lighter, and more power-efficient electronic systems. Beyond its reduced size, the device was far more reliable and economical, laying the foundation for the mass adoption of solid-state electronics.

The pioneering trio John Bardeen, William Shockley, and Walter Brattain – combined ingenuity made the transistor possible. The work of John Bardeen, William Shockley, and Walter Brattain was soon formalized by Shockley's theoretical analysis, published in his seminal paper, "The Theory of p-n Junctions in Semiconductors and p-n." This work not only explained the operational principles of the first bipolar transistor but also established a rigorous framework connecting semiconductor physics to practical circuit performance. For their efforts, Shockley and Brattain were awarded the Nobel Prize in Physics in 1956, recognizing the profound impact of their invention on both science and industry.

The collaboration between Bardeen, Brattain, and Shockley represents one of the most influential intersections of theory and experiment in the history of electronics. Although their personal and professional relationships were complex, their combined contributions formed a coherent intellectual progression: experimental observation, physical interpretation, and theoretical formalization. Bardeen and Brattain's experimental demonstrations revealed that semiconductor surfaces could support controllable charge modulation, while Shockley's subsequent theoretical work provided the mathematical framework needed to generalize these observations into a scalable device concept. This synthesis of empirical discovery and analytical modeling became a template for later developments in semiconductor device engineering.

Equally important was the institutional environment in which these advances occurred. Bell Laboratories provided an unusually fertile setting where physicists, chemists, and engineers collaborated closely, allowing ideas to move rapidly from fundamental physics to working hardware. The transistor emerged not as an isolated breakthrough but as the outcome of sustained inquiry into solid-state phenomena such as surface states, carrier transport, and junction behavior. This multidisciplinary approach would later become a defining feature of microelectronics research and remains essential in modern RF and microwave device development.

The historical significance of this period lies not only in the invention of the transistor itself, but also in the conceptual shift it introduced. Electronics transitioned from vacuum-based charge transport in macroscopic structures to solid-state conduction governed by quantum-mechanical principles. Concepts such as energy bands, carrier statistics, and recombination mechanisms became central to circuit understanding. These ideas underpin all later discussions of noise, gain, and linearity in semiconductor devices, making this era foundational for modern analog and RF theory.

Moreover, the early transistor era established the intimate connection between device physics and system-level performance. Limitations observed at the circuit level, such as noise, gain compression, or bandwidth constraints, can now be traced back to microscopic mechanisms, including carrier scattering, recombination, and junction capacitance. This cause-and-effect relationship between material properties and electrical behavior is a recurring theme throughout RF integrated circuit design and remains directly relevant when analyzing contemporary devices such as SiGe HBTs and advanced CMOS transistors.

The recognition awarded to the inventors in 1956 symbolized more than individual achievement; it marked the beginning of a technological trajectory that reshaped communication systems, computation, and sensing. From early discrete components to highly integrated RF systems-on-chip, the intellectual lineage of modern electronics can be traced directly to these foundational developments. Understanding this historical progression provides essential context for appreciating why modern transistor technologies behave as they do, and why their noise and frequency characteristics remain constrained by principles first uncovered during this formative period.

The original point-contact germanium transistor became the ancestor of today's high-speed silicon-germanium heterojunction bipolar transistors (SiGe HBTs) (Lambrechts and Sinha 2017)). These SiGe devices, first developed in 1951, were designed to push the operating frequency of bipolar junction transistor (BJTs) higher while maintaining the advantages of compact size and robust performance. By adjusting the material composition and junction profiles, engineers were able to dramatically increase the unity-gain frequency ($f_T$) and maximum oscillation frequency ($f_{MAX}$), enabling transistors suitable for RF and microwave applications.

Meanwhile, a parallel evolution in transistor technology gave rise to field-effect transistors (FETs). First conceptualized in 1953 and entering production in the 1960s, FETs offered an alternative amplification mechanism: rather than relying on both electrons and holes, as in bipolar devices, FETs use a single type of carrier flowing through a controlled channel. The channel's conductivity is modulated by a voltage applied to the gate, offering high input impedance and often lower noise than equivalent BJTs at certain frequencies. Variants such as junction FETs (JFETs), metal–semiconductor FETs (MESFETs), MOSFETs, and high-electron-mobility transistors (HEMTs) emerged over subsequent decades, each optimized for performance metrics such as noise figure, frequency response, and power efficiency.

Collectively, these advances in bipolar and field-effect devices laid the groundwork for modern LNAs and RF front ends. The transition from bulky, high-noise vacuum tubes to compact, high-speed solid-state devices allowed engineers to design receivers and transmitters with unprecedented sensitivity, bandwidth, and reliability. Noise performance, once dominated by shot noise in vacuum tubes, could now be carefully controlled through device choice, geometry, and biasing, enabling the next generation of high-frequency communication systems.

Research into electronic noise has historically been driven by the relentless pursuit of improved receiver sensitivity. As established earlier, the fundamental limitation to receiver performance is set by the noise generated within the active devices themselves. No matter how sophisticated the matching networks, filters, or antenna systems may be, the intrinsic noise of the amplifying element establishes the ultimate floor against which weak signals must compete. For this reason, the evolution of active devices has always gone hand in hand with attempts to reduce their noise figure (NF), making noise research central to the progress of both semiconductor physics and communication engineering.

Table 5.1 summarizes the steady improvement in the noise performance of active devices over successive technology generations. In the early days, spanning roughly 1940–1960, electronic vacuum tubes and the first BJTs typically exhibited noise

**TABLE 5.1**
**Evolution of the NF of active devices**

| Period | Active Device | Typical NF |
|---|---|---|
| 1940–1960 | Vacuum tube, BJT | > 3 dB |
| 1960–1980 | BJT, FET, MOSFET | 1–3 dB |
| 1980–2000 | GaAs FET | 0.5–1 dB |
| 2000–present | HEMT, pHEMT | < 0.2 dB |

figures greater than 3 dB. While such devices represented a significant improvement over earlier technologies, they still imposed clear limits on receiver sensitivity, especially in radio astronomy, radar, and satellite communications, where faint signals needed to be detected against cosmic or atmospheric background noise.

Between 1960 and 1980, progress in silicon processing and the development of FET and MOSFET devices allowed noise figures to be pushed into the range of 1–3 dB. This was already a breakthrough, as it brought electronic receivers closer to the theoretical thermal noise limit. The period from 1980 to 2000 saw the widespread adoption of GaAs FETs, which offered noise figures in the 0.5–1 dB range, enabling practical deployment of low-noise front ends for microwave and early satellite communication systems. More recently, beginning in the early 2000s, the introduction of HEMTs and their pseudomorphic variant (pHEMTs) has further reduced noise figures to below 0.2 dB. At these levels, active devices approach the fundamental quantum noise limit, and improvements are now incremental rather than revolutionary.

In parallel with these material and structural innovations, (SiGe HBTs have emerged as an attractive alternative to III–V compound semiconductors in many high-frequency applications. SiGe HBTs offer significant improvements in both $f_T$ and $f_{MAX}$ compared with conventional BJTs. These two figures of merit are central to transistor characterization: $f_T$ represents the frequency at which the small-signal current gain falls to unity, while $f_{MAX}$ reflects the highest frequency at which the device can sustain power gain. Together, they provide an effective measure of a transistor's intrinsic speed.

A key point is that improvements in $f_T$ are generally correlated with better noise performance and higher maximum operating frequency, regardless of whether the device is fabricated in silicon, silicon–germanium, or III–V materials. In practice, this means that by engineering device geometries, doping profiles, and heterojunction interfaces to increase $f_T$, designers can simultaneously enhance both the bandwidth and the noise figure of amplifiers. SiGe HBTs are especially valuable in this respect because they combine the high-speed and low-noise characteristics once thought to be the exclusive domain of III–V devices with the integration and cost advantages of CMOS-compatible processes.

Thus, the historical trajectory of noise research reflects not only incremental improvements in device physics but also a deeper convergence between materials science, circuit design, and system-level requirements. Each step forward in reducing the noise figure of active devices has directly expanded the horizons of communication systems, from long-distance telephony and broadcast radio to deep-space probes and high-capacity wireless networks.

The development of RF transistors following the first point-contact germanium device has been shaped not only by advances in circuit topologies but also by innovations in semiconductor materials. The following is an overview of the historical evolution of RF transistor technologies, highlighting how successive generations have leveraged different materials and device structures to extend speed, reduce noise, and improve power handling.

This technological evolution can be understood as a gradual expansion of the semiconductor design space, driven by the need to overcome fundamental trade-offs between speed, noise, breakdown voltage, and integration density. Early RF transistors were constrained primarily by material properties such as carrier mobility and achievable doping profiles, which directly limited transit frequency and gain. As device physics matured, successive generations introduced new material systems and structural refinements that systematically relaxed these constraints. Each transition represented a response to a specific bottleneck encountered in earlier technologies, whether that bottleneck was excessive base resistance, low breakdown voltage, poor thermal conductivity, or insufficient carrier velocity.

In the earliest generations, germanium and later silicon bipolar transistors defined the baseline for RF performance. Improvements in lithography, diffusion control, and junction engineering steadily pushed transition frequencies upward while reducing parasitic capacitances. The emergence of heterostructures marked a decisive conceptual shift: instead of relying on a single semiconductor, designers began combining materials with different bandgaps and transport properties to tailor carrier confinement and injection efficiency. This strategy enabled much higher current gain and reduced noise by suppressing unwanted recombination mechanisms, laying the groundwork for modern heterojunction bipolar transistors (HBTs).

As frequency targets moved from VHF and UHF into the microwave and mm-wave regimes, compound semiconductors became increasingly dominant. Materials such as GaAs and InP offered substantially higher electron mobility than silicon, allowing carriers to traverse the active region more quickly and thereby increasing both the unity-gain frequency and maximum oscillation frequency. These advantages made III–V technologies especially attractive for low-noise amplifiers (LNAs), mixers, and high-speed digital links, where transit-time limitations directly translate into performance ceilings. Over time, refinements in epitaxial growth and lithographic precision enabled the realization of sophisticated heterostructures with carefully engineered band profiles.

In parallel, wide-bandgap semiconductors emerged to address a different regime of operation, one dominated by high voltage, high power, and harsh environments. Materials such as silicon carbide and gallium nitride support much higher critical electric fields than silicon, allowing devices to operate at elevated voltages without breakdown. This capability, combined with high thermal conductivity and robustness, makes them particularly well-suited for radar transmitters, power amplifiers, and space or defense applications. Their adoption reflects a shift from purely frequency-driven optimization toward system-level considerations such as efficiency, reliability, and thermal management.

Across all these material systems, the introduction of heterojunctions represents a unifying theme. By deliberately engineering discontinuities in band structure, designers gain additional degrees of freedom to control carrier transport, confinement, and recombination. This approach underpins many of the most successful RF transistor families, including SiGe HBTs, InP HBTs, and various III–V FETs. The ability to shape electric fields and carrier distributions at the nanoscale enables improvements in noise figure, linearity, and gain that would be unattainable in homojunction devices.

Taken together, the historical progression of RF transistor technologies reflects an increasingly sophisticated interplay between materials science, device physics, and circuit requirements. Rather than a single linear evolution, it represents a branching landscape in which different material platforms coexist, each optimized for a particular operating regime. This diversity is a defining feature of modern RF engineering and directly informs the design choices encountered in contemporary LNAs, mixers, oscillators, and power stages.

Three broad categories of semiconductor materials now dominate RF transistor fabrication:

1. Group IV materials (Si and SiGe). Silicon remains the workhorse of the semiconductor industry, owing to its abundance, mature processing infrastructure, and excellent scalability. Germanium has historically played a role as the first transistor material and continues to be important in SiGe HBTs. The incorporation of germanium into silicon devices enables bandgap engineering and mobility enhancement, which translates into higher cutoff frequencies and improved noise performance compared with conventional silicon BJTs.
2. III–V compound semiconductors (GaAs, InP, and their alloys). These materials are formed by combining elements from columns III and V of the periodic table. The combinations may be binary (e.g., GaAs), ternary (e.g., AlGaAs), or quaternary (e.g., indium gallium arsenide [InGaAsP]). III–V semiconductors generally exhibit higher electron mobility compared with Si or Ge, which leads to superior carrier transport and high-frequency performance. For example, indium arsenide (InAs) offers electron mobility more than an order of magnitude higher than that of silicon, enabling devices with exceptional gain and speed. In practice, GaAs- and InP-based devices are widely used in low-noise microwave front ends, power amplifiers, and high-speed optical communication systems.
3. Wide-bandgap semiconductors (SiC, GaN, and diamond). Wide-bandgap materials offer advantages in high-voltage, high-temperature, and high-power applications. Their ability to sustain large electric fields makes them suitable for radar transmitters, satellite communications, and base-station power amplifiers. Beyond RF electronics, these materials are also foundational in optoelectronics, including light-emitting diodes and semiconductor lasers. Diamond, the only wide-bandgap material in Group IV, has been investigated

as an ultimate semiconductor for extreme power and frequency applications, though its fabrication challenges remain significant.

A further layer of complexity arises in heterojunction devices, which exploit bandgap differences between dissimilar materials to improve injection efficiency and reduce base resistance. Commonly encountered material stacks include GaAs/AlGaAs, GaAs/InGaP, InP/InGaAs, and Si/SiGe. By carefully engineering these heterostructures, device designers achieve better trade-offs between speed, linearity, and noise, enabling circuits that would be impossible with homojunction devices alone.

From a device-level perspective, RF transistors can be broadly classified into FETs, BJTs, and HBTs, as summarized in Table 5.2. Each family encompasses multiple subtypes and spans several semiconductor technologies:

- BJTs are traditionally realized in silicon and germanium. While limited in speed compared with modern alternatives, they remain important for analog and mixed-signal integration.
- HBTs are essentially improved BJTs that use heterojunctions to enhance carrier transport. Examples include GaAs HBTs fabricated with AlGaAs/GaAs or GaInP/GaAs layer stacks, and InP-based HBTs with InP/InGaAs/InP structures. The vertical flow of carriers in HBTs enables high current densities and excellent high-frequency performance.
- FETs include JFETs, MOSFETs, MESFETs, and HEMTs. Among these, pseudomorphic (pHEMT) and metamorphic (mHEMT) variants are especially prominent in low-noise and high-frequency applications. Unlike HBTs, FETs are lateral devices, with current flowing horizontally through the channel.

The distinction between vertical and lateral transistor architectures represents one of the most consequential structural divides in RF device engineering, with far-reaching implications for noise performance, power handling, scalability, and integration strategy. In vertical devices, such as BJTs and HBTs, charge carriers flow perpendicular to the wafer surface through a stack of epitaxially grown layers. This geometry allows precise control over doping profiles, junction depths, and electric-field distributions, making vertical devices particularly well suited to high-speed and low-noise operation. In contrast, lateral devices such as HEMTs

**TABLE 5.2**
**Principal categories of RF transistors and their material systems**

| Category | Examples | Material Systems |
|---|---|---|
| BJT | Si BJT | Group IV (Si, Ge) |
| HBT | GaAs HBT, InP HBT, SiGe HBT | GaAs/AlGaAs, GaAs/InGaP, InP/InGaAs, Si/SiGe |
| FET | JFET, MOSFET, MESFET, HEMT (pHEMT, mHEMT) | GaAs, GaN, InP, Si |

conduct current parallel to the substrate surface, typically within a two-dimensional electron gas (2DEG) formed at a heterointerface. While lateral transport enables extremely high carrier mobility and excellent high-frequency performance, it introduces different constraints related to breakdown voltage, scaling, and integration density.

From a noise perspective, vertical bipolar devices benefit from fundamentally favorable transport physics. Their operation relies on minority-carrier injection across a forward-biased emitter–base junction, where noise mechanisms are closely tied to base resistance, recombination, and shot noise associated with collector current. In early silicon BJTs, relatively high base resistance and limited control over doping profiles imposed a floor on achievable noise figure, especially at microwave frequencies. The introduction of heterojunctions fundamentally altered this situation. By replacing the homojunction base–emitter interface with a heterojunction, most notably in SiGe HBTs, engineers gained the ability to independently optimize carrier confinement, base resistance, and emitter efficiency.

In SiGe HBTs, the graded germanium profile in the base region creates a built-in electric field that accelerates carriers toward the collector, reducing base transit time while simultaneously lowering base resistance. This dual improvement directly reduces thermal and shot-noise contributions, leading to substantially lower minimum noise figures compared with conventional silicon BJTs. As a result, SiGe HBTs exhibit excellent noise performance over a broad frequency range, extending well into the mm-wave domain. Their performance often rivals, and in some regimes surpasses, that of III–V-based bipolar technologies, despite being fabricated in a silicon-compatible process.

The noise advantages of SiGe HBTs become especially evident when contrasted with traditional GaAs or InP bipolar devices. While III–V HBTs benefit from higher electron mobility, their material systems often suffer from higher base resistance, more complex processing, and limited integration density. These factors can offset intrinsic mobility advantages, particularly when system-level noise and manufacturability are considered together. Consequently, SiGe HBTs occupy a uniquely favorable position: they combine low noise, high cutoff frequency, and excellent uniformity with the economic and integration benefits of mainstream silicon fabrication. This combination has made them a dominant technology for RF front ends in wireless communications, radar, and high-speed data links.

It is also instructive to compare these bipolar devices with lateral field-effect technologies traditionally used for low-noise microwave applications, such as GaAs MESFETs, and HEMTs. Historically, MESFETs were regarded as the benchmark for low-noise amplification at microwave frequencies due to their high electron mobility and relatively simple biasing. However, advances in SiGe HBT design have narrowed and, in many frequency ranges, eliminated this advantage. Modern SiGe devices can achieve noise figures comparable to, or even better than, GaAs MESFETs while offering superior integration with digital circuitry, higher yield, and lower production cost. This convergence has played a major role in shifting commercial RFIC development away from discrete III–V technologies toward highly integrated silicon-based solutions.

The architectural distinction between vertical and lateral devices also influences power handling and scaling behavior. Vertical devices naturally support higher current densities and better thermal conduction paths into the substrate, making them well suited for moderate- to high-power operation in compact layouts. Lateral devices, by contrast, distribute electric fields along the surface and are often preferred when extremely high electron mobility or ultra-high-frequency operation is required. These complementary strengths explain why both architectures continue to coexist, each optimized for different corners of the RF design space.

In summary, the evolution from early silicon BJTs to modern SiGe HBTs illustrates how heterojunction engineering has reshaped the fundamental noise-speed trade-off in RF transistors. By enabling independent optimization of transport, resistance, and band structure, heterojunction devices achieve performance levels that were once accessible only to exotic compound semiconductors. At the same time, their compatibility with large-scale silicon integration makes them uniquely attractive for modern RF systems, where low noise, high frequency capability, and economic manufacturability must coexist.

A broader comparison across FET technologies further illustrates how material choice and device physics govern achievable noise performance in RF systems. Among FET-based devices, those built on indium phosphide (InP) substrates consistently demonstrate the lowest achievable noise figures, particularly at microwave and mm-wave frequencies. This advantage arises primarily from the exceptionally high electron mobility available in InP-based heterostructures, especially when combined with InGaAs channels. The resulting two-dimensional electron gas exhibits extremely low scattering rates, high carrier velocity, and reduced channel resistance, all of which contribute directly to superior low-noise behavior.

In InP HEMTs, the separation of the dopant layer from the conduction channel through modulation doping minimizes impurity scattering while maintaining a high carrier concentration. This structural feature enables very high transconductance at comparatively low bias currents, which in turn reduces equivalent input-referred noise. As frequency increases into the mm-wave regime, where parasitic resistances and capacitances dominate, the advantages of InP-based HEMTs become increasingly pronounced. Their combination of low access resistance, high cutoff frequency, and strong electrostatic control allows them to maintain low noise figures where many other transistor technologies begin to degrade rapidly.

By contrast, GaAs-based MESFETs and HEMTs occupy an intermediate position in the performance spectrum. Although their electron mobility is significantly higher than that of silicon-based MOSFETs, it remains lower than that achievable in InP-based structures. As a result, GaAs devices typically exhibit higher minimum noise figures than their InP counterparts, especially above tens of gigahertz. Nevertheless, GaAs technologies continue to be widely used in microwave front ends due to their maturity, reliability, and favorable balance between performance and manufacturing complexity. They remain attractive for applications such as LNAs, mixers, and driver stages where extreme noise minimization is not the sole design constraint.

When considered alongside bipolar technologies, an important pattern emerges. SiGe HBTs, despite relying on silicon substrates, achieve noise figures that rival or surpass many III–V FET technologies over a broad frequency range. This outcome underscores the effectiveness of heterojunction engineering in bipolar devices, where vertical transport, reduced base resistance, and optimized carrier injection collectively suppress noise generation. Consequently, SiGe HBTs have become the dominant solution in highly integrated RF and mixed-signal systems, particularly where low noise must coexist with dense digital integration and cost-sensitive manufacturing.

Taken together, these comparisons reinforce a central theme in RF device engineering: no single transistor technology universally outperforms all others. Instead, each occupies a distinct region of the design space defined by noise, frequency, power handling, cost, and integration constraints. InP HEMTs represent the benchmark for ultimate low-noise performance at the highest frequencies and are therefore indispensable in applications such as radio astronomy receivers, deep-space communication links, and ultra-sensitive instrumentation. SiGe HBTs offer an exceptional balance between noise performance and large-scale integration, making them the technology of choice for most commercial and defense RFICs. GaAs-based devices remain relevant where moderate noise performance, robustness, and proven fabrication platforms are sufficient.

This diversity of device technologies reflects a broader principle in RF system design: optimal performance emerges not from a single "best" device, but from a careful alignment between material properties, device physics, circuit architecture, and system-level requirements.

The noise and frequency performance of BJTs is strongly tied to their emitter–base doping ratio. For efficient carrier injection, the emitter must be doped much more heavily than the base. While this improves injection efficiency and current gain, it introduces two opposing penalties: a lightly doped base increases the base resistance, while a heavily doped emitter increases the base–emitter junction capacitance. Together, these parasitics limit both the noise performance and the unity-gain frequency, $f_T$. As a result, conventional BJTs typically struggle to operate much beyond a few gigahertz, with practical upper limits around 3–4 GHz.

The introduction of HBTs overcame these limitations by exploiting different semiconductor materials in the emitter and base regions. By choosing an emitter material with a wider bandgap than the base, HBTs achieve far more favorable injection efficiency without requiring extreme doping levels. This reduces base resistance and junction capacitance simultaneously, enabling devices with higher $f_T$, higher maximum oscillation frequency $f_{MAX}$, and much improved noise performance. Consequently, HBTs can extend reliable operation well into the mm-wave regime and have largely supplanted BJTs in modern RF and microwave ICs. For this reason, the discussion that follows will focus on HBTs rather than conventional BJTs.

A parallel story unfolds in the field of FETs. Among these, InP-based HEMTs stand out as the ultimate low-noise devices, combining ultrahigh electron mobility with excellent scalability to mm-wave and sub-THz frequencies. However, their specialized fabrication processes make them costly and less accessible outside of niche applications. In contrast, silicon MOSFETs, while not as intrinsically quiet as III–V devices, have achieved sub-1 dB noise figures in advanced CMOS nodes. Their ubiquity, integration density, and process accessibility make CMOS LNAs a practical and increasingly attractive choice, especially where integration and cost are as critical as raw noise performance.

In summary, while BJTs laid the foundation, it is HBTs, HEMTs, and MOSFETs that define the active-device landscape of today's low-noise RF design, each offering a different balance of performance, frequency range, and manufacturability.

## HEMTS

MESFETs were the preferred choice for microwave and high-speed applications before HEMTs. In MESFETs, current is controlled by a Schottky barrier gate, and electrons moving through a doped channel are subject to scattering due to ionized dopants, which reduces the mobility that they should have in a GaAs substrate. Therefore, to improve noise performance, gain, and frequency, the separation of the dopant channel from the electron transport channel is desired. This concept was invented in 1979 by Takashi Mimura, who was researching to improve MOSFET performance. HEMTs are based on quantum wells, an important and vast research topic in quantum physics that has led to two Nobel prizes in 2000. Electrons are trapped in quantum wells due to the bandgap differences of adjacent materials. The adjacent materials are made of thin semiconductor layers of different bandgaps, and when the width of the middle layer is comparable to the de Broglie wavelength of the carriers, the carriers are confined in a two-dimensional planar region with practically no thickness, at the heterojunction. Electrons trapped in the layer are referred to as two-dimensional electron gas (2DEG) because the middle layer is undoped and the scattering of electrons due to ionized impurities is prevented. Hence, many free electrons float in the layer, attaining very high mobilities.

The performance of HEMTs is closely related to the density of trapped electrons, to the quality of semiconductor materials, such as defects, impurities, heterojunction roughness, and to the strain caused by lattice mismatches. Therefore, techniques used for HEMT epitaxial growth are complex and require slow growth rates of a few atomic layers per second.

The cap layer is a heavily doped GaAs layer that reduces the ohmic resistance at the source and drain metal contacts. The donor layer below the cap layer is uniformly doped and fully depleted under normal conditions; it diffuses carriers in the undoped AlGaAs spacer layer. The spacer layer separates physically donor atoms from the channel and ensures that the 2D electron gas is confined in one layer. The heart of the device is the AlGaAs/GaAs heterojunction between the spacer layer and the buffer layer. The channel consists of a narrow bandgap material of high conductivity because of the large number of carriers trapped in the quantum well and the absence of impurities and defects in the GaAs buffer layer. The buffer layer is made of an undoped wide bandgap material that constitutes a barrier for electrons, reducing their intrusion from the channel into the buffer and, consequently, the substrate. The depletion region, controlled by the gate voltage, consists of the doped and undoped AlGaAs layers under the gate.

The addition of indium to the GaAs channel material significantly increases electron mobility, $f_T$, and decreases the noise. However, the lattice constant of InGaAs is not compatible with GaAs, and AlGaAs introduces strain in the InGaAs layer, resulting in imperfections and making the fabrication of InP HEMTs expensive. Such a device is referred to as pseudomorphic (PHEMT). Metamorphic HEMTs (MHEMTs) are an alternative to InP HEMTs because of their lower cost. Imperfections due to strain in the InGaAs layer are removed as the lattice constant of the buffer layer is tailored to the indium content by using a compositionally graded buffer layer, resulting in strain relaxation. The performance of InP MHEMTs grown on a GaAs substrate is comparable to the performance of devices grown on an InP substrate at a lower cost and with less fabrication complexity. HEMTs combine the physics of quantum confinement with advanced epitaxial growth to deliver extraordinary electron mobilities and record-low noise figures. From early GaAs-based structures to today's InP and metamorphic variants, each generation has pushed the limits of gain, frequency, and integration. While their fabrication remains challenging and costly, HEMTs continue to define the state of the art in low-noise and high-frequency transistor technology.

To fully appreciate their circuit-level impact, however, it is not enough to understand the device physics alone. The behavior of an HEMT in an RF front end is ultimately captured through its small-signal equivalent circuit and noise model, which translates the underlying material and structural properties into parameters directly useful for designers. The next subsection explores these models, laying the foundation for analyzing gain, noise figure, and bandwidth in practical low-noise amplifiers.

## SMALL-SIGNAL EQUIVALENT CIRCUIT AND NOISE MODEL

In LNA design, it is not enough to understand the physics of the transistor alone; designers also need compact models that predict how the device behaves at a given frequency and bias. The small-signal and noise models of HEMTs provide exactly this: they translate the underlying device physics into scattering parameters and noise parameters that can be directly used in circuit design.

For discrete transistors, the key noise parameters are the minimum noise figure ($F_{min}$), the equivalent noise resistance ($R_n$), and the optimum source admittance ($G_{sopt} + jB_{sopt}$). These four parameters, measured at a given frequency and bias point, are sufficient to estimate the noise performance of an amplifier. In practice, manufacturers often provide tabulated data or measured curves, which can be used directly in circuit simulations. However, relying only on tables has clear drawbacks: large datasets are required for multiple bias conditions, and the values cannot be easily scaled or extrapolated to higher frequencies. For this reason, compact analytical and empirical models are developed, offering more flexibility and insight.

Analytical models are rooted in the physics of the device. They allow us not only to predict operation under given conditions but also to explore how fabrication processes or material choices affect performance. This makes them powerful for optimization and for building large-signal models. The downside is computational cost: detailed physics-based models may slow down circuit simulators. Empirical models, by contrast, are faster because they rely on fitting coefficients to measured data. They are easier to implement in design tools, but they require extensive and sometimes difficult measurements, especially under demanding conditions such as cryogenic operation.

Increasingly, numerical simulation tools such as TCAD bridge the gap, enabling virtual device modeling with techniques like Monte Carlo simulation that can account for microscopic noise sources.

Historically, noise modeling in FETs started with Van der Ziel, who proposed a simple thermal noise current source at the device's output. However, this early model did not capture the complexity of microwave operation. Pucel, Haus, and Statz extended it by including channel diffusion noise in the velocity-saturated region of GaAs MESFETs, an important addition because the channel has both a linear and a saturated section at microwave frequencies. Although these models improved understanding, some coefficients could not be precisely determined from experiments. A more practical approach was introduced by Fukui, who developed a semiempirical model with equivalent circuit elements. This model connected noise parameters to transistor geometry, material properties, and parasitic elements, but it still fell short at very high frequencies because it ignored carrier transit time effects.

The most widely adopted approach today is the Pospieszalski model, introduced in 1989. This model represents the intrinsic noise of the HEMT with two frequency-independent parameters: the effective gate resistance temperature ($T_g$) and the drain conductance temperature ($T_d$). When $T_g$ and $T_d$ are known, the frequency, bias, and temperature dependence of the noise parameters can be predicted with good accuracy. In practice, $T_g$ can often be approximated by the physical device temperature, while $T_d$ depends quasi-linearly on the drain current.

At a fundamental level, noise in HEMTs originates from several physical mechanisms, including thermal noise, shot noise, diffusion noise, generation–recombination noise, and flicker (1/*f*) noise. However, for microwave and mm-wave operation, typically above approximately 1 MHz, only two mechanisms contribute significantly to the total noise: thermal noise associated with parasitic resistances and channel-related diffusion noise. As a result, practical noise modeling relies on a compact small-signal representation that captures only these dominant effects while remaining analytically tractable.

In this representation, the transistor is decomposed into extrinsic and intrinsic parts. The extrinsic network models access and contact resistances that arise from the physical layout and metallization. These include the gate resistance $r_g$, source resistance $r_s$, and drain resistance $r_d$, which appear in series with the intrinsic device. Each of these resistances generates thermal noise proportional to the ambient temperature $T_a$, reflecting the physical temperature of the device environment. Their noise contributions are commonly modeled as independent voltage or current noise sources associated with each resistive element.

The intrinsic portion of the device represents the active channel itself. At the input side, a resistance $r_{gs}$ is introduced to model losses associated with the gate–source region, including distributed gate effects and channel access phenomena. This element is assigned an effective noise temperature $T_g$, which may differ from the ambient temperature and serves as a fitting parameter that captures excess internal noise mechanisms. In parallel with this input path, the channel is characterized by an output conductance $g_{ds}$, representing finite drain–source conductance caused by carrier transport and channel-length modulation effects. The noise associated with this conductance is modeled as a drain current noise source and is described using an effective temperature $T_d$.

Together, these intrinsic elements form the core noise-generating region of the transistor. The channel noise associated with $g_{ds}$ is often referred to as diffusion noise and dominates at microwave and mm-wave frequencies. It reflects random fluctuations in carrier flow along the channel and is strongly dependent on bias conditions and device geometry. The intrinsic gate resistance $r_{gs}$, although sometimes hidden inside compact models, plays a critical role in determining the minimum achievable noise figure, particularly at high frequencies where gate losses become significant.

The complete noise representation, therefore, consists of a cascade of parasitic resistances at the terminals feeding an intrinsic transistor characterized by a noisy transconductance path and a finite output conductance. Each resistive element contributes thermal noise, while the channel contributes diffusion noise. These noise sources are not all referenced to the same temperature, reflecting the fact that different physical regions of the device exhibit different effective noise mechanisms. This separation of temperatures between $T_a$, $T_g$, and $T_d$ is a defining feature of the widely used Pospieszalski noise model.

This formulation provides a compact yet physically meaningful framework for analysing HEMT noise behavior. It enables the derivation of key noise parameters such as the minimum noise figure, the optimum source impedance, and the equivalent noise resistance, all of which are central to LNA design. Although simplified, the model captures the dominant physical mechanisms governing noise performance at microwave and mm-wave frequencies and therefore remains a standard reference in both academic research and industrial RF design practice.

The optimum source reactance is expressed as

$$X_{opt} = \frac{1}{\omega C_{gs}} \tag{5.1}$$

while the optimum source resistance is given by

$$R_{opt} = \sqrt{\left(\frac{f_T}{f}\right)^2 \frac{r_{gs}}{g_{ds}} \frac{T_g}{T_d} + r_{gs}^2} \tag{5.2}$$

and together, these form the complex optimum source impedance,

$$Z_{opt} = R_{opt} + jX_{opt} \tag{5.3}$$

which defines the source impedance that minimizes noise. The minimum achievable noise temperature is then defined as

$$T_{min} = 2\frac{f}{f_T}\sqrt{g_{ds}r_{gs}T_gT_d + \left(\frac{f_T}{f}\right)^2 r_{gs}^2 g_{ds}^2 T_d^2} + 2\left(\frac{f}{f_T}\right)^2 r_{gs}g_{ds}T_d \tag{5.4}$$

and the associated noise conductance is given by

$$g_n = \left(\frac{f}{f_T}\right)^2 \frac{g_{ds}T_d}{T_0} \tag{5.5}$$

where $T_0$ is the IEEE standard reference temperature (290 K). The noise resistance $R_n$, which quantifies how sensitive the device noise is to impedance mismatches, can be written as

$$R_n = \frac{T_g}{T_o}r_{gs} + \frac{T_d}{T_o}\frac{g_{ds}}{g_m^2}\left(1+\omega^2 C_{gs}^2 r_{gs}^2\right) \tag{5.6}$$

Finally, the unity gain frequency is defined by the familiar relation

$$f_T = \frac{g_m}{2\pi C_{gs}} \tag{5.7}$$

where $g_m$ is the device transconductance.

With these relations, the noise temperature of the transistor can be expressed as

$$T_n = T_{min} + \frac{g_n}{R_s}T_o\left|Z_s - Z_{opt}\right|^2 \tag{5.8}$$

where $Z_s$ is the source impedance. This form makes clear that the total noise temperature depends not only on the intrinsic minimum noise temperature but also on how closely the source impedance matches the optimum.

The intrinsic device performance is also shaped by geometry, particularly the ratio between channel length and depletion width. When the channel is long, its behavior can be approximated by

$$g_m \approx q\eta_s \frac{\mu_o}{L_g} \tag{5.9}$$

where $q$ is the electron charge, $\eta_s$ the electron concentration in the channel, $\mu_0$ the carrier mobility, and $L_g$ the gate length. However, as device dimensions shrink and the channel length becomes comparable to the depletion width, short-channel effects degrade performance. In this regime, the transconductance is better described as

$$g_m \approx \frac{\varepsilon_r \varepsilon_o v_e}{d_{bar}} \tag{5.10}$$

where $\varepsilon_0$ is the vacuum permittivity, $\varepsilon_r$ the relative permittivity of the spacer layer, $d_{bar}$ its thickness, and $v_e$ the carrier velocity.

From these expressions, the minimum possible noise temperature can be approximated as

$$T_{min} \approx 2\frac{f}{f_T}\sqrt{g_{ds} r_{gs} T_g T_d} \tag{5.11}$$

and, since $T_d$ scales approximately with $I_{ds}$ while $f_T$ is proportional to $g_m$, $T_{min}$ itself can be related to the bias conditions through

$$f\left(V_{ds}, I_{ds}\right) = \frac{\sqrt{I_{ds}}}{g_m} \tag{5.12}$$

This relationship emphasizes that both drain current and transconductance play central roles in setting the ultimate noise floor of the device.

Because of this bias dependence, the choice of operating point is critical. The RF output power, linearity, and noise performance of FETs and HEMTs are all strongly influenced by $V_{ds}$ and $I_{ds}$. For every given drain voltage $V_{ds}$, there exists an optimum drain current $I_{ds}$ that minimizes $T_{min}$. This condition defines the best bias point for achieving low-noise operation. Measurements confirm these trends: plots of $T_{min}$ (or equivalently $NF_{min}$) show a linear increase with frequency, and it has been demonstrated that the optimum low-noise bias occurs at $V_{ds}$ = 0.5 V and $I_{ds}$ = 100 mA/mm.

The minimum noise temperature $T_{min,}$ is not only a theoretical construct but has also been evaluated experimentally for different transistor technologies. For example, (Schleeh *et al.* 2013) shows modeled values of $T_{min}$ at 6 GHz and 10 K for an InGaAs/InAlAs/InP HEMT, comparing data extracted with and without consideration of gate current. It highlights the importance of including gate current effects when assessing noise behavior at cryogenic temperatures.

The bias dependence of noise parameters is further evident in GaN devices. (Dasgupta *et al.* 2015) shows how the minimum noise figure ($NF_{min}$) and the noise resistance ($R_n$) vary with the drain current at 9 GHz and $V_{DS}$ = 9 V for a 2 × 100 µm GaN device [36]. Both $NF_{min}$ and $R_n$ depend strongly on the operating bias, underlining the need for careful optimization.

The frequency dependence of the minimum noise figure is illustrated in (Dasgupta *et al.* 2015), where $NF_{min}$ is plotted as a function of frequency for a GaN device biased at $V_{GS}$ = -3.4 V and $V_{DS}$ = 5 V (Dasgupta *et al.* 2015). The nearly linear increase of $NF_{min}$ with frequency confirms earlier observations and provides practical guidelines for circuit design across different frequency ranges.

The transconductance $g_m$ and cutoff frequency $f_T$ are also bias-dependent and typically reach a maximum at an intermediate drain current, as illustrated in (Ture 2016). This behavior is reinforced in (Ture 2016), which shows peak values of gm and $f_T$ versus $V_{DS}$ for a GaN HEMT. Such plots are useful for identifying bias regions that simultaneously optimize gain and high-frequency response.

Noise and gain characteristics can also be assessed together. (Ture 2016) shows NF and gain as a function of $V_{DS}$ for an AlGaN/GaN HEMT, where the optimal noise and gain regions often coincide, simplifying bias selection.

The bias drain current that achieves the lowest noise can be determined by simulation. Because transistor parameters are closely linked to device geometry, it is also important to investigate how scaling affects noise behavior. Transconductance depends strongly on the gate-to-channel spacing, making gate scaling techniques attractive for noise optimization.

The correlation between transconductance and noise is made explicit in (Nsele *et al.* 2013), where both $g_m$ and $NF_{min}$ peak in the same region of $V_{DS}$. This trend is also seen for a 0.15 μm AlInN/GaN HEMT, where $NF_{min}$ and $g_m$ are plotted versus $V_{DS}$ at 12 GHz and 26 GHz under $V_{GS}$ = -1 V.

Finally, (Khalil *et al.* 2008) shows that in common-source HEMTs, the regions of maximum gain and minimum noise also overlap, confirming that bias optimization for low noise typically coincides with conditions that also deliver favourable gain.

To better understand how device geometry and biasing conditions affect the small-signal and noise behavior of HEMTs, it is useful to summarize the scaling trends of the key circuit parameters. Table 5.3 provides an overview of how small-signal

**TABLE 5.3**
**Small signal parameters versus bias voltage, frequency, and gate width**

| Parameters | Bias | Frequency | Gate Width |
|---|---|---|---|
| $L_g$ | Independent | Independent | Increase with *W* |
| $L_s$ | Independent | Independent | Increase with *W* |
| $L_d$ | Independent | Independent | Increase with *W* |
| $R_g$ | Independent | Independent | Proportional |
| $R_s$ | Independent | Independent | Inverse proportional |
| $R_d$ | Independent | Independent | Inverse proportional |
| $C_{pgs}$ | Independent | Independent | Increase with *W* |
| $C_{pgd}$ | Independent | Independent | Increase with *W* |
| $C_{pds}$ | Independent | Independent | Increase with *W* |
| $R_{gs}$ | Independent | Independent | Inverse proportional |
| $R_{ds}$ | Independent | Independent | Inverse proportional |
| $R_{gd}$ | Independent | Independent | Inverse proportional |
| $g_m$ | Dependent | Independent | Proportional |
| $C_{gs}$ | Dependent | Independent | Proportional |
| $C_{gd}$ | Dependent | Independent | Proportional |
| $\tau$ | Dependent | Independent | Increase with *W* |

*Source:* Hassan *et al.* (2019).

parameters vary with bias, frequency, and gate width. These relationships are not absolute values but general trends that guide both modeling and design optimization.

The table highlights a few important design insights. Extrinsic resistances such as $R_s$ and $R_d$ decrease with larger gate width, improving noise performance, while capacitances such as $C_{gs}$ and $C_{gd}$ scale proportionally with width and directly influence the cutoff frequency $f_T$. Similarly, the transconductance gm increases with width, though at the expense of higher capacitances. These trade-offs underline why device scaling must balance gain, bandwidth, and noise. The tabulated relationships, therefore, provide a practical reference when interpreting both the equivalent circuit models and the bias-dependent noise results presented earlier.

The tabulated relationships, therefore, provide a practical reference when interpreting both the equivalent circuit models and the bias-dependent noise results presented earlier.

A commonly used way to analyse the behavior of AlGaN/GaN HEMTs is through a detailed small-signal equivalent representation that separates extrinsic parasitic elements from the intrinsic device physics. Such a representation provides a structured framework for understanding how geometry, materials, and biasing conditions collectively determine gain, noise, and bandwidth. In this model, the external access regions between the intrinsic transistor and the measurement terminals are represented by parasitic resistances at the gate, source, and drain. These resistances originate from contact resistance, metallization resistance, and access regions formed during fabrication, and they play a critical role in determining high-frequency loss and thermal noise contributions.

Beyond the extrinsic elements, the intrinsic core of the AlGaN/GaN HEMT is described by a transconductance element that links the gate–source voltage to the drain current, together with capacitances that model charge storage and electrostatic coupling within the device. The most important intrinsic capacitances include the gate–source capacitance, the gate–drain capacitance, and the drain–source capacitance, each of which reflects physical overlap, fringing fields, and charge modulation in the two-dimensional electron gas (2DEG). These capacitances are strongly dependent on device geometry, particularly gate length and gate-to-channel separation, and they directly determine the cutoff frequency and high-frequency roll-off of the transistor.

The output characteristics of the device are captured through a finite drain–source conductance, which accounts for channel-length modulation and velocity saturation effects. This conductance introduces both gain reduction and additional noise, especially at high drain bias. In GaN-based HEMTs, the wide bandgap and strong polarization fields enable high breakdown voltages and large current densities, but they do not eliminate intrinsic output conductance. As a result, accurate modeling of this element remains essential for predicting realistic amplifier performance.

An important feature of this small-signal representation is its ability to connect physical scaling trends to circuit-level behavior. Reducing gate length shortens the carrier transit time across the channel, which directly increases the transconductance and improves the intrinsic cutoff frequency. At the same time, shorter gate lengths generally reduce intrinsic capacitances, particularly the gate–source and gate–drain components, thereby extending the usable frequency range. These effects collectively improve both gain and noise performance at microwave and mm-wave frequencies.

However, device scaling also introduces competing effects. As the gate length shrinks, access resistances tend to increase unless aggressive contact and layout optimizations are applied. Similarly, short-channel effects can increase output conductance and degrade voltage gain. The balance between reduced intrinsic capacitance and increased parasitic resistance, therefore, determines whether scaling produces a net improvement in radio-frequency performance. This trade-off is especially important in GaN technologies, where high electric fields and large carrier densities amplify both beneficial and adverse scaling effects.

Quantitatively, reductions in gate length are consistently associated with lower minimum noise figure and higher available gain at a fixed operating frequency. Shorter devices exhibit higher transconductance per unit width, which lowers the equivalent input-referred noise and improves impedance matching to typical source impedances. At the same time, the increased gain helps suppress the relative impact of parasitic noise sources located later in the signal path. These trends explain why aggressive gate-length scaling remains one of the most effective strategies for achieving low-noise operation in modern AlGaN/GaN HEMTs, particularly at frequencies around 10-20 GHz and beyond.

At a fixed frequency such as 12 GHz, shrinking the gate length leads to a monotonic improvement in both minimum noise figure and small-signal gain, provided that short-channel degradation is properly controlled. This behavior reflects the fundamental relationship between carrier transit time, transconductance, and noise generation. Devices with shorter gates operate closer to the ballistic transport regime, where carriers experience fewer scattering events, resulting in reduced excess noise and improved high-frequency response. Consequently, aggressive lithographic scaling has become a central design objective in GaN technology platforms aimed at LNAs and high-performance front ends.

The combined small-signal and noise modeling framework thus provides a powerful lens through which scaling trends can be interpreted. By linking physical dimensions to parasitic resistances, intrinsic capacitances, transconductance, and output conductance, the model enables designers to predict how changes in geometry translate into measurable performance metrics such as gain, cutoff frequency, and minimum noise figure. This approach also highlights why device optimization cannot be reduced to a single parameter: improvements in one aspect often introduce penalties elsewhere, requiring careful trade-offs at the technology and circuit levels.

In summary, the small-signal representation of AlGaN/GaN HEMTs serves not merely as a schematic abstraction but as a bridge between fabrication technology and RF system performance. It captures how scaling, material properties, and parasitic effects jointly shape noise and gain behaviur. By understanding these relationships, designers can better exploit the inherent advantages of wide-bandgap devices while managing their limitations, ultimately enabling high-performance, low-noise operation in demanding microwave and mm-wave applications.

Table 5.4 provides a compact summary of how $f_T$ and $f_{MAX}$ vary with gate length $L_g$. Both figures of merit increase strongly as the gate is shortened. These results

**TABLE 5.4**
**Impact of $L_g$ on $f_T$ and $f_{MAX}$**

| | $L_g$ (μm) | $W_g$ (μm) | $f_T$ (GHz) | $f_{MAX}$ (GHz) |
|---|---|---|---|---|
| #A | 0.15 | 2 × 75 | 41 | 70 |
| #B | 0.25 | 2 × 100 | 24 | 46 |
| #C | 0.70 | 2 × 100 | 15 | 25 |

highlight why device engineers push toward ever smaller gate dimensions, despite the fabrication challenges that arise at sub-0.1 μm scales.

The characteristics of HEMTs presented and discussed in this section allow us to propose a methodology for biasing HEMTs in ultra-low noise applications. From Table 5.3 and Table 5.4, $R_{opt}$ decreases with increasing gate length. Variations of $R_{opt}$ are more than proportional to variations in $L_g$, meaning that for a given operating frequency, $R_{opt}$ can be tuned close to the source resistance (typically 50 Ω) simply by adjusting the gate length. Once $R_{opt}$ is set, the HEMT bias point, defined by gate voltage and drain current $I_{DS}$, can be chosen to simultaneously minimize $NF_{min}$ and maximize gm.

In summary, the small-signal equivalent circuit and noise modeling framework, together with the scaling trends captured in figures and tables, provide a coherent picture of how geometry, biasing, and parasitics govern the noise performance of HEMTs. These insights form the foundation for practical LNA design, where device choice and bias optimization are inseparably linked.

## SI MOSFETS

While III–V devices such as HEMTs dominate the ultra-low-noise domain, silicon MOSFETs remain central to RF and mixed-signal circuit design. Their appeal lies not only in low cost and mature fabrication, but also in their compatibility with large-scale integration and system-on-chip solutions. Despite intrinsic challenges such as limited linearity and flicker noise, continued scaling and process improvements have made CMOS-based RF circuits viable in many practical applications. This section reviews the fundamentals of MOSFET operation and highlights their relevance in RF design.

## BACKGROUND

The widespread use of RF devices and the growth of the semiconductor industry have been driven by CMOS technology, which has become a contender for BiCMOS and group III–V semiconductors, owing to the low cost, high integration, and improved performance of CMOS devices. Such advantages allow for achieving mixed-mode and system-on-chip designs, which reduce the chip count and the pin count. MOSFETs were initially developed for digital applications and, despite their improved performance for analog applications, they achieve lower linearity and gain in comparison with bipolar transistors. But their main disadvantage comes from a high flicker noise, which is associated with their conduction nature: because the source–drain current flows near the gate interface, low-frequency noise is affected by carriers trapped at the Si-$SiO_2$ interface. Therefore, the performance of MOSFET-based VCOs, PLLs, mixers, and direct conversion receivers is limited by the phase noise resulting from transistors' flicker noise. Low-frequency noise in oscillators or mixers can be converted to higher-frequency noise and impair performance.

The operation of FETs, such as CMOS transistors and MOSFETs, is based on the modulation by the gate voltage of the cross-sectional area of the effective conduction channel between the drain and the source. For a JFET, this modulation results from a reverse-biased PN junction between the gate and the substrate. When the bias voltage is zero, JFETs are conducting; reduced conduction is achieved by extending the depletion region with the reverse bias. The type of FET used in integrated circuits is the MOSFET, which is a four-terminal device, where $t_{ox}$ is the oxide width, and $L$ and $Z$ are the channel length and width, respectively. In the literature, $L$ corresponds to $L_{eff}$, the effective length of the channel, and the transistor width is generally expressed by $W$. It has been claimed that MOSFET is the most abundant device made by mankind (Sah 1991).

The operating principles of MOSFETs are based on induced charges in the substrate at the interface between the oxide and the substrate. This interface is a plate of a two-terminal MOS capacitor. The analysis of charge regimes at the semiconductor interface and of voltage distribution between the insulator and the semiconductor allows us to investigate MOSFET characteristics such as the threshold voltage and the drain–source saturation voltage. Such analysis is found in (Ytterdal *et al.* 2003).

For the $n$-channel MOSFET, conduction between the drain and the source is established when a positive voltage is applied to the gate, separated from the p-substrate by the thin dielectric. The source and the substrate are at the same potential and are grounded. However, there is a potential drop in the top part of the substrate in the $x$-direction: the source–substrate PN junction is forward biased, and electrons are injected into the substrate. This injection happens mostly at the surface where the higher potential difference is higher and has an exponential relation with the gate voltage, as per the PN junction theory. In addition, the positive potential repels holes but attracts electrons from the substrate, creating, with the injection of electrons from the source, a thin conduction channel. As the gate voltage increases, the injected charge increases and finally screens the substrate from the gate, ending the exponential relationship. The screening of the substrate results from the change of charge repartition in the dielectric and in the substrate.

The capacitors $C_{ox}$ and $C_s$ form a voltage divider. As the gate–substrate voltage $V_{GB}$ increases, injected charges increase in the substrate capacitor, which increases $C_s$. Therefore, the potential $\psi_s$, which bias the source–substrate junction, become less dependent on $V_{GB}$.

More increase in the gate voltage is absorbed by more charge in the dielectric. This transition is gradual and is associated with the threshold voltage $V_T$. This dependence between the drain–source current $I_D$ and the gate voltage $V_{GS}$ is shown in Pulfrey (2010).

Such a device is referred to as an enhancement-mode device, in opposition to a depletion-mode device such as the JFET. Depending on the conduction channel, MOSFETs can be of N-channel type or -channel type. Each type can be in depletion mode or enhancement mode. For *n*-channel MOSFETs, charge transport is achieved by minority electrons in a P substrate, whereas for *p*-channel MOSFETs, the transport is performed by minority holes in an N substrate.

A conduction channel between the source and the drain is physically implanted in depletion-mode MOSFETs. For *n*-channel MOSFETs, this conduction channel is made of lightly doped *n*-type material at the surface of the substrate, between highly doped source and drain. Depletion-mode MOSFETs are turned on at zero gate voltage and can be used as load resistors. Enhancement mode MOSFETs are turned off at zero gate voltage, and conduction is only achieved when a voltage is applied to the gate. They are particularly used as switching devices. The circuit symbols of MOSFETs are shown in Figure 5.1 (Field effect transistors: a comprehensive overview from basic concepts to novel technologies by Valizadeh Pouya).

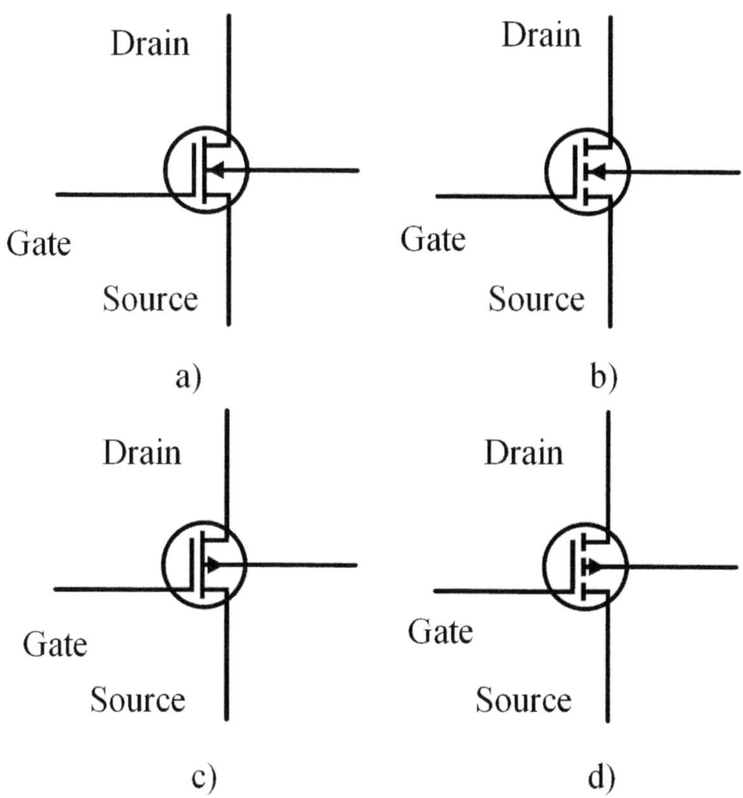


**FIGURE 5.1** Circuit symbols: (a) depletion mode n-channel MOSFET, (b) enhancement mode n-channel MOSFET, (c) depletion mode p-channel MOSFET, and (d) enhancement mode p-channel MOSFET.

In summary, MOSFETs form the backbone of CMOS technology and remain indispensable despite their noise limitations. Their versatility, with enhancement and depletion modes, *n*- and *p*-channel variants, makes them suitable across a wide range of analog, digital, and RF applications. The next section builds on this foundation by focusing on how MOSFET device physics and noise mechanisms translate into RF performance, particularly in the context of low-noise design.

## SMALL-SIGNAL EQUIVALENT CIRCUIT

In this section, we focus on the n-channel MOSFET operating in enhancement mode, which is the most used configuration in RF CMOS design. While a complementary *p*-channel device can be created by simply reversing the doping types in each region, the underlying operational principles remain analogous, so the discussion here applies generally to both device types.

Consider an *n*-channel MOSFET biased for normal enhancement-mode operation. The gate is positively biased relative to the source to induce an inversion layer of electrons at the semiconductor–oxide interface, forming a conduction channel. The drain is biased with a positive voltage relative to the source, creating an electric field along the channel. The density of mobile charge carriers per unit area near the surface, denoted as $Q_n$, determines the channel conductivity. As the drain–source voltage ($V_D$) increases, the potential along the substrate surface gradually decreases from source to drain, resulting in a nonuniform channel cross-section.

When $V_D$ is below the saturation voltage $V_{DS}$, the transistor operates in the linear (or triode) region. In this regime, the inversion channel extends continuously from source to drain, and the channel conductance, and hence the drain current $I_D$, is modulated primarily by the gate voltage $V_G$. The drain current increases approximately linearly with $V_D$, as the induced electrons flow through a nearly uniform channel.

As $V_D$ approaches $V_{DS}$, the channel near the drain begins to "pinch off." In other words, the channel cross-section narrows progressively toward the drain end due to the reduction of surface potential. At the pinch-off voltage, $V_D = V_{DS}$, the channel is fully constricted at the drain, and further increases in $V_D$ do not significantly increase the drain current $I_D$. The transistor enters saturation mode. In this mode, the current $I_D$ remains nearly constant, controlled mainly by $V_G$, while the pinched-off region sustains a strong electric field that accelerates electrons from the source-to-drain channel into the drain terminal. The resistance of this pinched-off region is typically negligible compared with the channel, so it does not impede the current flow.

The relationship between the drain current $I_D$, the drain–source voltage $V_D$, and the gate voltage $V_G$ can be conceptually visualized as a three-dimensional surface. At lower gate voltages, the channel forms slowly, and the corresponding $I_D$ is relatively small. As $V_G$ increases above the threshold voltage $V_T$, the inversion channel becomes stronger, leading to higher $I_D$. Below the saturation voltage, the surface forms a gently rising plane where $I_D$ increases with $V_D$. Beyond the saturation point, the surface flattens, representing constant $I_D$, illustrating the transition from linear to saturation operation. Dashed lines can be used in a conceptual diagram to indicate the onset of saturation for each $V_G$ level, highlighting how the drain current reaches a plateau even as $V_D$ increases further.

This textual description allows readers to mentally reconstruct the small-signal biasing and operating regions of the MOSFET without relying on graphical figures, while maintaining sufficient detail to follow subsequent derivations and equations. The discussion of pinch-off, surface potential, and channel modulation provides the necessary intuition for interpreting the drain current behavior as a function of both $V_D$ and $V_G$.

The total surface charge density $Q_s(x)$ at the semiconductor interface can be expressed as the sum of two contributions:

$$Q_s(x) = Q_n(x) + Q_b(x) \tag{5.13}$$

where $Q_b(x)$ is the fixed charge due to ionized acceptors in the depletion region, and $Q_n(x)$ is the mobile charge density in the inversion layer. At saturation, $Q_n(x)$ becomes negligible at the drain end compared to $Q_b(x)$.

At saturation, $Q_n(x) \ll Q_b(x)$ at the drain. The drain drift current $I_D$ is proportional to $Q_n(x)$ and to the velocity of the electron $v(x)$. This is expressed as

$$I_D = -WQ_n(x)v(x) \tag{5.14}$$

and for long-channel devices, the electric field in the channel is relatively weak, and the electron velocity is simply proportional to the mobility $\mu_n$ and the local electric field $E(x)$. At high fields, relevant for short-channel MOSFETs, this linear model no longer holds due to velocity saturation and other short-channel effects (Tsividis and McAndrew 2011). Assuming $E(x)$ is approximately constant along the channel, we may write:

$$I_D = -WQ_n(x)\mu_n E \tag{5.15}$$

where some considerations are required to evaluate $Q_n(x)$. The surface of the substrate is depleted of charges when the gate voltage is equal to $V_T$, and the inversion layer exists only when $V_G > V_T$. Therefore, when mobile charges are in the inversion layer, they are subject to the voltage $V_G - V_T - V(x)$. It is assumed that all mobile charges reside at the surface of the substrate. Hence, there is no potential drop across mobile charges in the $y$-direction, and $\psi_s$ applies to fixed ions only. Finally, because the inversion layer is depleted of holes, their charge can be neglected. A new expression of $I_D$ can therefore be derived

$$I_D = -WC_{ox}\left(V_G - V_T - V(x)\right)\mu_n \frac{dV(x)}{dx} \tag{5.16}$$

where $C_{ox}$ is the dielectric (oxide) capacitance per unit area. This formulation underpins the derivation of the MOSFET's current–voltage characteristics and provides the foundation for its equivalent small-signal models.

The integration of the channel yields the well-known expression for the drain current in the linear region (also called the triode region, by analogy with vacuum tubes):

$$I_D = \mu_n C_{ox} \frac{W}{L}\left[\left(V_G - V_T\right)V_D - \frac{V_D^2}{2}\right] \tag{5.17}$$

and at saturation, $V_D = V_{DS}$ and $V_{DS} = V_G - V_T$. When the drain voltage reaches this saturation condition, the device enters the active region. In this regime, the drain current no longer increases linearly with drain voltage but instead follows the so-called square-law dependence:

$$I_D = \frac{\mu_n C_{ox}}{2}\frac{W}{L}\left(V_G - V_T\right)^2 \tag{5.18}$$

where this quadratic dependence is characteristic of long-channel MOSFETs operating in saturation. Under these conditions, the transistor behaves as a current source with high output resistance. The threshold voltage itself is not a fixed constant but varies with the substrate bias $V_{SB}$. It can be expressed as:

$$V_T = V_{TO} + \gamma\left[\sqrt{2\phi_f + V_{SB}} - \sqrt{2\phi_f}\right] \tag{5.19}$$

where $V_{SB}$ is the substrate bias voltage, $V_{TO}$ is the threshold voltage when the source–substrate voltage $V_{SB}$ is zero, and $\phi_f$ is the Fermi level, usually about 0.3 V. Additionally,

$$\gamma = \frac{1}{C_{ox}}\sqrt{2q\epsilon N_A} \tag{5.20}$$

where $N_A$ is the doping density and $\epsilon$ is the permittivity of silicon. The oxide capacitance is given by

$$C_{ox} = \frac{\epsilon_{ox}}{t_{ox}} \tag{5.21}$$

where $\epsilon_{ox}$ is the permittivity of the dielectric (oxide).

The transconductance is the derivative of the drain current versus the gate voltage and is given by

$$g_m = \mu_n C_{ox} \frac{W}{L}\left(V_G - V_T\right) \tag{5.22}$$

and because the transistor is in saturation, it yields

$$g_m = \sqrt{2\mu_n C_{ox} \frac{W}{L} I_D} \tag{5.23}$$

The transconductance appears to be a monotonic function of $I_D$. Although the growth of $g_m$ decreases with $I_D$, $g_m$ never reaches a maximum. The growth of $g_m$ is limited because of parasitic source and drain resistances not included in the previous analysis. As the growth of $g_m$ decreases, $g_m$ is degraded and reaches a maximum. When the sum of the source and drain parasitic resistances $R_{SD}$ is considered (Subramanian *et al.* 2006), it is found that

$$I_D = \frac{\mu C_{ox} W}{2L} \frac{\left(V_G - V_T\right)^2}{1 + \frac{\mu C_{ox} W}{L}\left(V_G - V_T\right) R_{SD}} \tag{5.24}$$

The expression shows that $g_m$ is degraded because of $R_{SD}$. When an excitation is applied to the gate, charges are redistributed in the channel. The time required for this redistribution is modeled by the transconductance delay in multiplying $g_m$ by $exp(j\omega\tau)$, where $\tau \approx C_{gs}/5g_m$. The effects of the transconductance delay are mostly neglected for frequencies below $2\pi/\tau$ because they become important only for frequencies higher than $f_T$.

From previous paragraphs, long-channel MOSFETs in linear mode have two operating regimes in relation to the gate voltage: the weak inversion regime, also known as the sub-threshold regime, and the strong inversion regime. For the weak inversion regime, $V_G < V_T$. The boundary between the weak and strong inversion regimes, $V_{ON,}$ is arbitrary and defined by

$$V_{ON} = V_T + n\frac{kT}{q} \tag{5.25}$$

where $n$ is always greater than unity, typically about 3 or 4. The drain current corresponding to $V_{ON}$ is $I_{ON}$. The drain current in the weak inversion regime is exponential. The drain current is negligible but not zero, as sometimes presented in the literature.

$$I_D = I_{ON}\left[exp\left(\frac{q\left(V_G - V_T\right)}{nkT} - 1\right)\right] \tag{5.26}$$

In the strong inversion regime, the overdrive voltage, defined as $V_G - V_T$, is several times greater than $kT/q$. The drain current corresponds to the square law.

The equations above are basic and do not account for all effects that characterize MOSFETs. Such effects include the body-effect, "hot" electrons that create hole–electron pairs through impact ionization, hot electrons trapped in the oxide causing a gate current, non-quasi-static (NQS) effects, channel modulation, and short channel effects. Short channel refers to a high electric field in the channel. The electron velocity reaches a maximum of about $10^5$ m/s when the electric field in the channel is about $10^6$ V/m. At such a velocity, electrons are scattered by high-energy phonons, and their velocity becomes less dependent on the electric field, causing saturation of the drain current.

The small-signal model of the MOSFET, where the source and the substrate are not at the same potential. The substrate is viewed as another gate terminal at $v_{bs}$ and $g_{mb}$ is the body transconductance associated with $v_b$. In general, $v_{bs}$ is zero, which simplifies the model.

The expressions of parameters related to Figure 5.2 are provided next. The substrate transconductance $g_{mb}$ is given by

$$g_{mb} = \frac{\gamma}{2\sqrt{2\phi_f + V_{SB}}} g_m \tag{5.27}$$

and in triode mode, the gate–source and the gate–drain capacitances are given by

$$C_{gs} = C_{gd} = \frac{C_{ox}WL}{2} \tag{5.28}$$

and in saturation mode, the gate–drain capacitance is zero, while the gate–source capacitance is given by

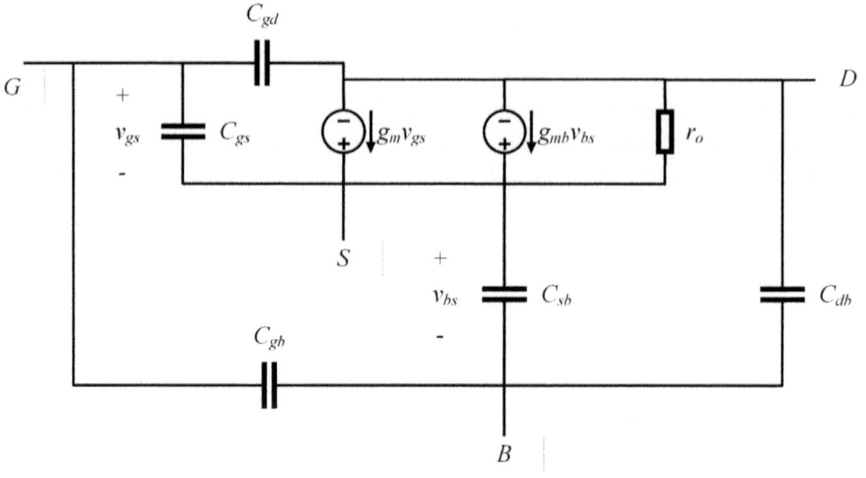


**FIGURE 5.2** Small-signal MOSFET equivalent circuit.

$$C_{sb} = \frac{C_{sb0}}{\left(1+\frac{V_{SB}}{\psi_0}\right)^{0.5}} \tag{5.29}$$

and

$$C_{sb} = \frac{C_{db0}}{\left(1+\frac{V_{DB}}{\psi_0}\right)^{0.5}} \tag{5.30}$$

where $C_{sb0}$ is the zero-bias source-bulk capacitance, $C_{db0}$ is the zero-bias drain-bulk capacitance, and $\psi_0$ is the built-in potential across the junction's source-bulk and drain-bulk. This potential is expressed as

$$\psi_0 = V_T \ln\frac{N_A N_D}{n_i^2} \tag{5.31}$$

where $V_T = kT/q \cong 26\ mV$ at 300 °K, $N_A$ and $N_D$ are the acceptors and donors doping densities in atoms/cm$^3$, and $n_i$ is the intrinsic carrier concentration of the silicon undoped semiconductor at 300 K.

The zero-bias source-bulk and drain-bulk capacitances, $C_{sb0}$ and $C_{db0}$, are approximated by

$$C_{sb0} = A_S\left(C_{j0}\right) + P_S\left(C_{jsw0}\right) \tag{5.32}$$

and

$$C_{db0} = A_D\left(C_{j0}\right) + P_D\left(C_{jsw0}\right) \tag{5.33}$$

where $C_{j0}$ is the source-bulk, drain-bulk junction capacitance at zero volts bias. $C_{jsw0}$ is the source, drain periphery capacitance at zero bias. $A_S$ and $P_S$ are the source and perimeter areas, respectively, and $A_D$ and $P_D$ are the drain and perimeter areas, respectively.

$$r_0 = \frac{1}{\lambda I_D} \tag{5.34}$$

where $\lambda$ is the channel-length modulation parameter.

The expression of the drain current when the transistor is in saturation and channel modulation effects are considered becomes

$$I_D = \frac{\mu_n C_{ox}}{2} \frac{W}{L} (V_G - V_T)^2 (1 + \lambda V_D) \tag{5.35}$$

and the unity gain frequency $f_T$ is given by

$$f_T = \frac{g_m}{2\pi (C_{gs} + C_{gd} + C_{gb})} \tag{5.36}$$

Other figures of merit are $f_{MAX}$ and $G_{MAX}$. The maximum oscillation frequency $f_{MAX}$ is defined when the power gain is unity, with $f_{MAX} \cong \sqrt{f_T / 8\pi R_g C_{gd}}$, where $R_g$ is the gate resistance. The gate resistance is discussed in the following subsections. The maximum power gain $G_{MAX}$ is the power gain of the transistor when its input and output impedances are matched to the source and load impedances, respectively. It can be shown that the maximum power gain is approximated by the expression: $G_{MAX} \cong f_T / 8\pi R_g C_{gd} f^2$.

## NOISE IN MOSFETS

Noise in MOSFETs arises from several physical mechanisms, each linked to either the intrinsic channel behavior or to extrinsic device elements such as contacts and parasitic resistances. The main sources can be grouped as follows:

- Thermal noise from extrinsic resistances: Parasitic resistances at the gate, source, and drain generate Johnson–Nyquist noise, just as in any passive resistor. Similarly, resistances associated with the substrate also contribute to thermal fluctuations.
- Channel thermal noise: This is the dominant noise mechanism in MOSFETs. Because the MOSFET operates as a voltage-controlled resistor, the channel itself has a resistive nature and produces thermal noise proportional to its conductance.
- Induced gate noise: The thermal noise in the channel is capacitively coupled to the gate via the gate-channel capacitance. This coupling generates a correlated noise current at the gate terminal, often referred to as induced gate noise.
- Flicker noise (1/*f* noise): At low frequencies, noise is dominated by flicker noise. In MOSFETs, this originates from trapping and de-trapping of carriers at defects located at the Si–$SiO_2$ interface. While its impact diminishes at microwave frequencies, flicker noise is critical in RF systems because it can up-convert into phase noise in oscillators or mixers.
- Gate shot noise: In deeply scaled technologies with ultrathin gate oxides, carriers can tunnel through the $SiO_2$ layer, giving rise to a gate leakage current. This leakage current exhibits shot noise, which can no longer be ignored in nanometer-scale MOSFETs.

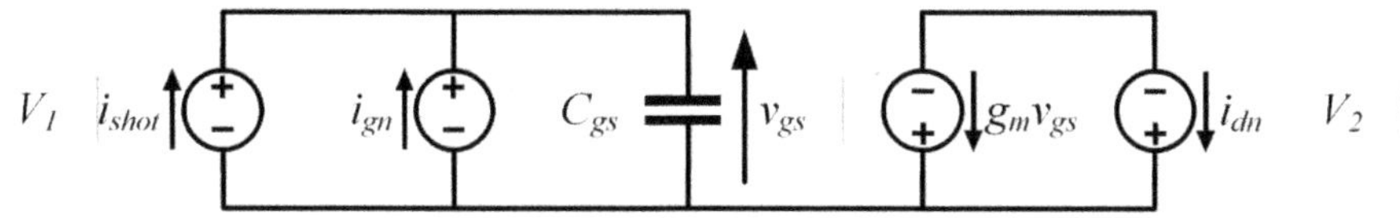


**FIGURE 5.3** Noise sources in the MOSFET.

For predictive modeling across technology nodes and over a broad range of operating conditions, noise models must remain grounded in these physical origins rather than being treated as purely empirical fitting parameters.

When the source and substrate are tied together, a common condition in integrated circuit design, the contribution from substrate thermal noise is effectively shorted out. Furthermore, if the thermal noise associated with gate, source, and drain parasitic resistances is neglected, the picture simplifies considerably. Under these assumptions, the dominant noise sources relevant at RF frequencies reduce to those shown in Figure 5.3.

This simplified model highlights the fact that channel thermal noise, induced gate noise, and flicker noise together set the fundamental noise floor of MOSFET-based RF circuits.

The shot noise is given by

$$i_{shot}^2 = 2qI_G\Delta f \tag{5.37}$$

where $I_G$ is the gate current, which is generally very small.

The gate noise is expressed by

$$i_{gm}^2 = \frac{4kT\,\delta\omega^2 C_{gs}^2}{5g_{d0}}\Delta f \tag{5.38}$$

where $g_{d0}$ is the drain–source conductance when $V_D$ is zero, and $V_G$ unaltered, $\omega$ is the angular operating frequency, and $\delta$ *is* the gate coefficient, which is equal to 4/3 for long-channel transistors. The value of the gate coefficient $\delta$ is twice $\gamma$, the channel noise coefficient, a parameter defined in the expression of the channel thermal noise.

The channel thermal noise is expressed as

$$i_{dm}^2 = 4kT\,\gamma g_{d0}\Delta f \tag{5.39}$$

where the channel noise coefficient $\gamma$ is unity when $V_D$ is zero, and $\gamma$ is equal to 2/3 for long-channel devices in saturation, which is the main operating regime of RFICs. For long-channel devices in the linear regime, $\gamma$ is comprised between 1 and 2/3. For short-channel devices, $\gamma$ is greater than 2/3 and varies with $V_G$.

The channel noise is independent of frequency but increases when the length of the channel decreases or when the gate voltage increases because the channel conductance increases. The relative contribution of the main types of noise to the MOSFET noise figure versus frequency for a 0.25 µm long-channel MOSFET is depicted in Scholten *et al.* (1999).

The gate-induced noise $i_{gn}^2$ and the channel thermal noise $i_{dm}^2$ are correlated because they both come from fluctuations in the channel. The normalized correlation coefficient is defined by

$$c = \frac{\overline{i_{gn} i_{dn}^*}}{\sqrt{\left(\overline{i_{gn}^2}\right)\left(\overline{i_{dn}^2}\right)}} \tag{5.40}$$

where the correlation coefficient is imaginary and equal to 0.395j for long-channel devices and comprised between 0.3j and 0.35j for short-channel devices (Manku 1999).

For short-channel transistors, $g_{d0}$ and $\gamma$ are complex functions of transistor parameters (Asgaran *et al.* 2004) because of short-channel effects. The use of the long-channel model for short-channel devices requires the adaptation of $\gamma$ and additional empirical parameters that are based on bias, geometry, and frequency. Some reported values of $\gamma$ for short-length devices corresponding to several technologies are depicted in Navid *et al.* (2007).

Navid *et al.* (2007) show that channel thermal noise is much larger in short-channel devices than predicted in long-channel transistors. When devices get smaller and the channel shorter, the aspect ratios of FETs are preserved during scaling. These ratios are defined as the ratios between the gate length and active vertical dimensions of the transistor, such as $t_{ox}$, the depth of the source and drain junctions $r_j$, the depletion depths of the source and the drain, $W_s$ and $W_d$, respectively. The aspect ratios are low for short-channel devices. The empirical expression below can be used to mark the transition from long channel to short channel (Ytterdal *et al.* 2003).

$$L < L_{min}\left(\mu m\right) = 0.4\left[r_j\left(\mu m\right)t_{ox}\left(\dot{A}\left(W_d - W_s\right)^2\left(\mu m^2\right)\right)\right]^{1/3} \tag{5.41}$$

The operation of the transistor is affected by short-channel effects when $L < L_{min,}$ and the actual length of the channel is not relevant. However, for short-channel devices, referred to as nanoscale transistors, the length of the channel is typically below 100 nm. The transition between long-channel and short-channel regimes can also be evaluated by the ratio $(V_G - V_T)/LE_{sat}$, where $E_{sat}$ is the electric field at which electron velocity is half of the velocity at low field. Short-channel effects are dominant when this ratio is large. Although $E_{sat}$ is related to the process, a typical value is $4 \times 10^6$ *V/m* (Lee 2004).

The effects of high electric fields on noise in MESFETs are not well known. Noise in short-channel devices is much higher than predicted, several factors higher (Abidi 2014), and even up to an order of magnitude for some cases. Theoretical studies and experimental results have led to contradictory explanations: in some literature, the additional noise was attributed to hot-carrier effects or to the substrate noise, which was not confirmed. It is understood that, when the length of the channel is inferior to the mean free path of electrons, carriers cease to be in quasi-thermal equilibrium with the lattice, and the thermal noise model is no longer fully applicable, as the scattering process is reduced. The flow of electrons is semi-ballistic and is also controlled by a potential barrier close to the source. Such a potential barrier creates shot noise. The current in the device is then limited by the channel resistance and the potential barrier. Short-channel MOSFETs are semi-ballistic devices in which every carrier injected from the source changes the electric field in the device until the carrier is absorbed by the drain. If the average time duration taken by a carrier from the source to reach the drain is less than the average time between consecutive injections, the injection probability of a new carrier is reduced until an equilibrium between the injection of electrons and their absorption is established. The PSD of this auto-regulated flow of electrons is less than $2qI_{DC,}$ and the process is referred to as partial shot noise suppression, a process well known in vacuum tubes (Carbone *et al.* 2022). In vacuum tubes, the anode–cathode current is limited by space charge when the potential between the anode and the cathode is low. The PSD of the noise is then given by $2K_SqI_A$, where $I_A$ is the current and $K_s$, known as the Fano factor, is less than unity. It is proposed in Navid *et al.* (2007) that noise in short-channel transistors should be characterized by $k_s$ rather than the channel noise coefficient $\gamma$ used for long-channel devices.

## NOISE PARAMETERS IN MOSFETS

Analytical expressions of noise parameters, such as noise resistance, minimum noise figure, and optimum source admittance, are required by the designer to predict the noise performance of MOSFET circuits before their fabrication. It is desired to express these parameters as functions of biasing and transistor geometry because the designer can set the biasing and the geometry to achieve, as closely as possible, the desired $NF_{min}$, $Y_{opt,}$ and $R_n$.

The noise models discussed so far are compact and include only intrinsic elements. Extrinsic elements such as source resistance, drain resistance, and gate resistance were neglected and ignored. However, at gigahertz frequencies, the gate resistance $R_g$ as well as the gate capacitance should be considered (Ytterdal *et al.* 2003). The intrinsic gate resistance consists mainly of the polysilicon resistance at low frequencies. At RF, the gate is affected by distributed transmission line effects and by NQS effects in the channel, and the gate resistance becomes dependent on the biasing. The resistance of the gate affects impedance matching, power gain, and impairs the noise figure. It should then be considered for both impedance matching and noise matching. In addition, the gate resistance reduces $f_{MAX}$. The expression of the gate resistance is given by

$$R_g = \frac{R_{gsh}W}{3Mn^2L} \tag{5.42}$$

where $R_{gsh}$ is the gate sheet resistance, $L$ and $W$ are the channel length and width, respectively, $n$ is the number of fingers, and $M = 1$ when the gate is connected at one

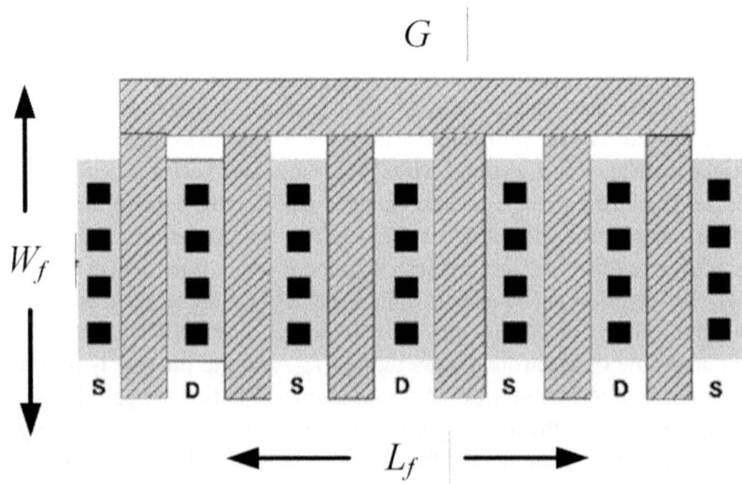


**FIGURE 5.4** MOSFET layout with multiple-gate-finger ($M = 4$, $n = 3$).

end, while $M = 4$ when the gate is connected at both ends. The factor 3 is derived from the distributed properties of the transistor gate. The layout of a multiple-gate MOSFET is depicted in Figure 5.4.

The expressions of noise parameters are derived from the work in Asgaran *et al.* (2004), which is based on the BSIM model in Ytterdal *et al.* (2003). In this work, the channel is divided into two regions: a region where the velocity of electrons is proportional to the electrostatic field in the channel and a region where the velocity is saturated. It is assumed that noise from the region with velocity saturation can be neglected because Ohm's law fails in this region, as the velocity of electrons is not affected by voltage fluctuations. Both regions are depicted in Asgaran *et al.* (2004). Although the model in Asgaran *et al.* (2004) is validated by measurements, it fails to predict the PSD when the drain–source voltage $V_D$ is zero. The voltage in the channel is given by $V_G - V_T - \alpha V(x)$ in this model. Hence, the charge per unit area becomes

$$Q_n(x) = -C_{ox}\left(V_G - V_T - \alpha V(x)\right) \tag{5.43}$$

where $\alpha$ is the bulk charge effect coefficient. The value of the bulk charge effect coefficient is 1.2. When the voltage $V_D$ is nonzero, the depletion region between the source and the drain is not uniform, and $V_T$ varies along the channel, even when the voltage between the substrate and the source is zero. It is assumed that $[\omega(C_{gs} + C_{gd})R_g]^2 \ll 1$, which is verified up to 10 GHz for a 0.18 μm process. Because $C_{gs}$, $C_{gd}$, and $R_g$ decrease with scaling, the inequality criterion will hold for frequencies higher than 10 GHz. The gate-induced noise is neglected in the determination of noise parameters. It is found that its inclusion improves only marginally the accuracy of $NF_{min}$ at 10 GHz, by less than 5%.

The expressions of noise parameters as functions of device characteristics and biasing are

$$R_n \approx R_g + \frac{\beta I_D}{g_m^2} \tag{5.44}$$

and

$$G_{opt} \approx \left(\frac{f}{f_T}\right)\frac{\sqrt{\beta R_g I_D}}{R_n} \tag{5.45}$$

and

$$B_{opt} \approx -\left(\frac{f}{f_T}\right)\frac{\sqrt{\beta I_D}}{g_m R_n} \tag{5.46}$$

and the minimum noise figure is defined by

$$NF_{min} \approx 1 + 2\left(\frac{f}{f_T}\right)\sqrt{\beta R_g I_D}\left[1 + \left(\frac{f}{f_T}\right)\sqrt{\beta R_g I_D}\right] \tag{5.47}$$

where

$$\beta = \frac{1}{V_{DS}} + \frac{\alpha^2 V_{DS}}{3\left(V_G - V_T\right)^2} \tag{5.48}$$

and

$$V_{DS} = \frac{E_C L\left(V_G - V_T\right)}{\alpha E_C L\left(V_G - V_T\right)}, \tag{5.49}$$

where $E_C$ is the critical electric field in the saturation region.

It is found that $R_n$ is independent of frequency, that $Z_{opt}$ varies as $1/f$, and that $NF_{min}$ increases linearly with $f$, the operating frequency. The noise parameters of MOSFETS with very short channels are given below, as functions for the transistor geometry.

$$R_n = R_g + \frac{W C_{ox} \mu_{eff}}{2 g_m^2}\frac{1}{L} \tag{5.50}$$

and where

$$G_{opt} \approx \left(\frac{f}{f_T}\right)\frac{\sqrt{WC_{ox}\mu_{eff}\left(V_G - V_T\right)R_g}}{R_n\sqrt{2}}\frac{1}{\sqrt{L}} \tag{5.51}$$

it is seen that

$$B_{opt} \approx -\left(\frac{f}{f_T}\right)\frac{\sqrt{WC_{ox}\mu_{eff}\left(V_G - V_T\right)}}{2g_m R_n}\frac{1}{\sqrt{L}} \tag{5.52}$$

and therefore

$$NF_{min} \approx 1+2\left[\left(\frac{f}{f_T}\right)^2\frac{WC_{ox}\mu_{eff}\left(V_G - V_T\right)R_g}{2L}+\left(\frac{f}{f_T}\right)\sqrt{\frac{WC_{ox}\mu_{eff}\left(V_G - V_T\right)R_g}{2L}}\right] \tag{5.53}$$

where $\mu_{eff}$ is the effective mobility, a function of the gate voltage.

Because $g_m$ and $f_T$ are proportional to $W/L$, $R_n$ and $NF_{min}$ decrease with $L$. However, for MOSFETs with extremely short channels, $g_m \approx WC_{ox}v_{sat}$ where $v_{sat}$ is the saturation velocity of electrons. In this case, $R_n$ and $NF_{min}$ are inversely related to the channel length.

The dependencies between the parameters of the MOSFET, referred to as figures of merit, and its geometry for scaling purposes are summarized in Table 5.5 (Manku 1999), where $r_i$ is the channel charging resistance and is in series with $C_{gs}$.

For the designer, the first questions arising consist of choosing the device size, selecting $V_G$ and $V_D$ for biasing, and setting the geometries of the transistor. In a

**TABLE 5.5**
**Scaling rules: dependence between figures of merit and MOSFET parameters**

| Parameters | Number of Fingers | Width |
|---|---|---|
| $g_m$ | $N_f$ | $W_f$ |
| $r_i$ | $1/N_i$ | $1/W_f$ |
| $R_g$ | $1/N_i$ | $W_f$ |
| $C_{gs}$, $C_{gd}$ | $N_i$ | $W_f$ |
| $r_d$, $r_s$ | $1/N_i$ | $1/W_f$ |
| $NF_{min}$ | Independent | $\sim W_f$ |
| $G_n$ | $N_i$ | $W_f$ |
| $R_{opt}$, $X_{opt}$ | $1/N_i$, $1/N_i$ | $\sim 1/W_f$, $1/W_f$ |
| $f_{MAX}$ | Independent | $\sim 1/W_f$ |
| $f_T$ | Independent | Independent |

design, the only variables that the designer can control are $V_G$, $V_D$ and $W$. When the transistor comprises several gates, $W = N_f \times W_f$, where $W_f$ is the width of each transistor and $N_f$ is the number of fingers.

The gate polysilicon resistance $R_g$ is related to the geometry of the transistor. A low $R_g$ is desired to achieve a low $NF_{min}$ and $R_g$. The gate resistance decreases when the gate electrode is connected to both ends of the polysilicon, when $L$ and $N_f$ increase or $W$ decreases.

For a given technology, the channel length is determined by the fabrication process and cannot be controlled by the designer. The overdrive voltage $V_G - V_T$ is then a function of the ratio $I_D/W$. Therefore, there is an optimum current per unit width ($I_D/W_{opt}$) where $NF_{min}$ is minimum. It was found in K. Yau *et al.* (2007) and Dickson *et al.* (2006) that the current densities corresponding to optimum $NF_{min}$, maximum $f_T$, and $f_{MAX}$ are approximately 0.15, 0.3, and 0.2 mA/μm, respectively, irrespective of foundries and technology nodes for n-channel MOSFETs. This is achieved when the ITRS constant field scaling guidelines, referring to the two-dimensional electric field profile, are applied during the fabrication. For recent technologies, $I_D/W_{opt}$ is comprised between 0.2 and 0.3 mA/μm$^2$. This change is due to strain engineering and due to the increase of the electric field between the gate and the channel. Generally, the optimum current can be determined by simulation of $NF_{min}$ versus $I_D/W$. In fact, low noise design techniques based on constant-current-density used for bipolar transistors can be used as well for MOSFETs (Voinigescu *et al.* 1997), but for MOSFETs, the current density is not related to the operating frequency, while it increases with frequency for BJTs and SiGe HBTs. In addition, these design techniques are independent of the semiconductor material (e.g., Si, SiGe, InP, GaAs, GaN, etc.). However, the sizing of MOSFETs and bipolar devices in low noise design differs in main ways. To understand the difficulty of simultaneous noise and power matching in MOSFETs, it is convenient to compare MOSFETs with BJTs. The optimum noise reactance $X_{opt}$ for a BJT is given by (Manku 1999)

$$X_{opt} \approx \frac{-X_{in}}{1+\left(f_T/f\right)^2 \beta^{-1}} \tag{5.54}$$

where $X_{in}$ is the input reactance of the BJT, $f$ is the operating frequency, and β is the DC gain.

When $\beta \gg (f_T/f)^2$, $X_{opt} \approx -X_{in}$, which is a condition for simultaneous noise and power matching for BJTs. This condition is generally realized when $f_T$ is three to five times $f$. Such a condition is realized because the base shot noise and the collector shot noise are not correlated and are not dependent on frequency. However, for MOSFETs, the induced gate noise and the channel thermal noise are correlated. In addition, while the channel noise is frequency independent, the induced gate noise varies with $f^2$, which makes it difficult to achieve simultaneous noise and power matching. It is shown in subsequent sections that $NF_{min}$ is independent of $W$ in BJTs, but $R_{opt}$ is inversely proportional to $W$. Therefore, $W$ can be adjusted to achieve $R_{opt}$. For MOSFETs, $NF_{min}$ increases with $W$, which constitutes a major difference in designing with MOSFETs in comparison with BJTs (Manku 1999). However, it was found that the effect of the induced gate noise on $X_{opt}$ is small, which makes it possible to simultaneously perform noise and power matching (Manku 1999). In fact, $X_{opt}$ and the imaginary part of the input impedance are within 15% for FETs and HBTs: impedance matching allows for matching at the same time the noise, with a 10% error margin in $X_{opt}$, when a series inductor is used at the input. $R_{opt}$ has an important feature that will be discussed later in this book, but that can already be highlighted in this paragraph, because of its importance in noise matching: $R_{opt}$ is invariant to lossless feedback and lossless impedance transformations.

The MOSFET with one finger can be viewed as a two-port network with $F_{min}$, $G_n$, and $Z_{opt}$ as noise parameters. If $N_i$ two-port networks are connected in parallel, the resulting noise factor is given by

$$F = F_{min} + N_i \frac{G_n}{R_S} \left| R_S - \frac{Z_{opt}}{N_i} \right|^2 \tag{5.55}$$

and because a MOSFET with $N_i$ fingers is equivalent to $N_i$ MOSFETs in parallel, the noise factor of a MOSFET with $N_i$ fingers is given above. It shows that $F_{min}$ is unchanged, hence $NF_{min}$, $R_{n,}$ and $Z_{opt}$ are divided by $N_i$, respectively. Because $R_{opt}$ scales as $1/N_i$, varying $N_i$ can be used to achieve $R_s = R_{opt}/N_i$ for noise matching.

The following methodology is proposed for biasing and scaling the MOSFET for minimum noise:

- The MOSFET technology with the shortest channel length is desired because $NF_{min}$ increases with $L$.
- An optimal finger width $W_f$ ensuring that the gate resistance $R_g$ and parasitic capacitances are balanced is first selected for the operating frequency. It is proposed to start with the smallest possible $W_f$.
- The optimum drain current density $I_d/W_f$ corresponding to the minimum $NF_{min}$ is determined.
- While the optimum drain current density and the optimal finger width are held constant, vary the number of gate fingers $N_i$ to achieve $R_S = \Re\{Z_{opt}\}$, where $R_s$ is the ohmic impedance of the source, typically 50 Ω.
- Verify if $g_m$ of $G_{MAX}$ is acceptable. If the area of the transistor is very large, the power consumption may be high, and the gain may be impaired because of the Miller effect. A trade-off between low noise, power consumption, and gain may be required.

This section has proposed, for short-channel devices, simple noise models with analytical expressions suitable for hand calculations of noise parameters. The intrinsic noise sources considered in these models are the channel noise, the gate-induced noise, and the gate thermal noise. These models, for which $\gamma = 2/3$ in saturation and $\gamma = 1$ in linear mode, include channel modulation effects and electron velocity saturation. For short-channel devices, the channel noise coefficient $\gamma$ used is greater than 2/3 and varies with the geometry and the biasing. Several other fitting coefficients dependent on geometry, bias, and frequency are used and determined by measurements. Models based on short-channel device physics are complex and still require fitting coefficients. The transport of charges in these models is based on semi-ballistic transport of electrons near the source, and electrons drift in the drain region, which results in shot and thermal noise. Cases for which $NF_{min}$ is underestimated by foundry models or where noise is not properly modeled by the manufacturer have been reported (Belostotski *et al.* 2006; Shen *et al.* 2017).

## SOI

Silicon-on-insulator (SOI) technology represents a major architectural evolution in CMOS device fabrication, motivated by the need to mitigate parasitic effects that increasingly limit performance as device dimensions shrink and operating frequencies rise. In its most basic form, SOI replaces the bulk silicon substrate beneath the active devices with a thin insulating layer, typically silicon dioxide, commonly referred to as the buried oxide (BOX). Above this insulating layer lies a thin silicon film in which the active devices are fabricated. This structural modification fundamentally alters the electrical environment of the transistor and has far-reaching implications for noise behavior, speed, power consumption, and isolation.

The motivation for SOI emerged from the growing dominance of parasitic capacitances and substrate-related losses in deeply scaled CMOS technologies. In bulk CMOS, the transistor channel resides directly on a conductive silicon substrate, which acts as an unintended participant in device operation. Electric fields extend into the substrate, charge is capacitively coupled into it, and minority carriers can be generated or collected in regions that are not part of the intended signal path. These effects lead to substrate noise coupling, latch-up susceptibility, increased junction capacitances, and degradation of high-frequency performance. SOI addresses these limitations by physically separating the active region from the bulk substrate, thereby breaking many of the parasitic paths that dominate at radio and mm-wave frequencies.

From a structural perspective, an SOI wafer consists of three main layers: a handle wafer that provides mechanical support, a buried oxide layer that provides electrical isolation, and a thin silicon device layer in which transistors are formed. Depending on the thickness of the silicon layer, SOI technologies are commonly classified as partially depleted (PD-SOI) or fully depleted (FD-SOI). In partially depleted SOI, the silicon film is thick enough that a neutral body region remains beneath the inversion channel, whereas in fully depleted SOI the silicon film is sufficiently thin that the entire body is depleted when the device is biased. This distinction has important consequences for electrostatics, noise behavior, and variability.

One of the most immediate benefits of SOI technology is the strong reduction in parasitic junction capacitances. Because the source and drain no longer form PN junctions with a conductive substrate, their capacitances are limited primarily to the device geometry and the BOX interface. This reduction directly improves high-frequency performance by lowering the total capacitive load seen at critical nodes, thereby increasing both the cutoff frequency $f_T$ and the maximum oscillation frequency $f_{MAX}$. From an RF design perspective, this translates into higher gain, wider bandwidth, and improved power efficiency compared to equivalent bulk CMOS implementations.

Another key advantage of SOI is improved isolation between neighboring devices. The buried oxide acts as an effective barrier against substrate coupling, suppressing the propagation of noise through the silicon bulk. This feature is particularly valuable in mixed-signal and RF systems, where digital switching noise can otherwise couple into sensitive analog front ends. In LNAs, oscillators, and mixers, reduced substrate coupling directly improves phase noise, noise figure, and spurious performance. As a result, SOI has become increasingly attractive for highly integrated RF transceivers, where digital, analog, and RF blocks must coexist on the same die.

SOI also mitigates several classical parasitic effects encountered in bulk CMOS. Latch-up, caused by parasitic bipolar action between pnp and npn structures in the substrate, is largely eliminated because the insulating layer interrupts the necessary current paths. Similarly, body effect is significantly reduced, especially in fully depleted SOI, because the body potential is either fixed or weakly coupled to external voltages. This leads to more stable threshold voltages and improved device matching, both of which are beneficial for analog and RF circuit design.

From a noise perspective, SOI technology offers both advantages and new considerations. The reduction in junction capacitances and substrate coupling generally lowers flicker noise coupling from the substrate and reduces modulation of channel charge by substrate potential fluctuations. This can lead to improved low-frequency noise performance, especially in circuits sensitive to $1/f$ noise up-conversion, such as oscillators and mixers. Furthermore, reduced parasitic capacitance lowers the effective noise contribution of resistive elements by reducing the frequency-dependent loading seen at sensitive nodes.

However, SOI devices are not entirely free of noise-related challenges. In partially depleted SOI, the floating body can accumulate charge due to impact ionization or capacitive coupling, leading to history-dependent behavior and excess low-frequency noise. This so-called "floating body effect" can introduce hysteresis, kinks in the output characteristics, and additional variability in analog circuits. Techniques such as body contacts, body ties, or the use of fully depleted SOI processes are commonly employed to mitigate these effects. Fully depleted SOI offers superior electrostatic control and reduced variability, making it increasingly attractive for RF and low-noise applications.

Thermal behavior is another important consideration in SOI technologies. The buried oxide has a much lower thermal conductivity than silicon, which can impede heat removal from the active region. As a result, self-heating effects can be more pronounced, especially in high-power or high-current-density applications. Elevated device temperature influences carrier mobility, threshold voltage, and noise, and must therefore be accounted for in RF design. Nonetheless, careful layout, thermal-aware biasing, and advanced packaging techniques can alleviate many of these concerns.

In the context of RF modeling, SOI MOSFETs require modified small-signal and noise models that account for the altered substrate coupling and body behavior.

Traditional bulk MOSFET models often assume a well-defined substrate node with fixed potential, whereas SOI devices introduce additional internal nodes associated with the body and buried oxide interfaces. These nodes can store charge and interact with the channel dynamically, influencing transconductance, output conductance, and noise sources. Consequently, compact models for SOI incorporate additional capacitances, resistances, and noise terms to capture floating-body effects, self-heating, and electrostatic coupling accurately.

Despite this added modeling complexity, SOI has proven highly attractive for RF system-on-chip implementations. The combination of high isolation reduced parasitic capacitances, improved linearity, and compatibility with advanced CMOS scaling has led to widespread adoption in commercial RF processes. Many modern RF front ends for wireless communication, including LNAs, mixers, and voltage-controlled oscillators, are implemented in SOI-based technologies. These processes strike a balance between the cost advantages of silicon and the performance traditionally associated with compound semiconductors.

In summary, SOI technology provides a structural and electrical environment that is particularly well suited to high-frequency and low-noise circuit design. By decoupling active devices from the bulk substrate, SOI reduces parasitic losses, suppresses unwanted coupling, and enhances scalability. While challenges such as self-heating and floating-body effects must be carefully managed, the overall benefits make SOI a key enabling technology in modern RF integrated circuits. Its role bridges the gap between conventional bulk CMOS and more exotic compound semiconductor platforms, offering a compelling compromise between performance, integration density, and manufacturing cost.

## SIGE HBT TECHNOLOGY

An extensive review of the SiGe HBT technology is provided in this section. The extent of the review is justified by the advantages of HBTs over Si-BJTs and MOSFETs. SiGe HBTs provide better performance than Si-BJTs in low noise design and provide high integration levels with a cost structure like standard BiCMOS processes. They consume less power than GaAs devices at much less cost. Although their noise performance is outperformed by HEMTs, SiGe HBT processes are less costly than HEMT processes. Like HEMTs, but contrary to Si-BJTs, they can operate at very low temperatures, which reduces noise. The basic equations describing Si-BJTs' operation are valid as well for SiGe HBTs. Hence, these equations are reviewed in this section.

The evolution of bipolar technology led in the 1990s to the emergence of SiGe HBTs. Heterojunction structures are semiconductor materials made from two or more semiconductors whose bandgaps differ, such as AlGaAs, InP, GaAs, InGaAsP, and SiGe. This evolution, which was rather a revolution, was difficult compared to other compound semiconductor technologies, such as AlGaAs/GaAs, that were already available. The difficulty comes from misfit dislocations at the interface between two semiconductors with relatively dissimilar lattice spacing. This was the case for Si and SiGe, contrary to Si and AlGaAs/GaAs.

The lattice constant of a SiGe crystal $Si_{1-x}Ge_x$, where $x$ is the fraction of Ge, is given by Vegard's rule:

$$a_{Si_{1-x}} = a_{Si} + x\left(a_{Ge} - a_{Si}\right) \tag{5.56}$$

which is approximately equal to

$$a_{Si_{1-x}} = 0.5431 + 0.0227x \tag{5.57}$$

where $a$ is the lattice constant of SiGe, $a_{Si}$ is the lattice constant of silicon, which is equal to 0.5431 nm, and $a_{Ge}$, the lattice constant of germanium, is equal to 0.5658 nm.

The lattice mismatch between Si and Ge is 4.2% at room temperature, and the total mismatch between a $Si_{1-x}Ge_x$ crystal and a Si substrate is expressed by the linear relationship

$$f = \frac{\left(a_{SiGe} - a_{Si}\right)}{a_{Si}} \tag{5.58}$$

which is approximately equal to $0.042x$.

It was found, from research carried out in the 1980s, that misfit dislocations could be avoided for thin SiGe layers and for moderate Ge concentrations, typically below 30%, and that, under these conditions, the epitaxial deposition of a thin layer of SiGe on a Si crystal induces stress. Hence, this epitaxial deposition allows the growth of SiGe on Si under strain and without misfit dislocations, which was a breakthrough in the emergence of SiGe HBTs. In fact, the theory that led to the development of SiGe HBTs was introduced much earlier in 1957 by Herbert Kroemer, who subsequently received a Nobel Prize in physics in 2000 (Banerjee *et al.* 2005). The SiGe layer is referred to as pseudomorphic when SiGe is under compressive strain.

SiGe HBTs are created by the incorporation of Ge atoms into the base crystal lattice. The atomic separation for Ge has been higher than for Si; a compressive strain is created in the silicon crystal, which leads to the reduction of the overall bandgap of the base material, the band gaps of Ge and Si being 0.66 eV and 1.12 eV, respectively. Electrons injected from the emitter are subject to a lower barrier and to an electrical accelerating field that increases with the concentration of Ge in the base area (Gao 2010). Compared to Si, the mobility of carriers and the speed of operation are increased.

For BJTs, a low base resistance is desired for high-frequency response, hence a high base doping. But a high current gain is also desired, which requires low base doping to reduce the injection of holes into the emitter for npn transistors. These requirements are contradictory for BJT but are resolved by HBTs because they have simultaneously low base resistance and high current gain.

Because the principles underlying the operation of BJTs and SiGe HBTs are fundamentally similar, it is helpful to revisit the construction of a standard BJT in its planar integrated form. Early point-contact transistors served as the progenitors of modern devices, but their fabrication relied on manual assembly of germanium contacts and could not be reliably scaled. With the advent of the planar process, pioneered in the late 1950s and early 1960s, it became possible to create all transistor regions on a single silicon substrate in a reproducible and lithographically defined manner.

In a typical planar BJT, the collector region forms the bulk of the substrate, providing mechanical support and serving as the main current-carrying layer. Above the collector, a lightly doped epitaxial layer may be grown to reduce collector resistance and improve breakdown voltage characteristics. Within this layer, a p-type base region is defined using precise ion implantation or diffusion, forming a thin layer through which electrons or holes travel. The base thickness is carefully controlled to optimize transistor gain, transit time, and noise performance.

The emitter, often heavily doped, is then diffused or implanted into the base, forming the junction responsible for charge injection. The resulting emitter–base junction determines the efficiency of carrier injection and strongly influences the current gain (beta) of the device. Metal contacts are deposited on the emitter, base, and collector regions to enable external connections. These contacts are patterned through photolithography to ensure alignment with the underlying semiconductor layers, minimizing parasitic resistances and capacitances.

Additional process steps define isolation regions around the transistor, preventing leakage currents to adjacent devices and enabling high-density integration of multiple transistors on a single chip. Passivation layers are applied to protect the surface from contamination and mechanical damage. The final device, although compact, contains all the essential junctions, regions, and contacts necessary to function as a reliable amplifier or switch.

Conceptually, the planar BJT can be visualized as a cross-sectional stack: starting from the substrate, the collector forms the base foundation, followed by the epitaxial layer, the p-type base region, and the n-type emitter, all topped with metal contacts and isolation structures. The device exhibits the classic behavior of current control through a minority carrier injection from the emitter into the base, with the collector current being a nearly linear function of the base current in the active region. Thermal and shot noise contributions arise from the junctions, while parasitic resistances and capacitances, defined by the physical layout, influence high-frequency performance.

The isolation between adjacent devices on a chip is a critical aspect of planar BJT design, ensuring that leakage currents do not degrade performance. In early implementations, isolation was achieved by diffusing p-type regions around each transistor and extending them from the front-end-of-line (FEOL) layers down to the substrate. These p-type guard regions form reverse-biased PN junctions when the substrate is connected to the most negative voltage in the circuit, effectively confining carriers to the intended device regions and preventing electrical cross talk.

Under the emitter and collector contacts, n+ regions are incorporated to reduce the resistance of the ohmic contacts, which is essential for both power efficiency and high-frequency operation. Early planar BJTs, used in operational amplifiers and TTL circuits, typically exhibited a transition frequency ($f_T$) limited to approximately 500

MHz. This limitation arose due to parasitic capacitances and resistances inherent in the original diffusion-based isolation and contact structures.

To overcome these constraints, a series of process innovations was introduced. Deep trench isolation replaced the surrounding p-diffusions, with trenches lined by oxide sidewalls and filled with polysilicon to enhance mechanical stability and electrical isolation. These trenches reduced parasitic base–emitter and base–collector capacitances, which are critical parameters that directly influence the transistor's high-frequency response. Additionally, polysilicon layers were introduced underneath the base contacts to lower base resistance, which improves current handling and reduces thermal noise. For the collector contact, connection to a buried n+ region within the collector layer further reduced series resistance, enabling faster charge transport through the device.

The combined effect of these enhancements, a double-polysilicon structure, deep trench isolation, and buried contacts, dramatically improved the device performance. The enhanced BJT could now achieve a transition frequency exceeding 30 GHz, making it suitable for high-frequency analog circuits and RF applications. These innovations also enabled the integration of BJTs with CMOS devices on the same die, giving rise to BiCMOS technology. In such mixed-technology circuits, BJTs serve as high-speed drivers capable of efficiently switching large capacitive loads, while CMOS devices provide low-power digital functionality.

Conceptually, this advanced BJT structure can be visualized as a multilayer stack: deep trenches surround the active regions for isolation, polysilicon layers reduce resistances, buried n+ regions enhance collector performance, and the p-type base is carefully controlled to optimize gain and bandwidth. Parasitic capacitances and resistances are minimized through these process enhancements, enabling superior high-frequency response.

To understand the interplay between transistor currents, noise sources, and physical parameters such as doping concentrations and base width, it is helpful to examine the behavior of charge carriers in an NPN BJT under typical operating conditions. In standard active-mode operation, the base–emitter junction is forward-biased, while the collector–base junction is reverse-biased. This biasing configuration allows electrons injected from the heavily doped n-type emitter to flow through the thin p-type base toward the collector, while only a small fraction recombines in the base. The forward bias at the base–emitter junction reduces the potential barrier, allowing efficient injection of carriers, whereas the reverse bias at the collector–base junction ensures rapid collection of these carriers with minimal recombination, contributing to the high current gain characteristic of BJTs.

The primary currents in the transistor can be conceptually described as follows: the emitter current ($I_E$) consists mainly of electrons injected into the base from the emitter, plus a small portion of holes flowing in the opposite direction. The base current ($I_B$) is primarily due to recombination of electrons within the base region, and it represents a small fraction of the total emitter current, typically 1–5% in a well-designed device. The collector current ($I_C$) is the electrons that successfully traverse the base without recombining; $I_C$ closely tracks $I_E$ minus $I_B$, with the relation $I_C \approx \alpha \times I_E$, where $\alpha$ is the common-base current gain, typically close to unity.

From a physical perspective, the flow of carriers is influenced by the doping profiles and the base width. A thinner base allows more electrons to reach the collector, improving current gain, while higher doping in the base can increase recombination, affecting both gain and noise characteristics. Thermal motion of carriers contributes to shot noise in the collector current, and fluctuations in recombination in the base contribute additional noise. The relative contributions of these mechanisms depend on both the geometry and the operating conditions of the transistor.

By describing these carrier flows and current paths in text, the reader can visualize the essential features typically shown in a schematic: the forward-biased base–emitter junction, the reverse-biased collector–base junction, the emitter injecting electrons, the base recombination, and the collection of electrons at the collector. This detailed textual description effectively replaces the illustrative schematic, providing a conceptual map of current distribution, carrier dynamics, and the origin of intrinsic transistor noise, all of which are central to understanding BJT behavior in analog and RF circuits.

At equilibrium, emitter electrons, which have reached the conduction band, are diffused into the base, leaving positive ions in the emitter and combining with holes in the base to create negative ions. The positive ions in the emitter and the negative ions in the base form a depletion region. This depletion region is thinner in the emitter because of the neutrality condition, the doping concentration in the emitter being higher than in the base. Equilibrium ceases when the device is biased, and electrons are injected from the emitter into the base, and holes are injected from the base into the emitter. Some electrons recombine with holes in the emitter–base depletion region, which results in the recombination current $I_{rg}$. The remaining holes recombine with electrons in the emitter if the emitter thickness is large compared to the hole diffusion length, giving rise to the current Ipe. If the emitter thickness is not large enough, some holes reach the metal contact and recombination is achieved at the contact, which generally occurs in high-speed BJTs.

The electrons from the emitter and in the base move by diffusion toward the collector because of the concentration gradient between the base and the collector. A small number of electrons injected into the base recombine with holes in the base, giving rise to the current $I_{rb}$, while the remaining electrons diffuse into the base–collector depletion region, giving rise to the current Inc. There is little recombination in the base–collector depletion region because of the high electric field across this junction, which is reverse-biased. This electric field drifts into the collector electrons that reach the base–collector depletion region.

Ohmic losses in the emitter, base, and collector regions and in the contacts create thermal noise. The crossing of the base–emitter potential barrier by base majority carriers (holes) from the base region is uncorrelated, and its distribution is Poissonian: it generates shot noise. Hence, the base shot noise is due to the base majority carriers. Similarly, the crossing of the emitter–base depletion region by emitter majority carriers (electrons) is also uncorrelated, with a Poissonian distribution, and generates shot noise. Therefore, the collector shot noise is due to the emitter's majority carriers.

## SIGE HBT LARGE-SIGNAL MODEL

A biased NPN BJT is presented in Figure 5.5, where the base–emitter junction is forward biased, and the collector–base junction is reverse biased. This model allows the calculation of direct currents and direct voltages in transistors.

The collector current is given by the expression

$$I_S = exp\frac{V_{BE}}{V_T} \tag{5.59}$$

where $I_S$ is a constant dependent on the transistor characteristics and is given by

$$I_S = \frac{qA\overline{D}_n n_i^2}{Q_B} \tag{5.60}$$

where $q$ is the electron charge, $A$ is the area of the emitter, and $Q_B$ is the number of doping atoms per unit of emitter area in the base. $\overline{D}_n$ is the average electron diffusion constant in the base, and $n_i$ is the silicon intrinsic carrier concentration. Typical values of $I_s$ are between $10^{-16}$ and $10^{-14}$ A (Gray *et al.* 2009).

As shown in Figure 5.5, the base current $I_B$ is given by $I_B = I_{rg} + I_{pe}$, the sum of the current due to the recombination of holes from the base and electrons from the emitter ($I_{rg}$), and the current due to the injection of holes from the base into the emitter ($I_{pe}$). The current $I_{pe}$ is inversely proportional to the emitter doping density. The emitter is highly doped, which reduces $I_{pe}$. The base current is given by

$$I_B = \frac{I_C}{\beta_F} \tag{5.61}$$

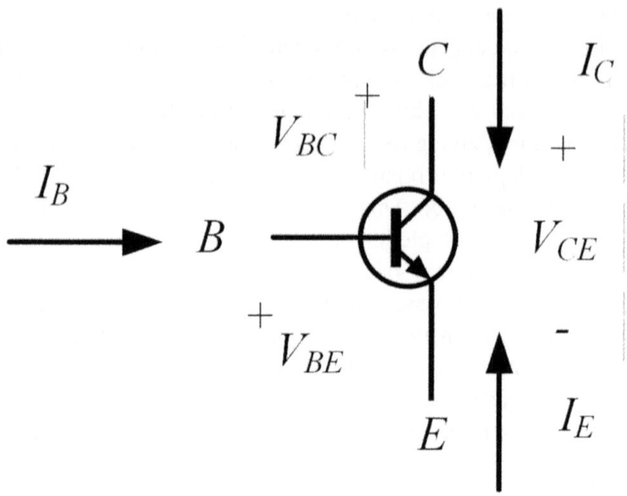


**FIGURE 5.5** Biased BJT.

where $\beta_F$ is the forward current gain, is related to the minority-carrier lifetime in the base $\tau_b$, to the diffusion constant of electrons in the base $D_n$, the diffusion constant of holes into the emitter $D_p$, the diffusion length of holes into the emitter $L_p$, the base width $W_B$, the base doping density $N_A$, and the emitter doping density $N_D$. The forward current gain is given by

$$\beta_F = \frac{1}{\dfrac{W_B^2}{2\tau_b D_n} + \dfrac{D_p}{D_n}\dfrac{W_B}{L_p}\dfrac{N_A}{N_D}} \tag{5.62}$$

Typical values of $\beta_F$ for integrated circuits are between 50 and 500 for npn transistors and between 10 and 100 for pnp transistors.

Because $I_B = I_E + I_C$, the collector and emitter currents are related by the equation

$$I_C = -\frac{\beta_F}{1+\beta_F} I_E \tag{5.63}$$

The large-signal model of an npn SiGe HBT is depicted in Figure 5.6, where $V_{BE}$ is between 0.6 and 0.7 V, owing to the exponential characteristic of $I_B$ as a function of $V_{BE}$. It can be noted that the collector current $I_C$ is independent of the collector voltage $V_{CE}$, an approximation that is not valid in real devices. The collector current increases slowly with $V_{CE}$, except for very small (transistor in saturation) or very large values of $V_{CE}$, as shown in the current–voltage characteristics of BJTs. The change of $I_C$ with $V_{CE}$ is referred to as the Early effect, which is associated with the Early voltage, a characterization parameter of BJTs. An increase in $V_{CE}$ increases $V_{CB}$

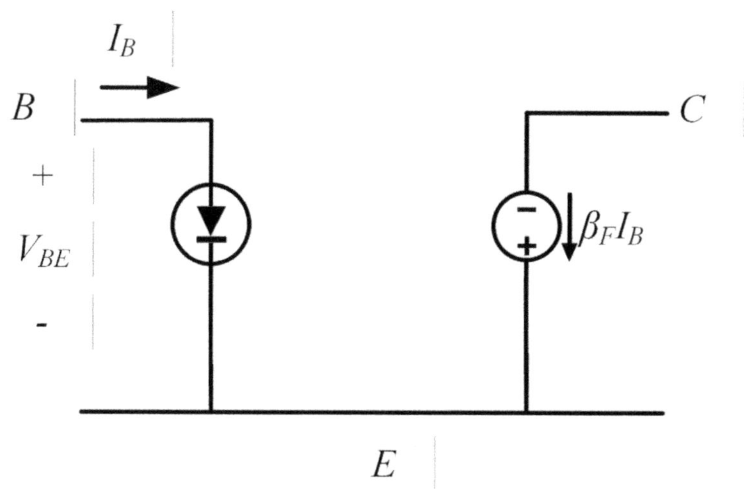


**FIGURE 5.6** Small-signal HBT model.

and the base–collector depletion region, which decreases the width of the base and increases the collector current. In addition, contrary to $I_{pe}$, $I_{rg}$ changes with $V_{CE}$, which affects the base current. However, because the main component of the base current is $I_{pe}$ in integrated circuits, $I_B$ is independent of $V_{CE}$ (Gray *et al.* 2009).

The typical value of $V_{CE}$ varies between 0.05 and 0.3 V when the transistor is in saturation, and the collector–base junction is forward biased. In saturation, a typical value of $\beta_F$ is 5. Because the gain of a transistor in saturation is very small, this operation mode is generally not used in analog circuits. Large values of $V_{CE}$ result in avalanche breakdown on the transistor junctions.

## SIGE HBT SMALL-SIGNAL MODEL

The simplified small-signal equivalent circuit of the SiGe HBT in the hybrid-π configuration is represented in Chen *et al.* (2006). The circuit is based on the VBIC model for vertical deep-trench isolated SiGe HBTs in the forward active region, an improvement of the SPICE Gummel–Poon (SGP) model widely used for bipolar transistors. Advanced compact models, such as the SGP and, more recently, the VBIC, Mextram, and HiCuM, do not lend themselves to simple analytical analysis. Simulation results from advanced and simple models are close in typical applications where current crowding effects, substrate effects, the dependence of the forward transit time, the base–collector capacitance, and the output conductance to bias conditions can be neglected.

The values of the elements of the intrinsic transistor are given by the following expressions. The transconductance is given when the base transit time is neglected.

$$g_m = \frac{qI_C}{kT} = \frac{I_C}{V_T} \tag{5.64}$$

and

$$r_{bc} = \beta_0 r_0 \tag{5.65}$$

where gain is $\beta_0$ is the small-signal current gain, and $r_0$ is the output resistance. The base–collector capacitance is given by

$$C_{bc} = \frac{C_{bc0}}{\left(1 - \frac{V_{CB}}{\psi_0}\right)^n} \tag{5.66}$$

where $C_{bc0}$ is the zero-bias base–collector depletion capacitance, $n$ is a real number between 0.2 and 0.5, $V_{CB}$ is the forward bias on the collector–base junction, and $\psi_0$ is the base–collector junction built-in potential.

The transistor's small-signal input resistance $r_{be}$ is given by

$$r_{be} = \frac{\beta_0}{g_m} \tag{5.67}$$

and the base–emitter capacitance is given by

$$C_{be} = C_b + 2C_{je0} \tag{5.68}$$

where $C_b$ is the base charging capacitance, $C_{je0}$ is the zero-bias base–emitter depletion capacitance. The base charging capacitance is given by

$$C_b = \tau_d g_m \tag{5.69}$$

where $\tau_d$ is the base transit time in the forward direction. The base transit time is the average time spent by minority carriers in the base to cross it. The typical values of $\tau_d$ vary between 10 and 500 ps for integrated NPN devices. The small-signal output resistance $r_o$ is related to the Early voltage by the relation

$$r_o = \frac{V_A}{I_C} \tag{5.70}$$

and the figure of merit $f_T$ is given by

$$f_T = \frac{1}{2\pi} \frac{g_m}{C_{be} + C_{bc}} \tag{5.71}$$

The base resistors model distributive effects and are not completely resistive components. The resistors can be lumped together by $r_b$, with $r_b = r_{b1} + r_{b2}$. The error deriving from the approximation of the base thermal noise by the noise due to $r_b$ is insignificant when the transistor is biased in the low injection regime and current crowding is insignificant (Cressler 1998). Low injection regime and insignificant current crowding are achieved for biases that are below $f_T$ and $f_{max}$, which is the general case. The transconductance is given by the expression $g_m = g_{m0} exp(-j\omega\tau_d) \approx (1-j\omega\tau_d)qI_C/kT$, where $g_{m0} \approx qI_C/kT$ and $\tau_d$ is the total base-to-collector delay time (Jun 2009). The delay time is of the order of picoseconds. Generally, $\omega\tau_d \ll 1$, hence $g_m \approx qI_C/kT$. The parasitic elements $C_{bcx}$, $C_{sub1}$, $C_{sub2}$, and $r_{sub}$ model substrate effects. For frequencies that are much lower than the maximum $f_T$, the extrinsic parasitic inductances and capacitances $L_b$, $L_e$, $L_c$, and $C_{bcp}$ can be neglected (Lee *et al.* 2005). The parasitic capacitor $C_{bcx}$ is associated with the base–collector–substrate junctions. For frequencies below mm-wave frequencies (e.g., 10 GHz), substrate couplings are not important and can be neglected (Pawlak *et al.* 2015). The small-signal equivalent circuit in Figure 5.7 is derived from the simplifications in this paragraph.

The model in Figure 5.7 includes some extrinsic components as well as base, emitter, collector, and substrate parasitic components that account for its high

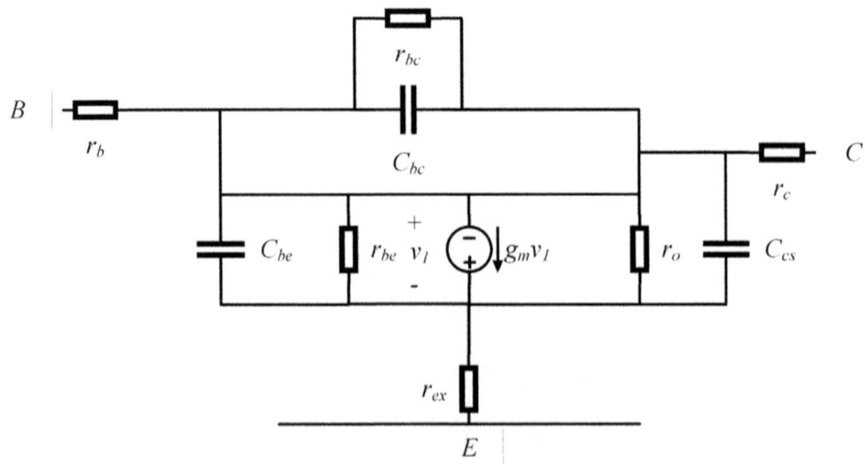


**FIGURE 5.7** Modified small-signal equivalent circuit of the SiGe HBT.

frequency response and noise performance. It is noted that the coupling between the collector and the substrate is represented by $C_{cs}$. The resistance $r_{bc}$ relates the modulation of the neutral base recombination current to the collector–base voltage and has an insignificant impact on transfer parameters and noise performance. The contribution of the collector parasitic resistance $r_c$ to the input noise is reduced, owing to the gain of the transistor. Hence, $r_c$ can be neglected. The output resistance $r_o$ can be neglected because the Early voltage is large for SiGe HBTs.

## *Y*- AND *Z*-PARAMETERS FOR THE CE AMPLIFIER

*Y*-parameters are expressed in terms of device parameters, where $r_0$ is neglected, as shown in Figure 5.8.

Calculation of $Y_{11}$:

$$Y_{11} = \left.\frac{i_1}{v_1}\right| v_2 = 0 \tag{5.72}$$

where

$$i_1 = v_b Y_{bc} + \frac{v_{be}}{Z_{be}} \tag{5.73}$$

and

$$v_1 = i_1 r_b + v_b \tag{5.74}$$

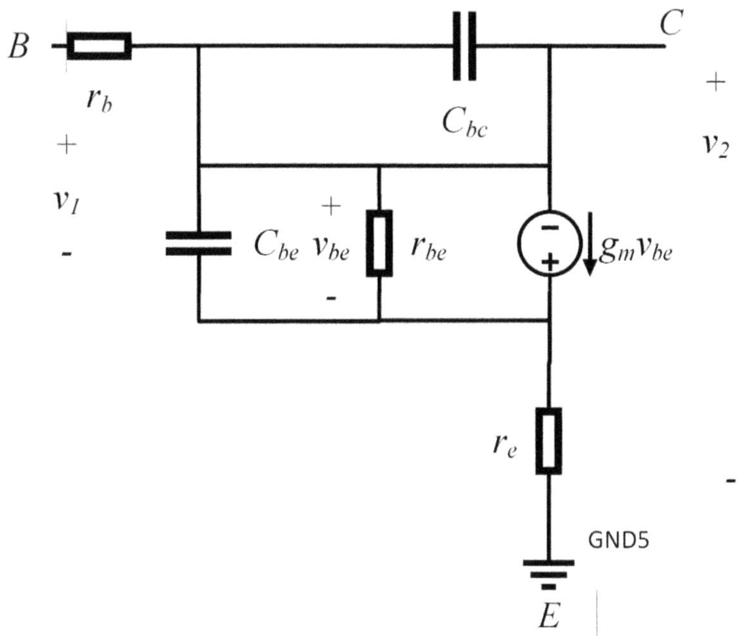


**FIGURE 5.8** Transistor equivalent circuit.

where $v_b$ is the voltage at node (b). Furthermore,

$$v_e = v_b \frac{r_e + Z_{bc} r_e g_m}{Z_{be} + r_e + Z_{bc} r_e g_m} \tag{5.75}$$

and

$$v_{be} = v_b \frac{Z_{be}}{Z_{be} + r_e + Z_{bc} r_e g_m} \tag{5.76}$$

and

$$Y_{11} = \frac{i_1}{v_1} = \frac{g_{be} + j\omega C_{be}\left(1 + r_e\left(g_{be} + j\omega C_{be} + g_m\right)\right)}{1 + r_b\left(g_{be} + j\omega C_{be} + j\omega C_{bc}\right) + g_m r_e\left(j\omega C_{bc} r_b + 1\right)} \tag{5.77}$$

where $g_{be} = 1/r_{be}$. $Y_{12}$ can be calculated as follows:

$$Y_{12} = \frac{i_1}{v_2}\bigg|_{v_1 = 0} \tag{5.78}$$

where

$$v_b = -i_1 r_b \tag{5.79}$$

and

$$i_1 = (v_b - v_2)Y_{bc} + \frac{v_{be}}{Z_{be}} \tag{5.80}$$

and

$$i_1\left(1 + r_b Y_{bc} + \frac{r_b}{Z_{be} + r_e + Z_{be} r_e g_m}\right) = -v_2 Y_{bc} \tag{5.81}$$

and

$$Y_{12} = \frac{i_1}{v_2} = \frac{-j\omega C_{bc}\left(1 + r_e\left(g_m + g_{be} + j\omega C_{be}\right)\right)}{1 + r_b\left(g_{be} + j\omega C_{be} + j\omega C_{bc}\right) + g_m r_e\left(j\omega C_{bc} r_b + 1\right)} \tag{5.82}$$

Similarly, $Y_{21}$ can be calculated as follows:

$$Y_{21} = \left.\frac{i_2}{v_1}\right| v_2 = 0 \tag{5.83}$$

where

$$v_e = v_b\left(j\omega C_{bc} + \frac{1}{Z_{be} + r_e + Z_{bc} r_e g_m}\right) \tag{5.84}$$

and

$$v_1 = v_b + i_1 r_b \tag{5.85}$$

and

$$v_b = \frac{v_1}{1 + r_b\left(j\omega C_{bc} + \frac{1}{Z_{be} + r_e + Z_{be} r_e g_m}\right)} \tag{5.86}$$

and

$$v_{be} = v_1 \frac{\dfrac{Z_{be}}{Z_{be} + r_e + Z_{be} r_e g_m}}{1 + r_b\left(j\omega C_{bc} + \dfrac{1}{Z_{be} + r_e + Z_{be} r_e g_m}\right)} \quad (5.87)$$

where

$$i_2 = g_m v_{be} - v_{be}\left(j\omega C_{bc}\right) \quad (5.88)$$

therefore

$$Y_{21} = \frac{i_2}{v_1} = \frac{g_m - j\omega C_{bc}}{1 + r_b\left(g_{be} + j\omega C_{be} + j\omega C_{bc}\right) + g_m r_e\left(j\omega C_{bc} r_b + 1\right)} \quad (5.89)$$

Finally, the calculation of $Y_{22}$ is performed:

$$Y_{22} = \frac{i_2}{v_2}\bigg|_{v_1 = 0} \quad (5.90)$$

where

$$i_2 = g_m v_{be} + \frac{v_b}{r_b} + \frac{v_{be}}{Z_{be}} \quad (5.91)$$

and

$$\frac{v_b - v_2}{Z_{bc}} + \frac{v_b}{r_b} + \frac{v_{be}}{Z_{be}} = 0 \quad (5.92)$$

such that

$$v_b = v_2 \frac{v_1\left(Z_{be} + \beta r_e\right)}{r_b\left(Z_{be} + \beta r_e\right) + Z_{bc}\left(r_b + Z_{be} + \beta r_e\right)} \quad (5.93)$$

where $\beta = 1 + Z_{be} g_m$.

$$i_2 = v_b\left[\left(\frac{Z_{be}}{Z_{be} + \beta r_e}\right)\left(g_m + \frac{1}{Z_{be}}\right) + \frac{1}{r_b}\right] \quad (5.94)$$

thus

$$Y_{22}=\frac{i_2}{v_2}=\frac{j\omega C_{bc}\left(1+\left(g_{be}+j\omega C_{be}+g_m\right)\left(r_e+r_b\right)\right)}{1+r_b\left(g_{be}+j\omega C_{be}+g_m\right)+g_m r_e\left(j\omega C_{bc} r_b+1\right)} \tag{5.95}$$

To calculate $Z_{11}$, a similar approach is followed:

$$Z_{11}=\left.\frac{v_1}{i_1}\right|_{i_2=0} \tag{5.96}$$

where

$$v_e=r_e\left(\frac{v_{be}}{Z_{be}}+g_m v_{be}\right) \tag{5.97}$$

and

$$v_{be}=v_b\left(\frac{1}{1+r_e\left(g_{be}+j\omega C_{be}+g_m\right)}\right) \tag{5.98}$$

and

$$v_1=v_b+i_1 r_b \tag{5.99}$$

and

$$i_1=g_m v_{be}+\frac{v_{be}}{Z_{be}} \tag{5.100}$$

such that

$$i_1=v_b\frac{g_m+g_{be}+j\omega C_{be}}{1+r_e\left(g_m+g_{be}+j\omega C_{be}\right)} \tag{5.101}$$

and

$$v_b=v_1\frac{1+r_e\left(g_m+g_{be}+j\omega C_{be}\right)}{1+\left(r_e+r_b\right)\left(g_m+g_{be}+j\omega C_{be}\right)} \tag{5.102}$$

therefore

$$Z_{11} = \frac{v_1}{i_1} = \frac{1 + (r_e + r_b)(g_m + g_{be} + j\omega C_{be})}{g_m + g_{be} + j\omega C_{be}} \tag{5.103}$$

The expressions derived above provide a complete small-signal *Y*- and *Z*-parameter description of the common-emitter transistor when parasitic resistances and capacitances are taken into account and the output resistance is neglected. These results explicitly show how transconductance, base and emitter resistances, and junction capacitances shape the frequency-dependent behavior of the two-port network. In particular, the coupling between input and output through the base–collector capacitance introduces feedback terms that directly affect $Y_{12}$ and $Y_{21}$, while the finite base resistance plays a central role in determining both the real and imaginary parts of the input impedance. The resulting formulations form the analytical foundation required for noise analysis, stability evaluation, and impedance matching in later sections. Although compact models and numerical simulators are typically used in practice, these closed-form expressions provide valuable physical insight into how device parameters influence gain, bandwidth, and noise performance in CE amplifiers, and they serve as a useful reference for understanding the behavior of bipolar and SiGe HBT devices at high frequencies.

## SIGE HBT NOISE MODEL

The high-frequency modeling of a transistor is not limited to its fundamental physical mechanisms because the effects of parasitic components must also be accounted for. Parasitics are represented in the extrinsic model, while the basic physical mechanism is modeled by the intrinsic model. The T-model provides a direct relation between model elements and the device physics, as shown in Sotoodeh *et al.* (2000) and Nallatamby *et al.* (2017).

Despite the correspondence between device physics and the components of the model, the extraction of $R_{be}$ and $C_{be}$ from measurements using the T-model can be difficult. The physical significance of the elements of the Π-model is ambiguous, but the model provides an accurate extraction of parameters such as $C_{be}$ and is widely used by designers and simulators. A one-to-one correspondence between the two models is found for frequencies for which the dependence of the components of the Π-model on frequency is not significant.

A comprehensive noise model of an HBT is shown in Figure 5.9 and comprises six intrinsic and extrinsic noise sources. The extrinsic noise sources are the emitter and collector thermal noise sources, and the extrinsic part of the base resistance $R_{bx}$ noise source. Because the base is sandwiched between the emitter and the collector, its temperature is generally higher than the temperature of the emitter and the collector; therefore, the noise due to the intrinsic part of the base resistance $R_{bi}$ accounts for most of the base thermal noise. The remaining intrinsic noise sources are the base and collector shot noises. Parasitic capacitors are $C_{pbe}$, $C_{pbc}$, and $C_{pce}$. Similarly, parasitic

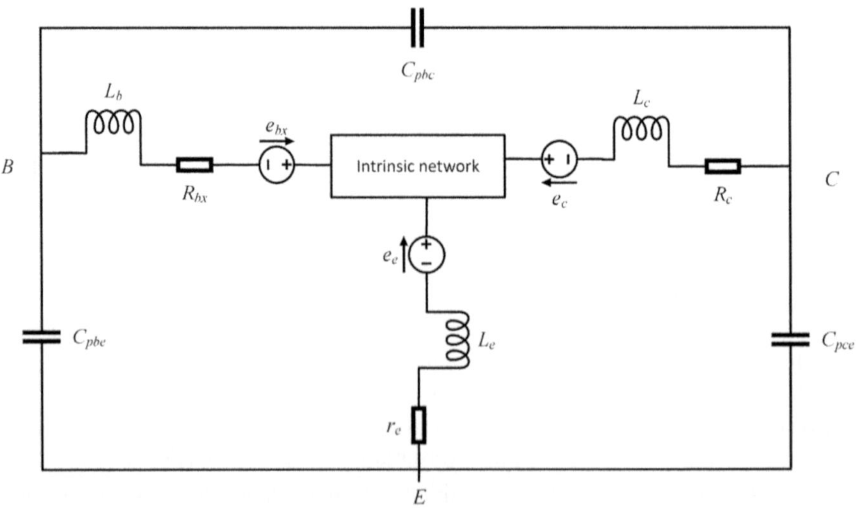


**FIGURE 5.9** Comprehensive noise model for HBTs.

inductors are $L_b$, $L_c$, and $L_e$. For analytical noise analysis, extrinsic noise sources and parasitic capacitors and inductors are generally omitted for simplicity.

The thermal noises $e_{bx}$, $e_e$, and $e_c$ are due to the base, emitter, and collector parasitic resistances.

The intrinsic noise model of an HBT is depicted in Figure 5.9. The types of noise sources and their qualitative contribution to the total noise are as follows:

- The shot noise $i_c$ is due to the collector direct current.
- The shot noise $i_{cr}$ is due to the base–collector reverse direct current and is not represented in Figure 5.1-9. The base–collector reverse current is small, and its shot noise can be disregarded.
- The shot noise $i_b$ is due to the base direct current.
- The thermal noise $e_{bi}$ is due to the base parasitic resistance. The resistances $R_{be}$ $(r_\pi)$, $R_{bc,}$ and $r_o$ do not correspond to physical resistors; they only model a physical phenomenon, which does not contribute to the transistor's intrinsic noise.

The thermal noise due to $R_{bi}$ and $R_e$ is not negligible. Hence, intrinsic and extrinsic thermal noise sources are used in the simplified comprehensive noise model in Figure 5.11, where a resistive source is connected to the input of the transistor.

The base resistance $R_b$ $(r_b)$ is relatively higher than the emitter and collector parasitic resistances shown above. The latter resistances are low owing to the heavy doping of the emitter and the buried nature of the collector layer. In addition, the collector parasitic resistance, in series with the load and all in parallel with the high output collector resistance $r_o$, is small and makes a small contribution to the output noise. Hence, the thermal noise due to the collector parasitic resistance can be neglected when referred to the input in view of the impact of the transistor gain. The noise due

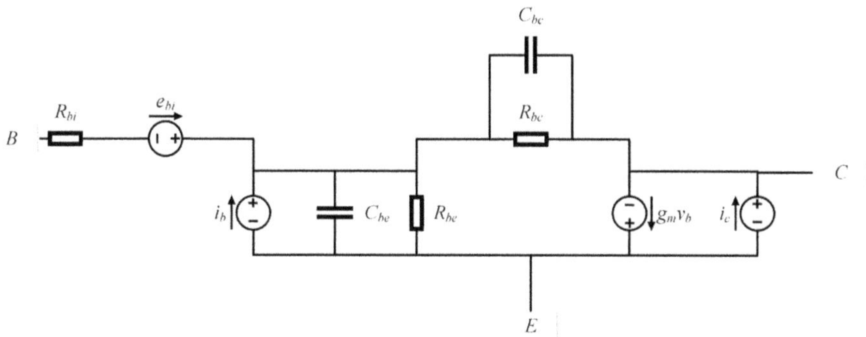


**FIGURE 5.10** Intrinsic noise.

to the emitter parasitic resistance *Re* is not affected by the transistor gain and should be accounted for. However, the calculation of input-referred noise leads to complex equations when the emitter resistance $R_e$ is accounted for. Noise calculations are simplified when $R_e$ is neglected, but only its thermal noise is considered. This approximation is valid when $R_e g_m << 1$, which has been verified (de Vreede *et al.* 1994). The thermal noise source due to $R_e$ is shifted to the transistor input through a Blakesley transformation. The thermal noise at the input can then be modeled according to the sum of $R_b$ ($r_b$) and $R_e$.

The crossing of the forward-biased base–emitter junction by electrons from the emitter and holes from the base is a random process that is at the origin of the base shot noise. However, the drift of the carriers through the reverse-biased base–collector junction is a deterministic rather than a random process, because the junction does not constitute a potential barrier. However, electrons crossing the base–collector junction carry a delayed noise component due to the crossing of the base from the emitter.

Generally, it is assumed that the collector and base shot noise are not correlated, which is valid at low frequencies (de Vreede *et al.* 1994). The correlation between the collector and base short noises is expressed by $\overline{i_b i_c^*} = 2qI_C\left[exp\left(j\omega\tau_n - 1\right)\right]$, where $\tau_n$, the noise transit time, is the transport time taken by short noise electrons from the base–emitter junction to reach the collector. The noise transit time is different from the total forward transit time, which is the delay between a change of voltage at the base and the time when electrons associated with the change arrive at the collector. The noise transit time is due to DC and does not include the time constant contributions due to the modulation of charge in the base area caused by a change in base voltage. At the operating frequency of the LNA, the noise transit time is assumed to be nil, as for impact, the effective noise due to the base and emitter DC is lower than the noise resulting from calculations. This can be shown in the expression of $F_{min}$ that considers $\tau_n$ in

$$F_{min} = 1 + \frac{1}{\beta} + \sqrt{\frac{2g_m R_n}{\beta} + \left[\frac{2R_n\left(\omega C_i\right)^2}{g_m} - \omega^2\left(C_{be} + C_{bc}\right)\tau_n R_n\right]\left(1 - \frac{1}{2g_m R_n}\right)} \tag{5.104}$$

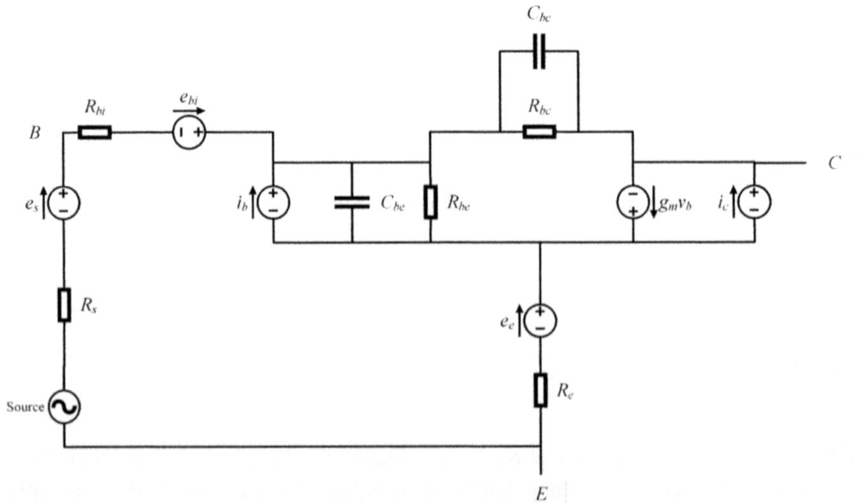


**FIGURE 5.11** Comprehensive noise model for HBTs.

The term in $\tau_n$ *is* small compared to other terms, and at low frequencies it is negligible. The impact of the correlation component on noise performance increases with $f/f_T$ (Yau 2006), which results in the reduction of $F_{min}$. At frequencies higher than 1/10 of the peak $f_T$, this term decreases $F_{min,}$ and a noise model with noise transit time is required. SPICE suffers from this limitation, producing a higher value of $F_{min}$ at high frequency (Sakalas *et al.* 2006).

Gummel–Poon, VBIC, Mextram, and Hicum transistor models assume that noise sources in BJT are of the thermal and shot types, and the base and collector shot noises are not correlated. For high-frequency applications and for high biasing currents, measurements diverge from model data. Models such as the Van Vliet and transport models consider the correlation between the base and collector shot noise. Semiempirical models allow accurate modeling of the base and collector shot noise, as well as the effects of high biasing current (Niu 2005).

## NOISE FIGURE OF THE CE SIGE HBT

The noise model is further simplified in Figure 5.12, where the emitter parasitic resistance ($R_e$) is not shown. The main sources of noise are represented where $v_b$ is the thermal noise associated with the base resistance $r_b$, $i_b$ is the base shot noise, and $i_c$ is the collector shot noise. The ABCD representation of the two-port noise model is also shown.

The noise power densities of sources in Figure 5.12 are given by the expressions

$$\overline{i_c^2} = 2qI_C \tag{5.105}$$

$$\overline{i_b^2} = 2qI_B \tag{5.106}$$

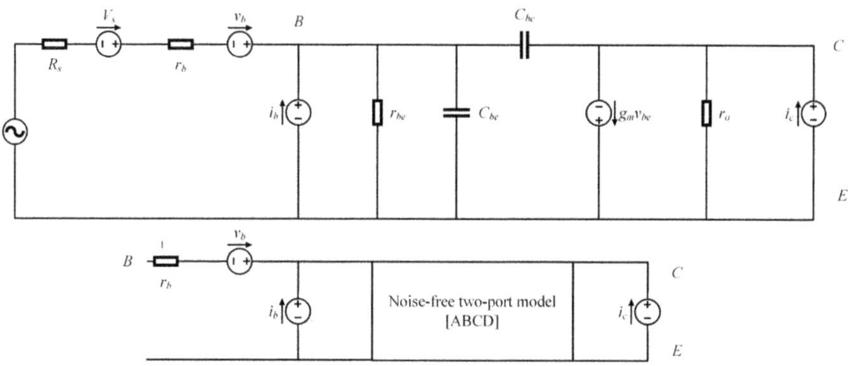


**FIGURE 5.12** Main noise sources of the CE SiGe HBT and the two-port model.

$$\overline{v_b^2} = 2kT\left(r_b + R_e\right) \tag{5.107}$$

$$\overline{v_s^2} = 4kTR_s \tag{5.108}$$

The ABCD representation of Figure 5.12 is first calculated because the matrix will be used to report to the input the collector shot noise source. In addition, an important property of the ABCD representation, which will be used in the calculations, is related to the ABCD representation of a cascade network, which is equal to the product of the ABCD matrix of each two-port network forming the cascade network. The base resistance $r_{b,}$ is neglected in the calculations of the ABCD matrix.

The chain matrix due to $r_{be}$ ($r_\pi$) at the input of the noiseless port is equal to

$$M_1 = \begin{bmatrix} A_1 & B_1 \\ C_1 & D_1 \end{bmatrix} = \begin{bmatrix} 1 & 0 \\ \dfrac{1}{r_\pi} & 1 \end{bmatrix} \tag{5.109}$$

The chain matrix due to $C_{bc}$ ($c_\pi$) at the input of the noiseless port is expressed by

$$M_2 = \begin{bmatrix} A_2 & B_2 \\ C_2 & D_2 \end{bmatrix} = \begin{bmatrix} 1 & 0 \\ \dfrac{1}{sC_\pi} & 1 \end{bmatrix} \tag{5.110}$$

where $s$ is the complex frequency, $s = j\omega f$.

The transmission matrix $M_3$ due to $C_\mu$, to $r_{o,}$ and the current source $g_m$ is derived from the results of Kirchhoff's current law (KCL) and Kirchhoff's voltage law (KVL), as shown in Figure 5.12.

$$M_3 = \begin{bmatrix} A_3 & B_3 \\ C_3 & D_3 \end{bmatrix} = \begin{bmatrix} sC_\mu\left(sC_\mu r_0 + 1\right) & 1 \\ sC_\mu r_0\left(sC_\mu - g_m\right) & sC_\mu - g_m \end{bmatrix} \tag{5.111}$$

and since the output resistance $r_o$ is very high and $g_m r_o >> 1$ and $sC_\mu r_o >> 1$, $M_3$ can then be approximated by

$$M_3 = \begin{bmatrix} \dfrac{sC_\mu}{sC_\mu - g_m} & \dfrac{1}{sC_\mu - g_m} \\ \dfrac{sC_\mu r_0}{sC_\mu - g_m} & \dfrac{sC_\mu}{sC_\mu - g_m} \end{bmatrix} \tag{5.112}$$

and the ABCD matrix $M$ is the product of $M_1$, $M_2$, and $M_3$, with $M = [M_1][M_2][M_3]$. Therefore

$$M = \begin{bmatrix} \dfrac{sC_\mu}{sC_\mu - g_m} & \dfrac{1}{sC_\mu - g_m} \\ \dfrac{\left(1 + sr_\pi C_\pi\right)sC_\mu}{\left(sC_\mu - g_m\right)r_\pi} + \dfrac{sC_\mu g_m}{sC_\mu - g_m} & \dfrac{1 + sr_\pi C_\pi}{\left(sC_\mu - g_m\right)r_\pi} + \dfrac{sC_\mu}{sC_\mu - g_m} \end{bmatrix} \tag{5.113}$$

The terms of the transmission matrix are $A$, $B$, $C$, and $D$, and are given by

$$A = \frac{sC_\mu}{sC_\mu - g_m} \tag{5.114}$$

$$B = \frac{1}{sC_\mu - g_m} \approx \frac{1}{-g_m} = -r_e, g_m \gg \omega C_\mu \tag{5.115}$$

and

$$C = \frac{sC_\mu\left(\dfrac{1}{r_\pi} + sC_\pi + g_m\right)}{sC_u - g_m} \tag{5.116}$$

and

$$D = \frac{1 + sr_\pi C_\pi}{\left(sC_\mu - g_m\right)r_\pi} + \frac{sC_\mu}{sC_\mu - g_m} = \frac{\dfrac{1}{r_\pi} + sC_\pi + sC_\mu}{sC_\mu - g_m} \tag{5.117}$$

and since $g_m \gg \omega C_\mu$, $D$ can be approximated by

$$D = -\frac{\frac{1}{r_\pi} + sC_\pi + sC_\mu}{g_m} = \frac{-1}{r_\pi g_m} - j\omega \frac{C_\pi + C_\mu}{g_m} \tag{5.118}$$

where it can be shown that

$$D = -\left(\frac{1}{\beta_0} + j\frac{f}{f_T}\right) \tag{5.119}$$

when it is considered that the small signal gain is $r_\pi g_m$, and the unity current gain frequency $f_T$ is given by

$$f_T = \frac{g_m}{2\pi\left(C_\pi + C_\mu\right)} \tag{5.120}$$

and

$$\beta_0 = \frac{\beta_F}{1 + j\beta_F\left(\frac{f}{f_T}\right)} \tag{5.121}$$

Since the transmission matrix $M$ is known, it is now possible to report to the input the shot noise contribution due to the collector current $I_C$. The noise of the power at the input is related to the square of $v_{ci}$ and $i_{ci}$.

$$v_{ci} = -Bi_c \tag{5.122}$$

and

$$i_{ci} = -Di_c \tag{5.123}$$

where the total intrinsic voltage noise of the transistor is given by

$$v_n = v_b + i_b\left(R_s + r_b\right) + Bi_c + \left(R_s + r_b\right)Di_c \tag{5.124}$$

such that

$$v_n = v_b + Bi_c + \left(R_s + r_b\right)\left(i_b + Di_c\right) \tag{5.125}$$

Therefore

$$v_n = v_b + r_e i_c + \left(R_s + r_b\right)\left(i_b + \left(\frac{1}{\beta_0} + j\frac{f}{f_T}\right)i_c\right) \tag{5.126}$$

and the noise factor $F$ is equal to the ratio of the total noise at the input to the source noise

$$F = \frac{\overline{v_s^2} + \overline{v_n^2}}{\overline{v_s^2}} = 1 + \frac{\overline{v_n^2}}{\overline{v_s^2}} \tag{5.127}$$

and the noise sources, $v_b$, $i_b$, and $i_c$ being uncorrelated, as discussed previously, the power density of the noise due to $v_n$ is expressed by

$$\overline{v_n^2} = \overline{v_b^2} + \overline{i_b^2}\left(R_s + r_b\right)^2 + \overline{i_c^2} r_e^2 + \overline{i_c^2}\left(R_s + r_b\right)^2 \frac{1}{\beta_0^2} + \overline{i_c^2}\left(R_s + r_b\right)^2 \frac{f^2}{f_T^2} \tag{5.128}$$

$$\begin{aligned}\overline{v_n^2} = {} & 4kT\left(r_b + R_e\right) + 2qI_B\left(R_s + r_b\right)^2 + 2qI_C r_e^2 + 2qI_C\left(R_s + r_b\right)^2 \frac{1}{\beta_0^2} \\ & + 2qI_C\left(R_s + r_b\right)^2 \frac{f^2}{f_T^2}\end{aligned} \tag{5.129}$$

$$\begin{aligned}F = {} & 1 + \frac{r_b + R_e}{R_s} + \frac{qI_C}{2kTR_s\beta_0}\left(R_s + r_b\right)^2 + \frac{qI_C}{2kTR_s} r_e^2 \\ & + \frac{qI_C}{2kTR_s}\left(R_s + r_b\right)^2 \frac{1}{\beta_0^2} + \frac{qI_C}{2kTR_s}\left(R_s + r_b\right)^2 \frac{f^2}{f_T^2}\end{aligned} \tag{5.130}$$

which can be simplified by introducing $r_e$, with $gm = qI_C/kT = 1/r_e$.

$$\begin{aligned}F = {} & 1 + \frac{r_b + R_e}{R_s} + \frac{1}{2r_eR_s\beta_0}\left(R_s + r_b\right)^2 + \frac{1}{2R_s} r_e + \frac{1}{2r_eR_s}\left(R_s + r_b\right)^2 \frac{1}{\beta_0^2} \\ & + \frac{1}{2r_eR_s}\left(R_s + r_b\right)^2 \frac{f^2}{f_T^2}\end{aligned} \tag{5.131}$$

and since $\beta_0 \gg 1$, we can write

$$F = 1 + \frac{r_b + R_e}{R_s} + \frac{1}{2r_eR_s\beta_0}\left(R_s + r_b\right)^2 + \frac{1}{2R_s} r_e + \frac{1}{2r_eR_s}\left(R_s + r_b\right)^2 \frac{f^2}{f_T^2} \tag{5.132}$$

which is equivalent to

$$F = 1 + \frac{r_b + R_e}{R_s} + \frac{g_m}{2R_s\beta_0}\left(R_s + r_b\right)^2 + \frac{1}{2g_m R_s} + \frac{g_m}{2R_s}\left(R_s + r_b\right)^2 \frac{f^2}{f_T^2} \tag{5.133}$$

The second term of the above equation is due to the base and emitter thermal noise sources; the third term is due to the base shot noise caused by the base DC bias, and the fourth and last terms are due to the shot noise contribution from the collector bias DC.

It shows that the noise factor increases with frequency, with the base and emitter parasitic resistances, but decreases when the DC gain and the unity frequency increase. At moderate frequencies, the noise factor is dominated by the base resistance, as shown by the second term. SiGe HBTs achieve better noise performance than BJTs owing to lower base resistance, higher DC gain, and higher $f_T$. In addition, HBTs are smaller than BJTs because of their higher current density. They are therefore subject to fewer parasitics. At high frequencies, the noise factor is dominated by the collector shot noise. The collector shot noise and $g_m$ are proportional to the collector bias current. Hence, when the bias current is low, the noise factor decreases until the gain reaches a maximum. When the peak gain has been reached, the noise factor increases with the collector current.

## RELATIONSHIPS BETWEEN *RN*, $F_{min}$, $Y_{opt}$, AND THE PARAMETERS OF A SIGE HBT

For the designer, it is important to evaluate $F_{min}$, $Y_{opt}$, and possibly the noise resistance $R_n$. Knowledge of $F_{min}$ allows derivation of the minimum possible noise that is achievable under optimum noise matching conditions, while $Y_{opt}$ allows derivation of the noise matching network that is required at the input. For this purpose, the noise parameters $F_{min}$, $R_n$, and $Y_{opt}$ can be related. $F_{min}$ can be derived from the correlation matrix elements using the equation

$$F_{min} = 1 + 2\left(Re\left\{C_{A12}\right\} + C_{A11}G_{sopt}\right) \tag{5.134}$$

The noise correlation matrix is calculated after reporting to the input the collector shot noise and deriving the elements of $C_A$. However, the determination of the correlation matrix can lead to cumbersome calculations. A desired approach entails using the $Y$ representation of the transistor circuit and converting the $C_Y$ noise correlation matrix into a $C_A$ correlation matrix. This approach simplifies calculations because the noise currents at the input and at the output are assumed to be uncorrelated at the operating frequency. Also, the $Y$ representation is preferred for nodal analysis. This approach is used in the following section for the calculation of two-port model noise parameters.

The transformation is done using the transfer function $T$, where $B$ and $D$ are elements of the chain matrix ABCD.

$$T = \begin{bmatrix} 0 & B \\ 1 & D \end{bmatrix} \tag{5.135}$$

The $Y$ representation of a SiGe HBT amplifier, when the base resistance $r_b$ is neglected, can also be used as:

$$M_1 = \begin{bmatrix} y_{11} & y_{12} \\ y_{21} & y_{22} \end{bmatrix} = \begin{bmatrix} \frac{1}{r_\pi} + s\left(C_\pi + C_\mu\right) & -sC_\mu \\ g_m - sC_\mu & sC_\mu \end{bmatrix} \tag{5.136}$$

The noise correlation matrix in admittance representation is not related to $r_b$ and is equal to

$$\begin{bmatrix} \overline{i_b i_b^*} & \overline{i_b i_c^*} \\ \overline{i_b i_c^*} & \overline{i_c i_c^*} \end{bmatrix} = 2kT \begin{bmatrix} \frac{g_m}{\beta} & 0 \\ 0 & g_m \end{bmatrix} = 2\Delta f \xi_Y = 4kT\Delta f C_Y \tag{5.137}$$

where $\xi_Y$ is the admittance correlation matrix, and $C_Y$ is the normalized admittance correlation matrix. Therefore

$$\xi_Y = 2kTC_Y \tag{5.138}$$

The ABCD matrix $M$ being the product of $M_1$, $M_2$, $M_3$, and $M_4$, $M = [M_1][M_2][M_3][M_4]$, with

$$M_1 = \begin{bmatrix} A_1 & B_1 \\ C_1 & D_1 \end{bmatrix} = \begin{bmatrix} 1 & r_b \\ 0 & 1 \end{bmatrix} \tag{5.139}$$

and

$$M_2 = \begin{bmatrix} A_2 & B_2 \\ C_2 & D_2 \end{bmatrix} = \begin{bmatrix} 1 & 0 \\ \frac{1}{r_\pi} & 1 \end{bmatrix} \tag{5.140}$$

and

$$M_3 = \begin{bmatrix} A_3 & B_3 \\ C_3 & D_3 \end{bmatrix} = \begin{bmatrix} 1 & 0 \\ sC_\pi & 1 \end{bmatrix} \tag{5.141}$$

and

$$M_4 = \begin{bmatrix} A_4 & B_4 \\ C_4 & D_4 \end{bmatrix} = \begin{bmatrix} \dfrac{sC_\mu}{sC_\mu - g_m} & \dfrac{1}{sC_\mu - g_m} \\ \dfrac{sC_\mu g_m}{sC_\mu - g_m} & \dfrac{sC_\mu}{sC_\mu - g_m} \end{bmatrix} \tag{5.142}$$

and when $r_b$ is not neglected, the elements of the chain representation are given by

$$C = \begin{bmatrix} \left\{ \dfrac{sC_\mu}{sC_\mu - g_m} + \dfrac{r_b sC_\mu \left( g_m + \dfrac{1}{r_\pi} + sC_\pi \right)}{sC_\mu - g_m} \right\} & \left\{ \dfrac{sC_\mu}{sC_\mu - g_m} + \dfrac{r_b \left( sC_\mu + \dfrac{1}{r_\pi} + sC_\pi \right)}{sC_\mu - g_m} \right\} \\ \dfrac{sC_\mu \left( g_m + \dfrac{1}{r_\pi} + sC_\pi \right)}{sC_\mu - g_m} & \dfrac{\left( sC_\mu + \dfrac{1}{r_\pi} + sC_\pi \right)}{sC_\mu - g_m} \end{bmatrix} \tag{5.143}$$

Using a $Y$ representation to chain representation conversion, the noise correlation matrix is given by

$$C_A = 2kY[T][C_Y][T]^+ \tag{5.144}$$

where

$$T = \begin{bmatrix} 0 & \left\{ \dfrac{sC_\mu}{sC_\mu - g_m} + \dfrac{r_b sC_\mu \left( g_m + \dfrac{1}{r_\pi} + sC_\pi \right)}{sC_\mu - g_m} \right\} \\ 1 & \dfrac{\left( sC_\mu + \dfrac{1}{r_\pi} + sC_\pi \right)}{sC_\mu - g_m} \end{bmatrix} \tag{5.145}$$

therefore

$$C_A = \begin{bmatrix} 0 & \left\{ \dfrac{1}{sC_\mu - g_m} + \dfrac{r_b\left(g_m + \dfrac{1}{r_\pi} + sC_\pi\right)}{sC_\mu - g_m} \right\} \\ 1 & \dfrac{\left(sC_\mu + \dfrac{1}{r_\pi} + sC_\pi\right)}{sC_\mu - g_m} \end{bmatrix} 2kT \begin{bmatrix} \dfrac{g_m}{\beta} & 0 \\ 0 & g_m \end{bmatrix} \times \begin{bmatrix} \left\{ \dfrac{1}{sC_\mu - g_m} + \dfrac{r_b\left(g_m + \dfrac{1}{r_\pi} + sC_\pi\right)}{sC_\mu - g_m} \right\}^* & \left\{ \dfrac{\left(sC_\mu + \dfrac{1}{r_\pi} + sC_\pi\right)}{sC_\mu - g_m} \right\}^* \end{bmatrix} \tag{5.146}$$

but, $C_A$ is also equal to

$$C_A = 2kT \begin{bmatrix} R_n & \left[\dfrac{F_{min} - 1}{2} - R_n Y_{sopt}^2\right] \\ \left[\dfrac{F_{min} - 1}{2} - R_n Y_{sopt}\right] & R_n \left|Y_{sopt}\right|^2 \end{bmatrix} \tag{5.147}$$

It can be shown that the noise parameters for the two-port noise model are given by

$$R_n = r_b\left(1 + \frac{1}{\beta}\right) + \frac{1}{2g_m} + \frac{g_m r_b^2}{2}\left(\frac{1}{\beta^2} + \frac{f^2}{f_T^2}\right) \tag{5.148}$$

and if $\beta \gg 1$ and $f_T \gg R_n$ becomes $r_b + 1/2g_m$.

$$G_{sopt} = \frac{\left[\dfrac{1}{\beta^2} + 2g_m r_b\left(1 + \dfrac{1}{\beta}\right)\left(\dfrac{1}{\beta^2} + \dfrac{f^2}{f_T^2}\right) + \left(g_m r_b\right)^2\left(\dfrac{1}{\beta^2} + \dfrac{f^2}{f_T^2}\right)^2\right]^{\frac{1}{2}}}{2r_b\left(1 + \dfrac{1}{\beta}\right) + \dfrac{1}{g_m} + g_m r_b^2\left(\dfrac{1}{\beta^2} + \dfrac{f^2}{f_T^2}\right)} \tag{5.149}$$

and

$$B_{sopt} = \frac{\dfrac{f}{f_T}}{2r_b\left(1 + \dfrac{1}{\beta}\right) + \dfrac{1}{g_m} + g_m r_b^2\left(\dfrac{1}{\beta^2} + \dfrac{f^2}{f_T^2}\right)} \tag{5.150}$$

with $Y_{sopt} = G_{sopt} + jB_{sopt}$

$$R_{sopt} = \frac{\left[2r_b\left(1+\frac{1}{\beta}\right)+\frac{1}{g_m}+g_m r_b^2\left(\frac{1}{\beta^2}+\frac{f^2}{f_T^2}\right)\right]\left[\frac{1}{\beta^2}+2g_m r_b\left(1+\frac{1}{\beta}\right)\left(\frac{1}{\beta^2}+\frac{f^2}{f_T^2}\right)+\left(g_m r_b\right)^2\left(\frac{1}{\beta^2}+\frac{f^2}{f_T^2}\right)^2\right]}{\frac{1}{\beta^2}+2g_m r_b\left(1+\frac{1}{\beta}\right)\left(\frac{1}{\beta^2}+\frac{f^2}{f_T^2}\right)+\left(g_m r_b\right)^2\left(\frac{1}{\beta^2}+\frac{f^2}{f_T^2}\right)^2+\frac{f^2}{f_T^2}} \tag{5.151}$$

thus

$$X_{sopt} = \frac{\frac{f}{f_T}\left[2r_b\left(1+\frac{1}{\beta}\right)+\frac{1}{g_m}+g_m r_b^2\left(\frac{1}{\beta^2}+\frac{f^2}{f_T^2}\right)\right]}{\frac{1}{\beta^2}+2g_m r_b\left(1+\frac{1}{\beta}\right)\left(\frac{1}{\beta^2}+\frac{f^2}{f_T^2}\right)+\left(g_m r_b\right)^2\left(\frac{1}{\beta^2}+\frac{f^2}{f_T^2}\right)^2+\frac{f^2}{f_T^2}} \tag{5.152}$$

and with $Z_{sopt} = 1/Y_{sopt} = R_{sopt} + jX_{sopt}$

$$F_{min} = \frac{1}{2g_m R_s}+\left(1+\frac{1}{\beta}\right)\left(1+\frac{r_b}{R_s}\right)+\frac{g_m}{2R_s}\left(r_b+R_s\right)^2\left(\frac{1}{\beta^2}+\frac{f^2}{f_T^2}\right) \tag{5.153}$$

The approach used to determine noise parameters assumed that $r_b$ by itself is not a noisy two-port network, and only the contribution of $r_b$ to the transfer function is considered in the calculations. A different approach that considers $r_b$ and the CE transistor circuit as two cascaded two-port networks has been investigated. The analytical expressions obtained for both methods for the noise resistance $R_n$ are close and provide the same results when $f_T \gg f$.

## TWO-PORT NOISE PARAMETERS OF THE CE SIGE HBT BASED ON ADMITTANCE PARAMETERS

The admittance parameters of the CE SiGe HBT can be determined accurately, considering the base and emitter parasitic resistances $r_b$ and $r_{e,}$ as well as the base–collector capacitance $C_\mu$. This approach allows the provision of accurate analytical expressions of the two-port noise parameters.

## CONCLUSION

This chapter has presented a comprehensive discussion of noise mechanisms, small-signal behavior, and scaling effects in modern transistor technologies used for RF and microwave circuit design. Beginning from device-level physical considerations and progressing toward two-port representations, the chapter has established a consistent framework that links microscopic noise sources to measurable circuit-level parameters. Although the detailed formulations differ among MOSFETs, BJTs, and SiGe HBTs, the underlying objective remains the same: to understand how device geometry, biasing, and material properties determine gain, bandwidth, and noise performance.

A central theme throughout the chapter is the role of intrinsic and extrinsic noise sources and their interaction with the small-signal parameters of the transistor. Thermal noise associated with resistive elements, shot noise related to carrier transport, and frequency-dependent noise mechanisms all contribute to the total input-referred noise. By expressing these contributions through two-port representations, it becomes possible to derive analytical relationships for the noise factor, minimum noise figure, optimal source admittance, and noise resistance. These quantities form the foundation of practical low-noise design methodologies.

For bipolar and SiGe HBTs, the analysis shows that noise performance is strongly influenced by base resistance, transconductance, current gain, and unity-gain frequency. At low and moderate frequencies, thermal noise associated with parasitic resistances typically dominates, whereas at higher frequencies the contribution of collector shot noise becomes increasingly significant. This behavior explains the characteristic increase of noise figure with frequency and highlights the importance of maximizing $f_T$ while minimizing parasitic elements. The dependence of noise on bias current further reveals the existence of an optimum operating region in which gain and noise performance are jointly optimized.

The chapter also emphasizes the importance of geometry and scaling. Device dimensions, finger multiplicity, and layout-dependent parasitics directly affect resistance, capacitance, and transconductance, and therefore strongly influence noise behavior. In MOSFETs, correlated channel and induced gate noise complicate simultaneous noise and power matching, and the minimum noise figure generally increases with device width. In contrast, bipolar and SiGe HBT devices benefit from higher transconductance efficiency and weaker noise correlation, allowing more flexible trade-offs between matching, gain, and noise. These differences explain why SiGe HBTs remain particularly attractive for low-noise and high-frequency applications.

The derivation of noise parameters such as $F_{min}$, $R_n$, and $Y_{opt}$ provides a compact and physically meaningful description of transistor noise behavior. These parameters enable the synthesis of input matching networks and allow direct comparison between devices and technologies. Their relative invariance under lossless transformations further justifies their widespread use in RF design practice. The analytical development also illustrates how two-port noise representations serve as a bridge between device physics and circuit-level design.

Although modern circuit design relies heavily on compact models and foundry-provided simulation data, analytical noise models retain significant value. They provide physical insight, allow rapid estimation of trends, and help identify the dominant mechanisms limiting performance. Discrepancies between modeled and measured noise behavior may arise from short-channel effects, high-field transport, or incomplete modeling of parasitic elements, emphasizing the need for careful interpretation of simulation results.

In summary, this chapter establishes a unified theoretical foundation for understanding noise in MOSFETs, BJTs, and SiGe HBTs. By combining device physics, small-signal modeling, and two-port noise analysis, it provides the conceptual tools needed to evaluate and optimize transistor performance for low-noise applications. The principles developed here form the basis for the circuit-level design approaches presented in subsequent chapters, particularly for LNAs and RF front-end systems operating from microwave to mm-wave frequencies.

## REFERENCES

Abidi, A.A., 2014. CMOS Microwave and Millimeter-Wave ICs: The Historical Background. In: *2014 IEEE International Symposium on Radio-Frequency Integration Technology*. IEEE, 1–5.

Asgaran, S., Deen, M.J., and Chen, C.-H., 2004. Analytical Modeling of MOSFETs Channel Noise and Noise Parameters. *IEEE Transactions on Electron Devices*, 51 (12), 2109–2114.

Banerjee, B., Venkataraman, S., Zhao, E., Lee, C.-H., Cressler, J.D., Laskar, J., El-Kareh, B., Balster, S., and Yasuda, H., 2005. Modeling of Broadband Noise in Complementary (npn + pnp) SiGe HBTs. In: *RFIC) Symposium, 2005. Digest of Papers. 2005 IEEE Radio Frequency integrated Circuits*. IEEE, 553–556.

Belostotski, L., Haslett, J.W., and Veidt, B., 2006. Wide-Band CMOS Low Noise Amplifier for Applications in Radio Astronomy. In: *2006 IEEE International Symposium on Circuits and Systems*. IEEE, 4.

Carbone, V., Telloni, D., Lepreti, F., and Vecchio, A., 2022. High-frequency Magnetic Fluctuations in Space Plasmas and the Role of Electron Landau Damping. *The Astrophysical Journal Letters*, 924 (2), L26.

Chen, H.-Y., Chen, K.-M., Huang, G.-W., and Chang, C.-Y., Small-Signal Modeling of SiGe HBTs Using Direct Parameter-Extraction Method. *IEEE Transactions on Electron Devices*, 53 (9), 2287–2295.

Cressler, J.D., 1998. SiGe HBT Technology: A New Contender for Si-Based RF and Microwave Circuit Applications. *IEEE Transactions on Microwave Theory and Techniques*, 46 (5), 572–589.

Dasgupta, A., Khandelwal, S., and Chauhan, Y.S., 2015. Surface Potential Based Modeling of Thermal Noise for HEMT Circuit Simulation. *IEEE Microwave and Wireless Components Letters*, 25 (6), 376–378.

de Vreede, L.C.N., de Graaff, H.C., Hurkx, G.A.M., Tauritz, J.L., and Baets, R.G.F., 1994. A Figure of Merit for the High-Frequency Noise Behavior of Bipolar Transistors. *IEEE Journal of Solid-State Circuits*, 29 (10), 1220–1226.

Dickson, T.O., Yau, K.H.K., Chalvatzis, T., Mangan, A.M., Laskin, E., Beerkens, R., Westergaard, P., Tazlauanu, M., Yang, M.-T., and Voinigescu, S.P., 2006. The Invariance of Characteristic Current Densities in Nanoscale MOSFETs and Its Impact on Algorithmic Design Methodologies and Design Porting of Si(Ge) (Bi)CMOS High-Speed Building Blocks. *IEEE Journal of Solid-State Circuits*, 41 (8), 1830–1845.

Gao, J., 2010. *Optoelectronic Integrated Circuit Design and Device Modeling*. Wiley.

Geddes, L.A., and Roeder, R.A., 2003. De Forest and the First Electrosurgical Unit. *IEEE Engineering in Medicine and Biology Magazine*, 22 (1), 84–87.

Gray, P.R., Hurst, P.J., and Meyer, R.G., 2009. *Analysis and Design of Analog Integrated Circuits*. 5th ed. Wiley.

Hassan, B., Cutiver, A., Rodriguez, C., Cozette, F., Soltani, A., and Maher, H., 2019. Scalable Small-Signal Modeling of AlGaN/GaN HEMT Based on Distributed Gate Resistance. 2019 IEEE BiCMOS and Compound Semiconductor Integrated Circuits and Technology Symposium (BCICTS), IEEE.

Jun, F., 2009. Small-Signal Model Parameter Extraction for Microwave SiGe HBTs Based on Y- and Z-Parameter Characterization. *Journal of Semiconductors*, 30 (8), 084005.

Khalil, I., Liero, A., Rudolph, M., Lossy, R., and Heinrich, W., 2008. GaN HEMT Potential for Low-Noise Highly Linear RF Applications. *IEEE Microwave and Wireless Components Letters*, 18 (9), 605–607.

Lambrechts, W., and Sinha, S., 2017. *SiGe-Based Re-engineering of Electronic Warfare Subsystems*. Springer International Publishing Imprint: Springer.

Lee, K., Choi, K., Kook, S.-H., Cho, D.-H., Park, K.-W., and Kim, B., 2005. Direct Parameter Extraction of SiGe HBTs for the VBIC Bipolar Compact Model. *IEEE Transactions on Electron Devices*, 52 (3), 375–384.

Lee, T.H., 2004. *The Design of CMOS Radio-Frequency Integrated Circuits*. Cambridge University Press.

Manku, T., 1999. Microwave Noise Modeling of CMOS Transistors. In: *Analog Circuit Design*, edited by W. Sansen, J. Huijsing, and R. van de Plassche. Springer US, 247–265.

Nallatamby, J.-C., Prigent, M., Cordeau, D., and Paillot, J.-M., 2017. Noise Sources Characterization and Its Compact Modeling Based on Physical Device Simulations for Designing Low Phase Noise Oscillators. In: *2017 Joint Conference of the European Frequency and Time Forum and IEEE International Frequency Control Symposium (EFTF/IFCS)*. IEEE, 175–175.

Navid, R., Navid, R., Lee, T.H., and Dutton, R.W., 2007. A Circuit-Based Noise Parameter Extraction Technique for MOSFETs. In: *2007 IEEE International Symposium on Circuits and Systems (ISCAS)*. IEEE, 3347–3350.

Niu, G., 2005. Noise in SiGe HBT RF Technology: Physics, Modeling, and Circuit Implications. *Proceedings of the IEEE*, 93 (9), 1583–1597.

Nsele, S.D., Escotte, L., Tartarin, J.-G., and Piotrowicz, S., 2013. Noise Characteristics of AlInN/GaN HEMTs at Microwave Frequencies. In: *2013 22nd International Conference on Noise and Fluctuations (ICNF)*. IEEE, 1–4.

Pawlak, A., Lehmann, S., Sakalas, P., Krause, J., Aufinger, K., Ardouin, B., and Schroter, M., 2015. SiGe HBT Modeling for mm-Wave Circuit Design. In: *2015 IEEE Bipolar/BiCMOS Circuits and Technology Meeting – BCTM*. IEEE, 149–156.

Pulfrey, D. L., 2010. *Understanding Modern Transistors and Diodes*. Cambridge University Press.

Sah, C.T.., 1991. *Fundamentals of Solid-State Electronics*. World Scientific.

Sakalas, P., Herricht, J., Chakravorty, A., and Schroter, M., 2006. Compact Modeling of High Frequency Correlated Noise in HBTs. In: *2006 Bipolar/BiCMOS Circuits and Technology Meeting*. IEEE, 1–4.

Schleeh, J., Rodilla, H., Wadefalk, N., Nilsson, P., and Grahn, J., 2013. Characterization and Modeling of Cryogenic Ultralow-Noise InP HEMTs. *IEEE Transactions on Electron Devices*, 60 (1), 206–212.

Scholten, A. J., Tiemeijer, L. F., Langevelde, R. van, Havens, R. J., Venezia, V. C., Zegers-van Duijnhoven, A. T. A., Vries, B. De, and D. Klaassen, B. M., 1999, Accurate Thermal Noise Model for Deep-Submicron CMOS. International Electron Devices Meeting (IEDM), IEEE. Washington, DC, USA.

Shen, Y., Cui, J., and Mohammadi, S., 2017. An Accurate Model for Predicting High Frequency Noise of Nanoscale NMOS SOI Transistors. *Solid-State Electronics*, 131, 45–52.

Sotoodeh, M., Sozzi, L., Vinay, A., Khalid, A.H., Hu, Z., Rezazadeh, A.A., and Menozzi, R., 2000. Stepping toward Standard Methods of Small-Signal Parameter Extraction for HBTs. *IEEE Transactions on Electron Devices*, 47 (6), 1139–1151.

Subramanian, V., Parvais, B., Borremans, J., Mercha, A., Linten, D., Wambacq, P., Loo, J., Dehan, M., Gustin, C., Collaert, N., Kubicek, S., Lander, R., Hooker, J., Cubaynes, F., Donnay, S., Jurczak, M., Groeseneken, G., Sansen, W., and Decoutere, S., 2006. Planar Bulk MOSFETs Versus FinFETs: An Analog/RF Perspective. *IEEE Transactions on Electron Devices*, 53 (12), 3071–3079.

Tsividis, Y., and McAndrew, C., 2011. *Operation and Modeling of the MOS Transistor*. Oxford University Press.

Ture, E., 2016. *GaN-Based Tri-Gate High Electron Mobility Transistors*. Albert-Ludwigs University of Freiburg.

Voinigescu, S.P., Maliepaard, M.C., Showell, J.L., Babcock, G.E., Marchesan, D., Schroter, M., Schvan, P., and Harame, D.L., 1997. A Scalable High-Frequency Noise Model for Bipolar Transistors with Application to Optimal Transistor Sizing for Low-Noise Amplifier Design. *IEEE Journal of Solid-State Circuits*, 32 (9), 1430–1439.

Yau, K. H. K., 2006. *Noise Modelling of Silicon Germanium Heterojunction Bipolar Transistors at Millimetre-Wave Frequencies*. University of Toronto.

Yau, K.H.K., Tang, K.K.W., Schvan, P., Chevalier, P., Sautreuil, B., and Voinigescu, S.P., 2007. The Invariance of the Noise Impedance in n-MOSFETs across Technology Nodes and Its Application to the Algorithmic Design of Tuned Low Noise Amplifiers. In: *2007 Topical Meeting on Silicon Monolithic Integrated Circuits in RF Systems*. IEEE, 245–248.

Ytterdal, T., Cheng, Y., and Fjeldly, T.A., 2003. *Device Modeling for Analog and RF CMOS Circuit Design*. Wiley.

# 6 Scaling and Integration of Ultra-Low-Noise RF CMOS

## INTRODUCTION

Low-noise amplifiers (LNAs) remain foundational to every modern wireless receiver, setting the ultimate sensitivity and signal integrity from which the entire front end must operate. As communication standards evolve toward mm-wave and eventually sub-THz bands, the performance of these LNAs is determined less by ideal transistor behavior and increasingly by the nuanced interaction between device physics, parasitic elements, and compact modeling accuracy. While early CMOS LNAs could be designed with simplified assumptions, moderate channel lengths, predictable gm scaling, and manageable matching constraints, nanoscale technologies require a more holistic methodology. Here, the transistor itself is no longer an isolated building block, but an electromagnetic structure with geometry-dependent impedances, layout-dependent coupling paths, and noise behavior that cannot be separated from process conditions or bias regime. As a result, compact models have grown significantly in complexity, and their parameter extraction demands rigorous measurement discipline, careful de-embedding, and validation across both small-signal and large-signal domains.

Simultaneously, device scaling continues to reshape what optimal means in front-end design. Short-channel effects, mobility degradation, and parasitic capacitances set new boundaries for gain, noise, and linearity, requiring circuit-level decisions to accommodate device-level physics rather than suppress them. FinFET, GAAFET, and emerging two-dimensional (2D) semiconductor structures further alter the balance, offering higher transconductance efficiency and lower capacitance, but only if modeling captures transport phenomena, substrate interactions, and non-quasi-static effects with sufficient fidelity. Ultimately, the success of next-generation RF systems depends not only on better transistors but also on the accuracy with which we can describe them, mathematically, physically, and in simulation. This chapter, therefore, examines modern parameter extraction workflows, the challenges of scaled CMOS LNA design, and the device technologies poised to carry CMOS into 6G and beyond.

## HISTORICAL ROLE OF CMOS IN RF AND MICROWAVE DESIGN

As indicated by Abidi (2014), fully integrated microwave integrated circuits (ICs) have been in various forms of development since the 1970s, and their later move to complex microwave and millimeter-wave (mm-wave) CMOS subsystems began significant activity throughout the late 1990s up to now. The work in Abidi (2014) (and to


DOI: 10.1201/9781003459385-6

an extent Lee 2007) offers a comprehensive viewpoint on the pre-CMOS experience where active microwave circuits were already placed on chip level, some even operating at 60 GHz. During this time, around 2014, mm-wave consumer devices were not yet considered mature, but research at the time showed that it would likely be a feasible technology in the future. Its approach, which varied significantly from the more typical radio frequency (RF) CMOS at the time, required researchers to experiment with the technology and find solutions for new problems that were not evident in the RF spectrum. From its early days, mm-wave research attracted attention from radar needs in World War II, and after that, when Bell Labs started looking for high-capacity trunk systems. The IMPATT diode was developed, a solid-state device that laid the foundation for silicon-based microwave, as well as mm-wave, integration programs such as Texas Instruments' solid-state phased array, called the Molecular Electronics for Radar Application program (MERA) (Hyltin 2000), developed around 1964. Arguably, this (and likely other programs) marked the start of (formal) development of silicon microwave (followed by mm-wave) ICs (McQuiddy *et al.* 1984, 1991; Abidi 2014), although no specific historical moment can really be viewed as the formalized starting point of the said research. Monolithic microwave integrated circuit (MMIC) technology was already under development since the 1950s, according to McQuiddy *et al.* (1984), but the push toward silicon (and gallium arsenide (GaAs)) marked a significant shift toward increasing the frequency of operation significantly. McQuiddy *et al.* provide an interesting and detailed historical perspective on MMICs, both in McQuiddy *et al.* (1991) and McQuiddy *et al.* (1984), and is highly recommended to understand and appreciate the activities around early high-frequency developments and the problems that materials and techniques presented at the time.

Around the 1970s, there was a significant push for both hybrid MICs and MMICs to evolve toward commercial research and applications, and not just operate in the realm of defense-driven programs. Up to this point, the Defense Advanced Research Projects Agency (DARPA) of the United States Department of Defense spearheaded most of the MMIC research, especially on GaAs, but there was a shift happening toward the commercialization of these high-frequency data transmission capabilities. In 1975 (with a publication in 1976 (Pengelly and Turner 1976)), a commercial company in the United Kingdom (UK), called Plessey, developed the first commercial GaAs MMIC, a single-stage and broadband amplifier, further encouraging DARPA to invest in programs at Texas Instruments, Raytheon, and General Electric to develop MMIC technology for S-band and X-band modules. The amplifier in Pengelly and Turner (1976) used lumped elements, integrated with the transistor, on semi-insulating GaAs, and was 1.8 × 1.2 mm in size. It had a gain of more than 4.5 dB over 7.5 to 11.5 GHz, and was installed into a low-parasitic package, giving it a practical advantage and commercial traction at the time (and being a major catalyst for DARPA's increased involvement during the late 1970s onward).

CMOS, by the 1990s, was still viewed primarily as a digital technology, and GaAs remained the dominant choice for high-frequency analog and RF implementations. However, the CMOS environment started to improve dramatically, and breakthroughs in deep submicron scaling enabled cutoff frequencies ($f_T$ and $f_{max}$) to rise dramatically, and this opened new possibilities in CMOS front-end design (Lee 2004). Books and papers on the topic started growing in popularity, and publications such as Razavi

(1997) became widely referenced and used. A large contributor to the research and development of high-frequency analog integrated solutions was wireless communication, specifically cellular (mobile) phones. At the time, GSM was used to transmit voice and data; however, the increased focus on data-oriented mobile experiences created commercial pressure to develop low-cost integrated solutions. CMOS was able to compete with GaAs, although having inferior noise/power performance, its scalability and cost were large drivers for its development. The integration density and cost of CMOS, compared to GaAs, where CMOS was able to integrate LNAs, mixers, oscillators (synthesizers), and baseband operation on a single die, overshadowed (Abou-Allam *et al.* 2000) any noise performance increases available in GaAs and other technologies. In effect, the focus shifted away from purely maximizing device performance towards maximizing system-level integration and manufacturability, and this marked the real turning point for CMOS as a serious contender in RF and, later, mm-wave applications. This momentum would carry into the 2000s, where the push toward higher frequencies, broadband data services, and eventually mm-wave systems created an entirely new set of challenges and opportunities for CMOS technology.

During the early 2000s, CMOS shifted from a potential candidate for consumer RF integration, such as Bluetooth, Wi-Fi, GSM (and EDGE), followed by UMTS, to the dominant go-to technology with immense potential for future development. Its low-cost and dense solutions easily outweigh GaAs and most commercial products (Lee 2004), except for high-end and/or military applications, where GaAs was still superior in its noise performance (Razavi 1997). Another breakthrough in CMOS development was the system-on-chip (SoC) for RF applications, a technique to integrate RF front ends with baseband and digital control circuits (Chun-Lin Ko *et al.* n.d.). The mid-2000s (Floyd *et al.* 2005) is known for its focused development toward sub-100 nm CMOS scaling and base-level doping in technologies such as silicon-germanium (SiGe) to increase $f_T$ and $f_{max}$ into the 100– 200 GHz range. Research into applications in mm-wave, especially analog front ends (Razavi 2006), was commonplace at the time, and the field was spearheaded by numerous publications and mm-wave CMOS designs being published at the time. Dr. Ali Hajimir and his team at Caltech University were among the leaders in the research field, specifically with their invention of the distributed active transformer (DAT). This device enabled the integration of RF CMOS power amplifiers, required and suitable for cellular phones, also later used to demonstrate radar-on-chip in silicon technology. Although numerous other institutions and researchers developed groundbreaking technologies at the time, related to tight CMOS integration, the entire research and development community worked toward novel and commercially viable solutions, with mm-wave and THz applications in mind. The direction of the late 2000s and early 2010s moved toward better commercialization, increased SoC integration, and novel applications such as phased arrays, beam forming, and eventually low-cost radar and wireless standards.

In 2013, Rappaport *et al.* (2013) presented a highly cited and recommended article on mm-wave technology in 5G telecommunications. 5G became a dominant driver of mm-wave (and THz) technology from the mid-2010s, and still is, toward 6G communications. Rappaport *et al.* (2013) showed via measurements at 28 GHz and 38 GHz that with steerable, directional antennas at both base station and handset, line-of-sight path loss approaches free-space levels, and even non-LOS links are viable with careful beamforming. The paper also emphasized that while delay spread (multipath dispersion) in NLOS conditions at mm-wave is larger and more variable, it remains within bounds that allow for practical system design, again provided antenna gain and steering are exploited. Researchers at Caltech, industrial collaboration with companies like Infineon, NXP, Texas Instruments, and Analog Devices; conference proceedings; patents from Qualcomm/Broadcom; and work by Princeton University under K. Sengupta all further contributed to mm-wave ICs, automotive radar, 60 GHz consumer electronics, and THz/sub-THz CMOS research to reach a technology maturity as can be witnessed in modern technology. Further evidence of this maturity can be seen in doctoral work (Forsberg 2018) that not only surveyed mm-wave CMOS transmitters but also presented a complete set of fabricated designs, ranging from 65 nm PAs to low-power voltage-controlled oscillators and phase-locked loops operating in the 30–60 GHz bands, demonstrating that efficiency, phase-noise, and integration challenges were being solved in practice, not just theory. Key papers on mm-wave CMOS during this time focused on improving amplifier efficiency and linearity (Park *et al.* 2016), designing oscillators and VCOs, and modeling components for high-frequency operation. Significant work was done on CMOS processes such as 65 nm and 130 nm, developing advanced techniques, including Doherty power amplifiers, and applying port voltage/current optimization and *T*-embedding for high-efficiency oscillators.

Figure 6.1 visually depicts the timeline and key inflection points of CMOS and III–V RF integration from the 1950s to the 2020s.

Shown in Figure 6.1, the *x*-axis represents the years between the 1950s and the 2020s, as discussed in this section. The *y*-axis represents the operational frequency of

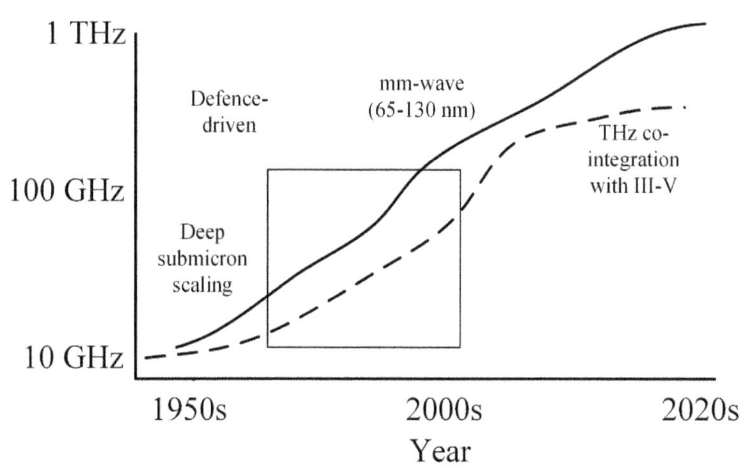


**FIGURE 6.1** Timeline and key inflection points of CMOS and III–V RF integration between 1950s and 2020s.

transistors, extending from the MHz range to the THz range, on a logarithmic scale. The solid line represents CMOS evolution, and the dashed line represents III–Vs. The shaded (overlapping) region can be referred to as a competitive regime, a time where CMOS and III–V frequency capabilities overlapped, around the 30–150 GHz. From Figure 6.1, it is shown that around the 1950s and 1960s, early MMICs (GaAs, hybrid circuits, and the DARPA MERA program) were key research areas. In the mid-1970s, the first commercial MMICs (Plessey, UK) were produced. In the 1990s, CMOS deep submicron scaling led to a significant rise in $f_T$, and by the 2000s, SoC integration in commercial applications such as Bluetooth and Wi-Fi was commonplace. During the 2010s, mm-wave CMOS in the 65–130 nm nodes were popular, leading to significant advances in radar-on-chip and 5G prototypes. In the 2020s, 6G and THz communication are dominant research areas, with nano-ridge engineering co-integration with III–Vs being explored.

## Why CMOS Is Central to Integrated LNAs vs III–V Solutions

Co-integration of CMOS and III–V solutions (Zota *et al.* 2019) remains a strong contender for RF and mm-wave front ends, especially when low noise and high-power solutions are required. However, the balance and trade-offs between integration complexity, overall cost, and performance remain a challenging balance to get just right. CMOS amplifiers, especially LNAs, which are considered the interface to the real world, are ubiquitous, low-cost, and have many general-purpose commercial applications. At its core, the cost of CMOS remains a factor that cannot be overlooked, even if performance is considered, depending on the figure of merit, worse than III–V (GaAs and GaN, for example) integrated devices. III–V materials also exhibit additional performance gains when operated cryogenically (Zota *et al.* 2019), but again, the cost involved in realizing these solutions is difficult to justify in many cases. For military applications, cryogenically cooled III–V active devices remain a popular solution, as their noise performance remains superior to CMOS implementation. Its increased cost is often justified in these applications as state-of-the-art performance is typically a high priority.

Still, for consumer-grade (wireless) systems, the sheer momentum of CMOS (with reference to Moore's law (Moore 1998)) has been difficult to ignore. As mm-wave wireless networks for 5G and 6G are developed, numerous operational specifications must be satisfied, including operating frequency, power consumption, efficiency, and integration density. As consumer electronics also shrink in physical size (mobile phones, Internet of Things (IoT)), the operating requirements, along with the integration density, remain imperative. A proposed route is CMOS-compatible III–V co-integration, where the high-powered GaN-on-Si devices, or, depending on the application, the very high speed InP-based heterojunction bipolar transistors (HBTs) are combined – typically on enlarged (compared to typical dies) silicon substrates. Nano-ridge engineering (NRE) (Yudistira *et al.* 2025) has demonstrated progress in this field by depositing III–V material into narrow oxide trenches on silicon wafers. This process traps relaxation defects, and the continuous growth from the trench pattern results in high-quality III–V nano-ridges with engineered shapes and increased material volume (Yudistira *et al.* 2025). Engineering progress such as these NREs demonstrates that there is still a possibility to truly hybridize III–V and CMOS for future 5G and 6G solutions toward SoC RF integration, albeit still at a higher cost than traditional CMOS.

Apart from the typical performance metrics and integration cost, designing for CMOS and III–V also has different aspects to consider, and developing an IC on these two technologies differs quite significantly. Considering LNA design and specifically looking at component-level physics, CMOS circuits (or components) (if designed for MOSFETs) often require clever engineering to extract the maximum amount of transconductance ($g_m$) per unit bias current (OKTYABRSKY *et al.* 2008; del Alamo *et al.* 2015), as this can be limited in CMOS (when compared to III–Vs – HEMTs and HBTs). III–Vs inherently exhibit higher electron mobility, which translates to higher transconductance for the same bias current. In LNA design, gain is one of the most important performance metrics to consider, together with noise, of course, but amplifying a weak incoming signal to something usable and easy to manipulate is a critical feature of any LNA, especially since the interface is directly affected by the environment and can be extremely weak and noisy. On noise, at baseband, CMOS flicker noise (1/*f*) is more pronounced than in III–Vs due to lower channel noise and higher output impedance (Maity and Pandit 2021). High-powered III–V transistors, and as the name suggests, having higher breakdown voltages and higher bias voltages, often demonstrate better linearity and fewer limitations in operating headroom, when compared to CMOS transistors (at comparable operating conditions). Moving away slightly from the component-level physics, the topologies of CMOS LNAs compared to III–V LNAs are also different. In CMOS, it is well known that the inductive source degeneration common-source (CS) LNA topology is a tried, tested, and trusted topology, even for high-frequency applications. This CS topology sets the input matching up and lowers the noise figure (NF) of the circuit. However, it requires degeneration to achieve proper 50 Ω matching. It has been shown that III–V LNAs operate reliably in CS (HEMT) or common-emitter (CE) (HBT) configurations with passive matching. As a result, these LNAs require less degeneration to get a 50 Ω match due to their inherently high transconductance. Furthermore, as CMOS amplifiers are often cascaded to increase isolation and stability, it is well known that the voltage headroom in these circuits can be limited (often seen in 3.3, 2.5, and 1.8 V technology nodes), whereas high-powered III–V technologies by definition have more voltage headroom and therefore fewer limitations when stacking or cascading amplification stages. Device parasitics also play an important role in high-frequency application-specific integrated circuits (ASICs) and, generally, III–Vs have lower (lower is better) parasitic values, especially in terms of their parasitic capacitance – although comparing these values requires ensuring that they are compared under similar conditions, and in some circumstances, CMOS can also have "better" parasitic performance. In most cases, the maturity of silicon-based CMOS and its predictability, along with its internal material properties, which are often considered as a "perfect" material, especially when compared to III–Vs, still count as a major advantage for traditional CMOS LNAs. Its SoC capabilities, where III–Vs often require multi-chip modules (MCMs), also count heavily in its favor for

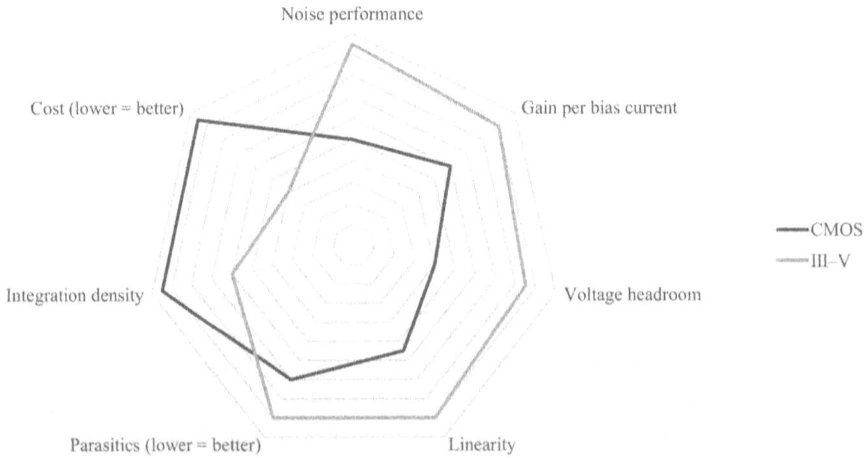


**FIGURE 6.2** Comparison of CMOS versus III–V LNA design trade-offs.

commercial applications. Figure 6.2 is a visual representation of LNA design trade-offs in CMOS and III–V technology.

Figure 6.2, a radar chart, shows the normalized contributions/comparison between the trade-offs in LNA design, for CMOS and III–Vs. It highlights the parameters such as noise performance, gain per bias current, therefore gm efficiency, voltage headroom, linearity, parasitics, integration density, and cost. From this figure, III–V technology is typically superior in noise performance, gain per bias current, voltage headroom, linearity, and parasitics. However, CMOS research has shown that some notable enhancements and mitigations significantly increase CMOS performance, typically at the expense of other trade-offs, which are application-specific. The radar chart is also not weighted; therefore, the advantages that CMOS offer in cost and integration density typically outweigh some of the other performance parameters, again, especially true if application specificity is considered.

## Target Metrics: NF, Gain, $f_t$, $f_{max}$, Power Consumption

LNA performance metrics are well known and briefly discussed in the previous paragraphs. These metrics are the cornerstone of any good LNA, and researchers and engineers always strive to design amplifiers that minimize trade-offs and deliver on the required performance specifications. But, in CMOS scaling, these performance metrics evolve. Designers are increasingly required to consider and balance trade-offs when choosing a technology node and/or application. NF is the most unforgiving when designing LNAs, and in CMOS, NF is often considered as "good-enough" rather than best in class. In amplifier applications, smaller devices that result from technology scaling often benefit from the decreased size as higher $f_T$ and better noise matching (especially at mm-wave) are observed. Thus, there is generally an inverse relationship between $f_T$ and physical size (specifically, gate length) for FETs (Poulain *et al.* 2011).

Gain is off course one of the critical, but often misunderstood, performance metrics of any LNA. In general, designers usually want *more* (Choi *et al.* 2020) of it without considering all factors, such as for what purpose and at what cost, in terms of trade-offs. In a practical LNA, the available gain is set first by the device physics (principally transconductance per unit bias current) and second by the impedance environment that the transistor sees. In CMOS, this interplay forces designers into a set of recurring selections. At the device level, gain is proportional to $g_m$ and to how effectively that transconductance is translated into a voltage at the output node. The traditional approaches to increase that conversion, cascading to raise the intrinsic output resistance (Zhang *et al.* 2020), multi-stage amplification, or using transformers and resonant loads to create highly effective impedances, are essentially the same across technologies. CMOS, however, differs in that $g_m$ per ampere is typically lower than for high-mobility III–V devices, so achieving a large gain in a single stage is more difficult. The implication is twofold: first, CMOS designs often rely on careful matching networks (sometimes active matching) and multi-stage topologies to steadily increase gain, and second, the designer must accept trade-offs in area, power, or bandwidth to achieve the required or desired gain. Adding amplification stages raises total gain but also increases sensitivity to instability, mismatch, and accumulated noise from subsequent stages (importantly, as a design rule, the first stage dominates the noise budget). Using resonant or high-$Q$ loads (Pascht *et al.* 2001) gives high gain at a narrow bandwidth, but that gain disappears out of band. Transformers and interstage matching networks can be used to transform impedances so the transistor operates in its optimal region; at mm-wave, on-chip transformers or transmission-line matching become attractive because discrete lumped inductors become impractical. So, if gain is chased, one pays with noise, bandwidth, stability, or power; the art is then in choosing the right trade-off for the specific application, showing that there is no single rule to follow, and each case must be looked at on individual merit.

$f_T$ and $f_{max}$ are normally quoted device parameters; these numbers accurately indicate, respectively, the frequency at which current gain decreases to unity and the maximum frequency at which a positive power gain can be obtained. However, $f_T/f_{max}$ numbers alone are not sufficient to predict complete circuit performance unless they are quoted in context with the application and additional circuit (and system) parameters. First, CMOS scaling has driven $f_T$ significantly higher, and as a result, circuit and system operation and microwave, and mm-wave have become practical and common in modern applications (Qian *et al.* 2023). However, as a rule of thumb, circuits should not be designed to operate right at $f_T$. In practice, a comfortable design margin is used: designers typically target operation at some fraction of $f_T$, approximately 1/5 to 1/10 of $f_T$, depending on the topology and how aggressive the matching is. Pushing the operating frequency too close to $f_T$ leaves little headroom for matching losses, packaging parasitics, or model inaccuracies at high frequency. $f_{max}$ is often a better single-number predictor for both power and oscillator feasibility because it considers parasitic resistances and feedback effects. A device can have a high $f_T$ yet a modest $f_{max}$ if parasitic gate or source resistances are large, and those parasitics have a much more prominent effect at mm-wave. For this reason, layout, contact resistance, and interconnect length are important to consider and fundamentally set the usable portion of $f_T$. Also, at mm-wave, the "loss budget" of matching networks is challenging as on-chip metal losses, finite $Q$ of transmission lines, and via resistance limit the usable margin between the intrinsic gain of the transistor and the gain requirements of the system. Figure 6.3 shows the evolution of CMOS LNA performance constraints with technology scaling. While transistor speed ($f_T$, $f_{max}$) continues to improve, headroom and achievable gain per stage decrease, and NF improvements begin to saturate. The resulting design space shifts from "device-limited" to "layout- and system-limited" performance in advanced nodes.

Figure 6.3, a radar (spider) chart, shows the normalized key metrics in LNA design as they evolve across technology nodes, therefore a function of scaling. It highlights the nodes 130, 65 28, and 7 nm. Each metric (NF, gain, $f_T/f_{max}$, power efficiency, and voltage headroom) shows a non-monotonic trend; therefore, some improve while others degrade across node evolution. These trends show that, for example, voltage headroom decreases as nodes scale down, while power efficiency increases. As $f_T/f_{max}$ increases, a significant driver of node scaling, gain will decrease, and NF will degrade.

Power consumption is another primary system constraint that strongly influences topology choices, device sizing, and bias strategy. Technology scaling reduces supply voltages and therefore the available voltage headroom per device; this constrains the ability to stack devices, limits the range of linearization and degeneration techniques, and reduces the margin available for analog tuning. In practice, power budgeting for an LNA proceeds from system requirements (battery life, thermal budget, and

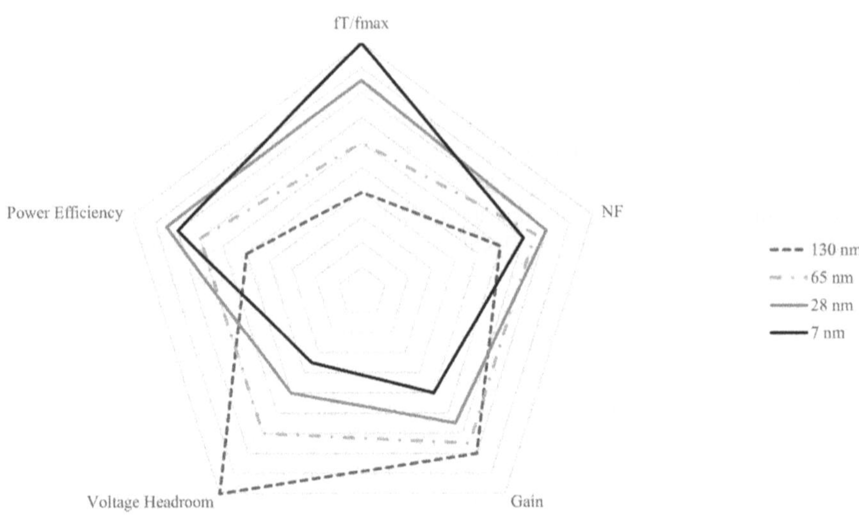


**FIGURE 6.3** Comparison of CMOS versus III–V LNA design trade-offs.

per-element power in arrays) and is used to set a maximum allowable drain current per unit width and a maximum supply voltage. Design techniques commonly applied to respect a tight power budget include $g_m/I_d$-based biasing to maximize transconductance per unit current (Kobal *et al.* 2022; Lai *et al.* 2023), current reuse, careful device sizing to achieve a target $g_m$ without excessive parasitics, and use of passive gain (transformers and resonant loads) to trade DC power for stored energy in the RF network. In arrayed systems (phased arrays, massive MIMO), per-element power becomes a dominant factor in architecture trade-offs: a modestly higher NF per element can be acceptable if array aperture or beamforming recovers system sensitivity at a far lower aggregate power cost.

Noise behavior in CMOS LNAs is determined by multiple, interacting mechanisms: channel thermal noise, induced gate noise, gate-resistance thermal noise, flicker (1/*f*) noise at low offset frequencies, and parasitic coupling to the substrate. The designer's objective is to minimize the contribution of these sources within the first stage (or within the dominant stage in the chain). Two practical consequences follow. First, the input matching network is treated as a noise-matching problem rather than a pure power match: the optimal source impedance for minimum noise is usually complex and differs from 50 Ω, so low-loss impedance transformation is required to reconcile on-chip device optimality with the external environment. Second, the noise contribution of the transistor is bias dependent: there is typically a bias point ($V_{gs}$, $I_{ds}$) that minimizes $T_{min}$ for a given device size and geometry; identifying and biasing to that point is fundamental to good LNA design. For CMOS, gate resistance and induced gate noise often set practical limits; careful multi-finger layout, gate-poly shielding, and low-resistance contacts are therefore critical to achieving the NFs predicted by models.

## FUNDAMENTALS OF RF CMOS DEVICE SCALING

This section moves from system targets to transistor realities. It summarizes the empirical trends observed as CMOS moved from micron to deep-submicron nodes and the practical consequences for RF behavior, explains the short-channel phenomena that alter transconductance, capacitances and noise, follows how $f_T$ and $f_{max}$ evolve with scaling and why those figures must be interpreted in context, and discusses the perennial tension between digital logic scaling and analog/RF requirements (device geometry, supply voltage, parasitic values, and modeling). Each subsection links measurable device quantities back to the LNA design choices you have just read about so that the reader can translate process-level numbers into concrete circuit decisions.

### SCALING TRENDS FROM MICRON TO NANOMETER NODES

CMOS scaling to smaller nodes, therefore smaller component sizes, has been accelerated by production (commercial and military) and research efforts, predicted decades ago by Moore's law (Moore 1998) but slowed down in modern history due to physical limitations, as has been written about extensively in the literature. An interesting outlook is presented in Borkar (2006), where CMOS scaling and evolution are and were driven by three tenets, namely:

1. Gain
2. Signal-to-noise ratio
3. Scalability

LNA performance and scalability can in many ways be similarly linked to its evolution, albeit not necessarily in the same order. According to, for example, Schwierz and Liou (2020), the gate length scaling steamroller tactics slowed down after around 2005 and levelled off at around 10–12 nm. The less aggressive scaling as would be required to maintain Moore's law can be attributed to the high cost of production-ready extreme ultraviolet (EUV) lithography tools, costing around $380 million to $400 million in 2025 (Tarasov 2025), equipment that is only affordable by companies such as Intel, as well as technical (physics-based) limitations, such as issues in suppressing short-channel effects (when the channel length is comparable to the depletion layer) in sub-10 nm gate MOSFETs (Tsuchiya *et al.* 1998). The three-dimensional nature of FinFET architecture, introduced in the 2012–2015 period (for mass production), has allowed researchers to continue scaling and reduce transistor footprint, without needing to physically shorten its gate of traditional planar transistors. Even though many complex designs were ported to 22, 16, and 14 nm, getting the same SoC performance from this technology is not a given and is dependent on various trade-offs (Banna 2015) and limitations (Saini and K Rana 2011). As shown in Saini and Rana (2011), the ultimate scaling of FinFETs is limited by electrostatic integrity, with drain-induced barrier lowering (DIBL) and subthreshold swing degrading sharply once the $L / T_{fin}$ ratio falls below approximately 1.5. While triple-gate FinFETs allow slightly more aggressive scaling than double-gate variants, short-channel effects and mechanical stability constraints on fin height and thickness ultimately restrict further dimensional reduction. Figure 6.4 is a three-dimensional representation of the FinFET structure.

CMOS scaling has greatly benefited digital logic, but analog and RF front ends compete with various additional parameters, where scaling in some circumstances can negatively impact their performance. As device dimensions decrease, the intrinsic gain ($g_m r_o$) and voltage headroom also decrease. Both these parameters directly impact the achievable LNA NF and linearity. Therefore, especially at high frequencies (mm-wave and beyond), LNA designers must consider the paradox where a higher $f_T$ or $f_{max}$ often results in poorer analog performance due to reduced intrinsic gain and increased flicker noise density (Yasuda *et al.* 2008; Srinivasan 2011; da Silva *et al.* 2023).

Furthermore, in transistor-level scaling, layout-dependent parasitic parameters become increasingly dominant, most significantly in high-frequency applications in nanometer nodes. These parasitics complicate and limit input matching (Moran Guizan *et al.* 2024) and gain flatness in wideband LNA (and power amplifier) designs. Essentially, the reduction and limitation of input voltage, combined with the smaller gate capacitances internally and inherently limit the achievable transconductance.

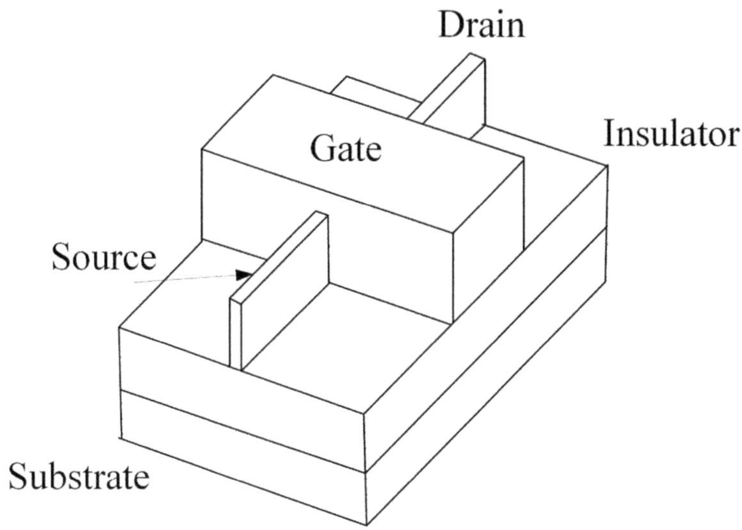


**FIGURE 6.4** The three-dimensional representation of the FinFET structure.

This forces designers to again rely on techniques such as inductive degeneration and transformer-based matching to maintain the LNA performance that is achievable in larger nodes (Hu and Chi 2023).

The CMOS versus III–V decision is not binary; it is an engineering judgment driven by system-level constraints. III–V materials (GaAs, InP, and GaN) retain clear device-level advantages for LNA applications: higher carrier mobility (Dewey *et al.* 2008), larger intrinsic transconductance per unit bias, greater breakdown voltages and voltage headroom, and, in many process points, lower intrinsic channel noise. These material properties translate directly into circuit-level benefits, higher single-stage gain, wider linear operating range, and, where biasing and cooling permit, superior noise performance. Thus, III–V remains the technology of choice for narrowband, performance-critical applications such as radio-astronomy front ends, specialized radar receivers, and certain defense systems where the incremental cost can be justified.

Conversely, CMOS offers a distinctive set of system advantages that frequently override per-device performance metrics. CMOS permits very large production volumes, dense functional integration, and low marginal cost. It allows LNAs, mixers, VCOs, and the requisite digital control and baseband functions to co-exist on a single die, reducing assembly steps, interconnect loss, and system cost while simplifying thermal and power management in mobile and consumer products. Heterogeneous integration techniques (monolithic III–V on Si, 2.5D/3D stacking (Dewey *et al.* 2008), nano-ridge engineering, and related methods) seek to combine III–V device physics with silicon manufacturing economy; these approaches are promising but introduce additional process risk, yield, and supply-chain complexity and therefore appear first in premium or specialized programs.

Accordingly, the appropriate technology is determined by the specification set. Where the absolute minimum NF, maximum linearity, and widest headroom are primary and cost is secondary, III–V technologies are typically most appropriate. Where integration density, per-unit cost, power budget, and tight co-integration with digital subsystems dominate, CMOS is normally preferable, subject to the well-known trade-offs (multi-stage gain strategies, aggressive impedance transformation, meticulous layout to control gate resistance and induced-gate noise, and conservative operation below raw device $f_T$). Hybrid solutions (for example, silicon LNAs with a co-packaged (Seifried *et al.* 2018; Masselink *et al.* 2023) III–V cold stage, or III–V islands on a silicon interposer) represent practical compromises that will become more common as heterogeneous manufacturing matures.

The remainder of this chapter shifts attention from the technology choice to the transistor realities that follow from it. The following sections relate process-level trends to circuit consequences: how scaling modifies transconductance, intrinsic gain, and capacitances; how short-channel phenomena alter noise and linearity; how $f_T$ and $f_{max}$ should be interpreted within matching and loss budgets; and where digital-driven process choices complicate analog design. These discussions provide practical rules of thumb and quantitative intuition to support process and topology decisions for a given NF, gain, and power budget. Figure 6.5 illustrates the relationship between the gate length-to-fin thickness ratio ($L/T_{fin}$) and two critical short-channel indicators, namely DIBL and subthreshold swing (SS), for advanced FinFET devices.

Figure 6.5 presents the characteristic degradation in electrostatic control as the physical dimensions of a FinFET are reduced. The figure shows DIBL and SS as functions of the geometric ratio $L/T_{fin}$, which describes the proportion between the effective gate length and the fin body thickness. As this ratio decreases, both DIBL and SS increase, illustrating the onset of short-channel effects. At large ratios (around

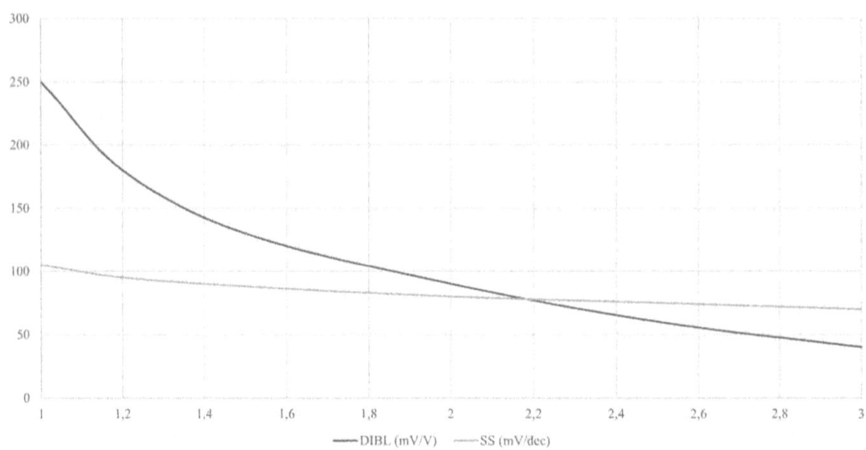


**FIGURE 6.5** Short-channel behavior and electrostatic control limits in advanced FinFET geometries. Dependence of drain-induced barrier lowering and subthreshold swing on the $L/T_{fin}$ ratio.

$L/T_{fin}$ = 3), electrostatic integrity is high: the gate exerts strong control over the channel potential, resulting in low DIBL (approximately 40 mV/V) and near-ideal subthreshold swing (approximately 70 mV/dec). In this region, the device exhibits good isolation between drain and source, predictable threshold voltage behavior, and minimal leakage current.

As the ratio is reduced to around 2.0, DIBL increases to approximately 90 mV/V, and SS rises to 80 mV/dec. This transition marks the point at which the drain's electric field begins to significantly penetrate the channel, partially collapsing the potential barrier. The gate's ability to fully deplete the fin is weakened, resulting in increased leakage and degraded switching behavior. Once $L/T_{fin}$ approaches 1.5, the degradation becomes more evident, with DIBL exceeding 130 mV/V and SS reaching close to 90 mV/dec. This ratio is often cited as the practical lower limit for FinFET design, since further dimensional reduction leads to exponential worsening of electrostatic behavior.

Below $L/T_{fin} \approx 1.5$, the device effectively loses its ideal gate control. At $L/T_{fin}$ = 1.2 and 1.0, DIBL values of 180 mV/V and 250 mV/V are observed, respectively, and the subthreshold swing rises beyond 100 mV/dec. These values correspond to a substantial deterioration in the subthreshold characteristics of the device. The drain and source depletion regions overlap significantly, the barrier height is reduced even in the off state, and the threshold voltage becomes increasingly sensitive to drain bias. This condition not only increases static power consumption through leakage current but also introduces variability in threshold voltage, transconductance, and gain, each of which directly impacts LNA design predictability at high frequencies.

The physical implication of this relationship is that FinFET scaling encounters a practical electrostatic limit governed not by lithography but by geometry. Even if shorter gates are theoretically possible, once the channel is no longer well-controlled by the gate, further scaling degrades rather than improves analog and RF performance. For RF circuits such as LNAs, this degradation manifests as reduced intrinsic gain ($g_m \cdot r_o$), poorer linearity, and increased device noise, despite nominal improvements in $f_T$ and $f_{max}$. In this context, Figure 6.4 demonstrates that aggressive scaling into sub-10 nm regions yields diminishing returns for analog front ends, reinforcing the need for careful design trade-offs that balance transistor speed with electrostatic stability.

## Short-Channel Effects and Their Implications for Noise

As transistor scaling in CMOS (and III–Vs) reaches channel lengths of less than approximately 1 μm, second-order effects, which were ignored in devices with longer channel lengths (so, approximately above 1 μm), become increasingly important to consider, understand, and mitigate. MOSFETs that fall in this category of channel length are called short-channel (Wong 2018) devices, and some of the second-order effects evident in these devices are velocity saturation, mobility degradation, threshold voltage variations ($V_T$ roll-off and DIBL), oxide and avalanche breakdown, and hot carrier effects.

In velocity saturation (Taylor 1984), the carrier drift velocity reaches a maximum, referred to as the saturation velocity, and this occurs specifically at high electric fields (common in short-channel MOSFETs) and when it reaches its critical electric field, $E_C$. The ideally linear relationship (following Ohm's law) between velocity and electric field disappears, and electrons start colliding (Yamasaki and Hirayama 1986) with the crystal lattice, called phonon scattering, and their velocity is limited to a constant value, $v_{sat}$. As a result, the drain current saturates earlier than expected and loses its dependence on the drain–source voltage, and the relationship between the drain current and gate-source voltage can become more linear. In Figure 6.6, the velocity saturation effect on the VI (voltage-current, which also depends on surface mobility) characteristics of a MOSFET is shown.

If short-channel devices are implemented, it necessitates alternative and arguably more complex modeling approaches, for example, velocity saturated models, as the standard square law model for long-channel devices no longer accurately describes the behavior of the device. In LNAs, particularly, velocity saturation severely limits the performance of the amplifier, limiting the transconductance and cutoff frequency, both critical performance metrics for an LNA.

For a long-channel NMOS, the threshold voltage is given by the body–effect equation as

$$V_T = V_{T0} + \gamma\left(\sqrt{|-2\phi_F + V_{SB}|} - \sqrt{|-2\phi_F|}\right) \tag{6.1}$$

where $V_{T0}$ is the zero-bias threshold voltage, when the body (substrate) and source are at the same potential ($V_{SB} = 0$), $\gamma$ is the body–effect coefficient, also called the body–bias coefficient, which measures how strongly $V_T$ changes with $V_{SB}$ (thus, larger substrate doping increases this parameter), $\phi_F$ is the Fermi potential, and $V_{SB}$ is the

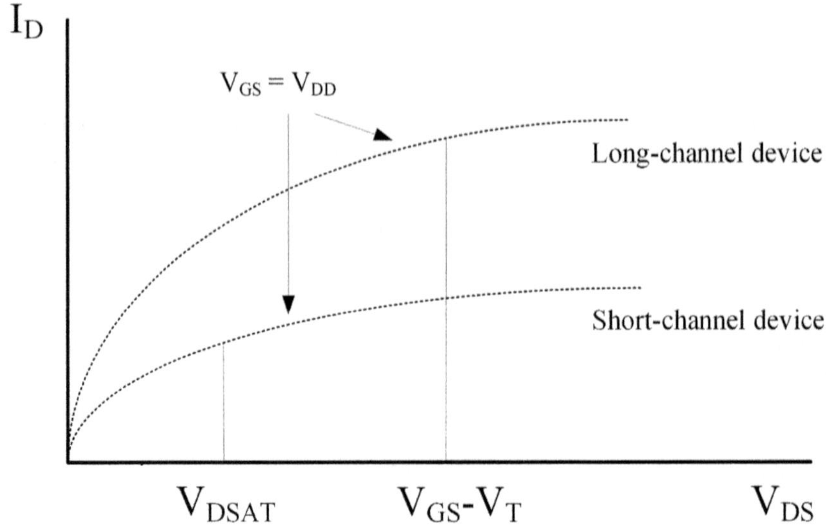


**FIGURE 6.6** Short-channel devices display an extended saturation region due to velocity saturation.

source-to-body potential, so the bias between the source and the substrate. From (6.1), it is noticeable that the threshold voltage is only a function of the technology and the applied substrate potential. However, in short-channel devices, this model becomes inaccurate, and the threshold voltage becomes a function of the length and drain–source voltage, and (6.1) becomes

$$V_T\left(L, V_{SB}, V_{DS}\right) = V_{T0} + \gamma\left(\sqrt{\left|-2\phi_F + V_{SB}\right|} - \sqrt{\left|-2\phi_F\right|}\right) - \eta_1 e^{-\frac{L}{\lambda}} - \eta_2 V_{DS} \quad (6.2)$$

where $\eta_1$is a fitting parameter (a few hundred mV), $L$ is the channel length, is the DIBL coefficient (in the range of 0.05–0.2 V/V in modern CMOS nodes), and $\lambda$ is the characteristic length (Masuda *et al.* 1979) of the device electrostatics, given by

$$\lambda = \sqrt{\frac{\varepsilon_{Si}}{\varepsilon_{ox}} t_{ox} x_d} \quad (6.3)$$

where $\varepsilon_{Si}$ and $\varepsilon_{ox}$ are the permittivity of silicon, and the device oxide, respectively, $t_{ox}$ is the oxide thickness, and $x_d$ is the depletion width. As the channel length decreases, the exponential and drain-dependent terms in (6.2) reduce the threshold voltage, leading to threshold roll-off DIBL. These effects collectively degrade the electrostatic integrity of the channel, resulting in increased subthreshold leakage, degraded transconductance efficiency, and higher flicker noise, key concerns in low-noise analog and RF circuits. The combined influence of velocity saturation and short-channel threshold modulation thus directly constrains achievable gain and noise performance in scaled LNAs, motivating closer examination of how $f_T$ and $\mathrm{f}_{\mathrm{max}}$ evolve with continued device scaling.

## Evolution of $F_T$ and $F_{MAX}$ with Scaling

To continue the evolution of $f_T$ and $f_{max}$ with scaling, it is important to recall what each of these terms means, specifically in the context of LNAs. These terms are usually evaluated based on transistor measurements performed at low GHz frequencies and extrapolated to a range of several hundred GHz (Rimmelspacher *et al.* 2019). Firstly, the transit frequency, $f_T$, is the frequency (in Hz) where the short-circuit current gain drops down to unity. This parameter is also primarily dependent on the transconductance, $g_m$, and the total gate capacitance, $C_{gg}$, of the transistor. The maximum oscillation frequency, $f_{max}$, is the frequency (again in Hz) where the power gain drops down to unity and is typically considered a more practical limit for RF circuits since it inherently includes parasitics, most notably the gate resistance ($R_g$), drain resistance ($R_d$), and the feedback capacitance ($C_{gd}$). In LNAs, the maximum oscillation frequency is also more representative of the usable gain-bandwidth product (GBP) than the transit frequency; however, both parameters trend together with scaling.

As scaling evolves, so do these terms, and various internal and external parameters also change, or become more or less apparent and important to consider. As a result, several trade-offs should be identified and understood and weighed against each other, depending on the application. Typically, as the transistor gate length $L$ decreases, the transconductance $g_m$ will increase, since a higher carrier velocity and stronger gate control up to velocity saturation is observed. However, the gate–source and gate–drain capacitances proportionally increase due to shorter channels and higher fringing fields (Kumar *et al.* 2006; Pahwa *et al.* 2022). Fringing fields arise from the electric field lines that extend beyond the intended gate-channel overlap region, particularly at the gate edges where the field "fringes" into the source and drain extensions. These fields introduce additional parasitic capacitances, called internal fringe capacitances, that couple the gate to the source and drain, effectively modifying the surface potential and lowering the threshold voltage as the gate-dielectric permittivity increases. As a result, $f_T$ will initially rise (with scaling), but the improvement will saturate (or in some cases be negated) as velocity saturation, parasitic resistance, and DIBL start dominating during circuit operation. Therefore, while earlier CMOS nodes exhibited close to linear improvement in $f_T$ with gate length reduction (an approximately $1/L$ ratio), modern deep-submicron and FinFET technologies deviate quite significantly from this trend, primarily as transport shifts from mobility-limited to velocity-limited regimes.

In terms of actual performance of both CMOS and III–V transistors, specifically related to $f_T$ and $f_{max}$, the values have followed trends, and at some risk of listing values that might seem outdated if referenced in the future, it might still be worthwhile to have some indication of these trends. Modern CMOS transistors, in the 5 nm and lower nodes, can reach $f_T$'s of around 300 to 400 GHz, where $f_{max}$ is somewhat lower (due to the inclusion of parasitics), at around 200 to 300 GHz. Conversely, III–Vs, specifically InP and GaN HEMTs (~50 nm nodes), can have $f_T$'s upward of 500 GHz, with $f_{max}$ higher than 1 THz already reported in 2007 (Lai *et al.* 2007), and amplification at 1 THz using 25-nm InP high electron mobility transistors already reported in 2015 (Mei *et al.* 2015). In 2025, Plouchart *et al.* (2005) reported a 243 GHz $f_T$ and 208 GHz $f_{max}$ for a 90 nm SOI CMOS SoC with low-power mm-wave digital and RF circuit capability.

## Trade-Off between Digital Logic Scaling vs Analog/RF Requirements

To avoid sounding repetitive and referring to Moore's law, which has been mentioned in a few places in this chapter, a slightly different approach is presented here. In this paragraph, the trade-off between digital logic scaling (Davari *et al.* 1995) and analog (RF and mm-wave) requirements is presented, since the fundamental idea of transistor scaling affects these two disciplines differently. LNAs fall into the second category of analog RF or mm-wave scaling and therefore experience the same advantages and limitations of scaling.

Digital scaling is driven by a narrower set of priorities (Allam *et al.* n.d.) when compared to analog scaling. These priorities primarily include transistor density, switching energy, and logic speed. To the advantage of digital evolution, foundry process scaling and evolution directly reflect these priorities. In contrast, analog (Spessot *et al.* 2020) front ends are governed by a different (and larger) set of constraints. These include voltage headroom, intrinsic gain, passive quality factors, parasitic effects, and low-frequency and flicker noise. Unfortunately, CMOS device scaling limits some of these parameters and therefore largely induces stricter constraints on analog design in smaller nodes. As a result, in some cases, digital logic blocks can be largely reused in smaller nodes, and re-designing and re-evaluating integrated analog LNA topologies and layout strategies are often required. In some cases, moving to a smaller node might not be feasible, depending on the application and performance requirements. For example, designing the readout integrated circuit for an infrared sensor array requires per-pixel analog processing. Due to the small currents being generated by the pixels and the high likelihood of noise, choosing a transistor technology has several limitations. Although smaller nodes allow for more transistors per pixel, their impact on noise, specifically flicker noise, can be a limiting factor, resulting in larger nodes being more ideal.

Commercially viable mainstream CMOS digital logic, like FinFETs and gate-all-around (GAA) (Barraud *et al.* 2017; Hemantha *et al.* 2024), requires additional processing enhancements that are RF-focused, such as RF-SOI, FDSOI, RF-CMOS, and technologies such as silicon-germanium (SiGe) BiCMOS (Lambrechts and Sinha 2017). These process enhancements specifically focus on parasitic effects, substrate losses, and offer alternatives such as body biasing, thick top metals, and high-resistivity substrates.

While digital logic benefits directly from standard foundry scaling, analog and RF circuits rely on process variants and modeling refinements that deviate significantly from the digital baseline. Foundries often develop specialized process extensions or RF corners of an existing logic node to meet these needs, incorporating options such as thicker gate oxides, multiple threshold devices, dedicated passive layers, or high-resistivity substrates. These variants trade density and switching speed for improved voltage handling, linearity, and matching, attributes that are essential for analog or mm-wave design but secondary for digital systems. For instance, RF-SOI and SiGe BiCMOS technologies often employ optimized isolation structures and backend metal stacks to suppress substrate coupling and enhance Q-factors of passives, directly addressing the limitations introduced by aggressive digital scaling.

Equally critical is how these process enhancements are captured and represented in the process design kit (PDK). While digital design flows can rely on compact SPICE models optimized for switching behavior, RF and analog circuits require much more detailed modeling of parasitic capacitances, distributed effects, and temperature dependencies. Modern PDKs, therefore, include frequency-dependent models, noise parameter extractions, and scalable layout-dependent effects (LDE) models that accurately represent the physical behavior of transistors and passives up to mm-wave frequencies. This modeling accuracy is vital since even small deviations between simulated and measured $S$-parameters can lead to significant mismatches in gain, NF, or stability at mm-wave frequencies. Consequently, the continual refinement of RF-focused PDKs has become a cornerstone of enabling reliable design migration to advanced technology nodes.

The most immediate and tangible consequence of digital scaling is the steady reduction of supply voltage, a change that propagates through topology choices, bias strategies, and matching network design. We therefore begin with a focused examination of voltage headroom and its consequences for LNA topologies before moving on to intrinsic gain erosion, parasitic penalties, and the practical process selections that matter to RF designers.

## PROCESS TECHNOLOGY AND VARIANTS FOR RF CMOS

In RF and mm-wave design, the choice of process technology does not merely set device dimensions or supply voltage; it defines the attainable NF, linearity range, matching constraints, and ultimately the scalability of the entire front-end architecture. As CMOS has migrated into the tens-of-nanometer domain, device behavior has become increasingly sensitive to substrate properties, junction isolation, and parasitic conduction paths that were negligible in earlier nodes. Process choice, therefore, becomes a first-order design variable rather than an implementation detail, influencing everything from compact model accuracy and stability margins to achievable gain at mm-wave frequencies. RF CMOS technologies today span multiple structural variants – bulk, SOI, FinFET, GAA nanosheet, and emerging 2D semiconductor hybrids – each introducing unique trade-offs in cost, substrate loss, fmax, and integration density. Before these distinctions are explored in detail, it is useful to establish the baseline behavior of the most pervasive technology: Bulk CMOS, and how its limitations motivated the development of SOI-based and later fully depleted multigate alternatives.

### BULK CMOS VS SOI CMOS

CMOS processes have traditionally used bulk silicon substrates, where the active devices share a conductive substrate that also supports interconnect and passives. While this configuration is ideal for cost and thermal management, it introduces substantial parasitic coupling paths at RF frequencies. The alternative, silicon-on-insulator (SOI), introduces a buried oxide (BOX) layer (Miura-Mattausch *et al.* 2014) that electrically isolates the active region from the substrate, reducing loss and improving passive quality factors. These structural distinctions make the choice between bulk and SOI CMOS particularly significant for RF designers. In bulk CMOS, substrate coupling and eddy currents reduce the $Q$-factor of passive components like the inductors, and therefore increase signal loss, proportional to the operating frequency. Through using the BOX layer, SOI mitigates these losses. Figure 6.7 is a simplified representation of the BOX layer in a SOI substrate, showing where the electrical isolation occurs.

To indicate how SOI can improve passive device $Q$-factor, it is feasible to first look at the traditional definition of inductor quality factor, being

$$Q = \frac{\omega L}{R_{eq}} \tag{6.4}$$

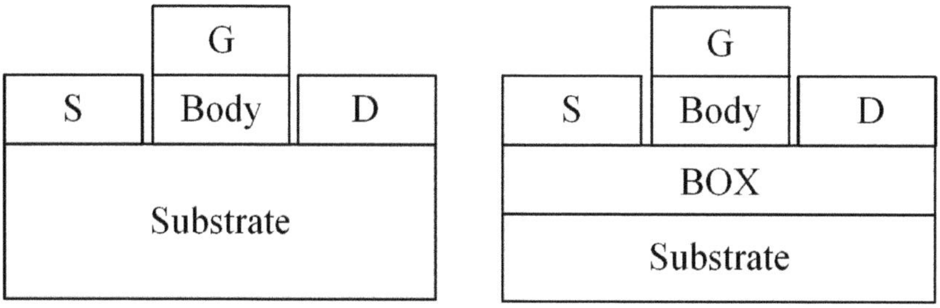


**FIGURE 6.7** Structural comparison between bulk CMOS (left) and SOI CMOS (right) technologies. In bulk CMOS (with a typical substrate resistivity of around 10 to 20 Ω•cm), transistors are built directly on a conductive silicon substrate, allowing substrate coupling and limiting passive *Q*-factors. In SOI CMOS (with a substrate resistivity of around 1 kΩ•cm), a BOX layer isolates the active silicon film from the substrate, reducing parasitic capacitance, improving isolation, and enhancing high-frequency performance.

where $L$ is the inductance, $\omega$ is the angular frequency, and $R_{eq}$ is the effective series loss resistance seen by the inductor. This equivalent resistance considers all loss mechanisms as seen by the inductor. These contributions are shown by

$$R_{eq} = R_{metal} + R_{diel} + R_{sub} \tag{6.5}$$

such that

$$\frac{1}{Q} = \frac{R_{metal}}{\omega L} + \frac{R_{diel}}{\omega L} + \frac{R_{sub}}{\omega L} \tag{6.6}$$

and therefore

$$\frac{1}{Q} = \frac{1}{Q_{metal}} + \frac{1}{Q_{diel}} + \frac{1}{Q_{sub}} \tag{6.7}$$

and each of these terms can be expressed in simplified physical quantities.

First, the metal loss, also known as the ohmic or skin effect loss, increases at high frequencies, primarily due to the skin effect, and

$$R_{metal} \approx R_s \cdot \frac{l}{\omega_{eff}} \tag{6.8}$$

with

$$R_s \approx \frac{1}{\sigma_{metal} \delta_{skin}} \tag{6.9}$$

where $l$ is the conductor length, $\omega_{eff}$ is the effective conduction width, $\sigma_{metal}$ is the metal conductivity, and $\delta_{skin}$ is the skin depth, defined as

$$\delta_{skin} = \sqrt{\frac{2}{\omega\mu\sigma_{metal}}} \tag{6.10}$$

and, as a result, the $Q$-factor of the metal can be written as

$$Q_{metal} \cong \frac{\omega L}{R_s\left(\frac{l}{\omega_{eff}}\right)} \tag{6.11}$$

Second, the dielectric loss (oxide and passive components) manifests through the loss tangent and passive capacitance that stores energy with a dissipative loss, and a practical expression for this Q-factor is

$$Q_{diel} \cong \frac{1}{\omega C_p \tan\delta \cdot R_{eq,diel}} \tag{6.12}$$

which can be simplified to

$$Q_{diel} \approx \frac{1}{\omega C_p \tan\delta} \cdot \frac{1}{\omega L} \tag{6.13}$$

and interpreted as the dielectric loss contribution to $1/Q$.

Finally, the substrate loss (eddy currents and induced current loss) is the crucial term to determine the effective quality of the passive components, and the dominant difference between bulk and SOI CMOS. A simple and commonly used approximation that is sufficient to show the parameter dependence is given by representing the substrate conductance ($G_{sub}$), which is the inverse of the substrate resistance ($R_{sub}$), as a function of the coupling capacitance ($C_c$), such that

$$G_{sub,eff} \approx \omega^2 C_c^2 R_{sub} \tag{6.14}$$

This term can be converted to an equivalent series resistance seen by the inductor, given by

$$R_{sub}^{(eq)} \approx \frac{1}{\omega^2 L C_c^2 R_{sub}} \tag{6.15}$$

and therefore, the substrate contribution to $Q$ can be written as

$$Q_{sub} \approx \frac{\omega L}{R_{sub}^{(eq)}} = \omega^3 L^2 C_c^2 R_{sub} \tag{6.16}$$

and some interpretations on these parameters can be implied. To do this, the terms are first combined:

$$\frac{1}{Q} = \frac{R_{metal}}{\omega L} + \omega C_p \tan\delta + \frac{1}{\omega^3 L^2 C_c^2 R_{sub}} \tag{6.17}$$

From this combined representation, it can be noted that $Q_{sub}$ increases with $R_{sub}$, the inverse of the substrate conductivity, and it decreases (falls) with the coupling capacitance, $C_c$. Therefore, if the substrate resistivity is increased, so is the $Q$-factor of the substrate. If we reduce $C_c$, for example, by inserting a BOX layer or moving a metal layer further away from the substrate, the $Q$-factor of the substrate is also increased, since $C_c$ appears squared in the denominator of the dissipation. Thus, the BOX layer affects these parameters in two ways, namely, it reduces the coupling to the conductive substrate ($C_c$ reduces as the BOX increases separation and lowers fringe coupling), and it electrically isolates the device so that the effective substrate conductance seen by the inductor is significantly smaller, since the substrate resistance is larger.

As mentioned earlier, the primary function of the BOX layer is to provide electrical isolation for the active semiconductors that are created on top of the silicon substrate, and this technique creates significant performance advantages over conventional bulk wafers at the device level and for LNAs. These performance advantages include:

- A reduction in parasitic capacitance (and therefore the body effect) that allows for faster switching speeds and lowers power consumption. Traditional bulk substrates and active devices have junctions that are directly connected to the underlying substrate, and this is where the higher unwanted parasitic capacitance is created, something that the BOX layer eliminates. As a result, LNAs have better linearity and matching if the parasitic capacitance is reduced.
- Latch-up is significantly reduced because the electrical isolation of the BOX layer minimizes the likelihood of a destructive short circuit between the power rails and the ground connection.
- For sensitive applications in space and military electronics, this layer leads to an improvement in radiation hardness by suppressing undesirable short-channel effects that can degrade device performance.
- An overall improvement in RF performance in high-performance RF circuits such as 5G, 6G, and Wi-Fi.

- In microelectromechanical systems (MEMS) fabrication, the BOX layer isolation serves as an inherent etch-stop that enables and simplifies the creation of complex suspended (or released) structures.

Despite the advantages that SOI offers over bulk CMOS, the substrate isolation, lower parasitics, and higher *Q*-factors, when it comes to thermal management, SOI has some less-intuitive disadvantages, and bulk CMOS outperforms it in terms of heat spreading and thermal conduction. This becomes increasingly important at mm-wave and sub-THz operation and can have dramatic effects on LNA performance. At these higher operating frequencies, power densities and dynamic losses increase significantly and are most notable in LNAs, PAs (Koga and Kurita 2019), and oscillators. While SOI's BOX layer provides excellent electrical isolation, it simultaneously acts as a thermal insulator. The BOX layer has a thermal conductivity that is roughly one hundred times lower than that of pure silicon. Silicon has a thermal conductivity of approximately 1.4 W/m•K, whereas the isolation layer, typically constructed of $SiO_2$, has a thermal conductivity, on average, of around 0.014 W/m•K. This essentially means that inside the active device, the heat that is being generated cannot easily dissipate downward into the silicon substrate and becomes trapped within the thin silicon film right above the BOX layer. Conversely, bulk silicon active devices are manufactured directly on the silicon substrate, and this provides a continuous thermal path for heat dissipation (and spreading). At high RF and mm-wave frequencies, in constant operation such as communication front ends, this characteristic becomes especially significant and critical for several reasons, namely:

- The dynamic power dissipation increases, as a function of frequency, leading to switching losses, gate charging currents, and interconnect losses.
- Key device parameters such as transconductance and threshold voltage are internally and inherently altered due to self-heating, which often reduces gain and increases NF.
- The local temperature in SOI can rise significantly above the ambient temperature, even during modest power consumption, and this can lead to thermal runaway in some circumstances, effectively increasing cost and complexity if additional cooling is required.
- Temperature gradients within the fin, or channel, can modulate carrier mobility and velocity saturation, effectively impacting the $f_T$ and $f_{max}$ of the active devices, particularly at mm-wave, where small changes in mobility can shift the gain and phase significantly.

Researchers and designers have identified several strategies to mitigate and reduce these effects, where such strategies include creating a thinned BOX layer, or using high thermal conductivity oxides such as $Si_3N_4$ or AlN to improve heat flow, including backside thermal vias, wafer thinning, or applying strict thermal-aware layout to space devices apart, distributing power stages, and using large top-metal heatsinks.

Another area where bulk CMOS still outperforms SOI is in cost and yield. SOI is used for high-frequency and low-noise applications, for example, in 5G smartphone

**TABLE 6.1**
**The comparative features of bulk CMOS and SOI and their impact on RF and LNAs**

| Feature | Bulk CMOS | SOI CMOS | Impact on RF |
|---|---|---|---|
| Substrate isolation | Poor | Excellent | Improved passive Q-factor and reduced crosstalk |
| Self-heating | Low | High | Impacts linearity and matching |
| Cost | Low | Higher | Limitations on volume use |
| Body effect | Strong | Weak | Better voltage threshold stability |
| RF use cases | Consumer SoC | RF front-end | Performance-driven choice |

front ends, but bulk CMOS remains a more cost-effective solution with extremely high integration density. Table 6.1 provides a brief overview of the features of bulk CMOS and SOI, and their impact on RF design and applications such as LNAs.

While SOI CMOS offers substantial advantages in substrate isolation, parasitic reduction, and overall RF performance, bulk CMOS remains highly relevant for cost-sensitive, thermally demanding, and densely integrated applications. The trade-off between electrical isolation and thermal conduction continues to define the choice of substrate technology, with modern RF CMOS design often balancing these factors according to target frequency, power level, and integration requirements.

## RF-Optimized Process Nodes

In parallel with the scaling of digital CMOS, foundries have introduced process variants optimized for analog and RF applications, often referred to as RF CMOS options. These processes retain the same core transistor design rules as the base logic node but introduce enhancements targeted at high-frequency and mixed-signal operation. Such enhancements include

- Thicker top metal layers (Rotella *et al.* 2003; Guo Lihui and Jie 2004) for low-loss inductors
- High-$Q$ metal-insulator-metal (MIM) capacitors
- Optimized varactors
- Device layout options with reduced gate resistance and parasitic capacitance

The goal is to extend the usable frequency range of CMOS into the mm-wave regime while preserving the cost and integration density advantages of mainstream logic nodes. These RF-optimized nodes are specifically designed to accommodate high-frequency applications like 5G, WLAN, positioning, vehicle-to-everything (V2x) communication, IoT, and other communication and consumer devices. These advanced RF capabilities are often categorized into three distinct categories, namely:

1. RF CMOS that supports high-voltage, optoelectronic, and high-temperature devices
2. RF integrated passive devices engineered through an HR-Si substrate and thick copper (Cu) metallization to achieve low-loss RF passive devices with high quality factors
3. BiCMOS for mixed high-precision analog RF designs (reviewed in the next section)

The specific implementation of these enhancements varies across foundries and process generations, but the general trend is consistent: to maintain the digital design ecosystem of mainstream CMOS while incrementally improving its analog and RF performance through process-level add-ons. Table 6.2 summarizes several representative process variants and their corresponding high-frequency figures of merit. These examples highlight how foundries extend standard logic technologies to accommodate RF front-end integration without resorting to fundamentally different device structures.

One of the most impactful enhancements, and often the simplest to implement and use, is the inclusion of a thicker top-level metal layer, sometimes referred to as the ultra-thick-metal (UTM) layer (Kim *et al.* 2024). This thicker metal layer can typically be a few microns of Cu metallization and plays an important role in reducing the conductor loss in on-chip passive components, such as inductors and transmission lines. The ohmic component, $R_{metal}$, is inversely proportional to the effective cross-sectional area of the conductor and, therefore, increasing the thickness of the conductor inherently reduces the series resistance and improves the $Q$-factor of the passive device, which also positively impacts the phase linearity. In practice, foundries often designate this thick metal layer as the *RF metal*, or *Top Metal 10* layer (depending on the offered number of metal layers), and the thickness is often around 2–4 µm. This is significantly thicker than the traditional, or standard, metal layers, which are typically around 0.7 –1 µm back-end-of-line (BEOL) thick (therefore, after all processing steps, as received by the designer). This enhancement by itself can improve the $Q$-factor for passive devices from the 5–10 range, all the way up to the 20–40 range, depending on the frequency of operation (typically measured around 5 – 40 GHz (RF to mm-wave). Additionally, the inherent increase in separation between the metal layer and the substrate also decreases the parasitic capacitance $C_c$, which further reduces and mitigates internal losses. Beyond the use of inductors, a thicker

**TABLE 6.2**
**Process variants and technology nodes that accommodate RF design**

| Node | Base process | RF variant | $f_T$ (GHz) | $f_{max}$ (GHz) | Comments |
|---|---|---|---|---|---|
| 65 nm | Logic CMOS | RFCMOS | 150 | 200 | Added top metal, MIM cap |
| 28 nm | LP CMOS | 28 RF | 250 | 320 | SOI or FD-SOI variant |
| 22 nm | FD-SOI | 22 FDX-RF | 300 | 380 | Back-bias capability |

top metal layer can also improve the overall current-handling capacity; therefore, designers are encouraged to use this layer for high-current paths, and it also improves the thermal robustness of interconnects in high-power RF paths (driver stages of PAs or LNA matching networks). Finally, another advantage of the thicker top metal is that it naturally provides a solid foundation for integrating on-chip antennas and baluns in mm-wave front ends, since series resistance is lower and surface planarity is high. It should be noted that thick metal layers, such as UTM, have certain drawbacks and disadvantages, including:

- Increased process complexity (and therefore cost).
- Higher mechanical stress on the dielectric stack can cause cracking, delamination, or even wafer warping.
- Reduced yield due to longer chemical-mechanical polishing (CMP) times.
- Increases in inter-metal capacitance, which can be mitigated through design strategies.
- Routing limitations and design rule restrictions through dedicated RF routing layers.
- Thermomechanical expansion due to mismatched thermal expansion coefficients between thick copper and the surrounding dielectrics.

Therefore, while the thick top metal layer provides clear improvements in $Q$-factor and passive efficiency, these benefits come at the expense of increased fabrication complexity, thermal stress management, and layout constraints, all of which must be considered during high-frequency circuit design and integration phases, which might increase the complexity and impact the time frames.

Another effective and essential design tool in enhancing RF-CMOS applications is the integration and use of high-$Q$ MIM capacitors (Jenei *et al.* 2001; Wenger *et al.* 2006; Mu *et al.* 2018), a component (feature) that is specified by most foundries in their RF-enhanced processes. High-$Q$ MIM capacitors are simply capacitors that are fabricated using two or more metal plates that are separated by a thin, but high-quality, dielectric layer (such as $SiN_x$ or $Al_2O_3$). Furthermore, these capacitors are typically positioned in the upper interconnect layers to minimize parasitic coupling to the Si substrate. Its quality factor stems from the known equation in RF design, given by

$$Q_C = \frac{1}{\omega R_{ESR} C} \tag{6.18}$$

where $R_{ESR}$ is the equivalent series resistance, $C$ is the capacitance, and $\omega$ is again the angular frequency. It is also known that in RF design, a higher $Q$ corresponds to lower dissipative loss, and this in turn affects filter selectivity, LNA input matching, and VCO phase noise. In MIM capacitors, the dielectric layer is specifically designed to balance the capacitance density (in fF/μm$^2$) and loss tangent (tan $\delta$), where the capacitive density in RF CMOS optimized nodes is typically higher than in conventional CMOS (Wenger *et al.* 2006). Furthermore, these capacitors offer better linearity, higher stability, compact tuning networks, and enhanced reliability; however, some additional considerations include limited breakdown voltage, increased process complexity (and therefore cost), and potentially higher coupling to adjacent metal layers if not laid out correctly.

Another passive component often used in RF design, especially in VCOs, tunable filters, and adaptive matching networks, and that can be enhanced by foundries for RF applications, is variable capacitors, or varactors. These capacitors are voltage-dependent (Pedersen 2001) and are typically implemented in MOS capacitors in traditional (or standard) CMOS processes. However, at high frequencies, MOS capacitors are known to suffer from poor linearity (and limitations in $Q$-factor) and significant parasitic resistance (and capacitance) that severely degrade phase noise performance and tuning range. In RF-enhanced CMOS processes, these limitations are addressed by developed and dedicated varactor structures that take advantage of process-level refinements that aim to improve the $Q$-factor, and therefore also the tuning efficiency. One such process-level refinement is the use of accumulation-mode MOS varactors (Dunwell and Frank 2007) (AMOS), which improves linearity and quality by eliminating inversion charge delay (since it isimplemented in accumulation mode as opposed to inversion mode) and minimizing channel resistance. Since the inherent varactor $Q$-factor is inversely proportional to the series resistance ($R_S$), as described by

$$Q_{var} = \frac{1}{\omega R_S C_{var}}, \tag{6.19}$$

advanced RF nodes implement silicide gate structures, low-resistance source and drain extensions, and shallow trench isolation (STI) to reduce the series resistance – this is especially effective at mm-wave, where this resistance significantly impacts both linearity and phase noise. Additionally, to achieve further enhancements of varactors in RF CMOS nodes, HR-Si or SOI substrates can also be used, as they reduce parasitic substrate coupling and improve C-V linearity. It should also be noted that although these improvements are ideal in RF design, some caveats and trade-offs exist, including a smaller tuning range and voltage headroom limitations.

Beyond process-level enhancements, a substantial portion of RF performance in CMOS technologies is ultimately determined by layout optimization, a discipline that bridges device physics, electromagnetics, and practical circuit implementation. For advanced nodes, layout is as important as device selection, since parasitic capacitances, resistances, and inductive couplings introduced through the layout can dominate the high-frequency response of the circuit. The physical arrangement of transistors, interconnects, and passives determines how effectively the theoretical device performance translates into circuit-level metrics such as gain, NF, and linearity.

Optimized layout strategies typically focus on minimizing gate resistance, source/drain parasitics, substrate coupling, and interconnect loss, while maximizing symmetry and matching between differential signal paths. However, at mm-wave and sub-THz frequencies, the complexity increases dramatically as every routing choice, via placement, and ground return path can influence performance. As a result, layout design for RF CMOS has evolved into a specialized domain that blends analytical modeling, 3D electromagnetic simulation, and empirical design heuristics – where artificial intelligence can even play a role.

Although many of these techniques are process-dependent and often proprietary to specific foundries, the underlying principles are consistent: shorter and wider gate fingers reduce resistance, symmetrical routing improves balance and isolation, and careful ground shielding reduces substrate noise coupling. Advanced EDA tools now integrate layout-aware extraction, enabling designers to predict and compensate for parasitic effects early in the design flow. In essence, while transistor scaling and process enhancements have extended CMOS operation into the mm-wave regime, it is layout experience and expertise (refined through iterative tuning) that ultimately enable these circuits to achieve the simulated levels of performance. The art of RF layout thus remains a defining factor in the practical realization of low-noise, high-gain, and power-efficient front-end circuits in nanoscale CMOS technologies.

These RF-optimized process nodes therefore bridge the gap between mainstream logic CMOS and dedicated compound-semiconductor processes. They enable monolithic integration of analog, digital, and RF blocks, offering sufficient frequency headroom for front-end circuits up to the lower mm-wave bands while maintaining the economies of scale of CMOS manufacturing. Even though there exist numerous techniques and strategies to enhance CMOS for RF applications, certain performance limitations that are inherent to CMOS cannot be mitigated or ignored. These limitations include constraints in

- Transconductance efficiency
- Breakdown voltage
- High-frequency gain

and applications that require superior performance in these regimes, and operate at ultra-low noise, high linearity, or deep in the mm-wave spectrum, still require additional enhancements that cannot necessarily be provided by traditional CMOS. In certain circumstances, combining bipolar and MOS transistors has added benefits, and the use of BiCMOS process nodes becomes a requirement.

## CMOS vs BiCMOS

Traditional CMOS uses only metal–oxide semiconductor field-effect transistors (MOSFETs) that have low static power consumption and very high integration density, ideal for scaling, and used in a variety of applications. Traditional CMOS, if not enhanced for RF or mm-wave, is typically ideal for digital logic where millions, or billions, of transistors perform digital operations at high speeds, typically in the low-GHz range. BiCMOS is a combination of CMOS and bipolar junction transistors (BJTs) that leverage the benefits of both transistor types to achieve higher speed and improved analogue (Brodic 1991), RF, and mm-wave performance (Lambrechts and

Sinha 2017) when compared to standard CMOS nodes – all while aiming to retain the low-power performance of the CMOS integration/parts while driving applications with high-speed and higher current-driving capabilities. Even back in 1989, Watanabe *et al.* (1989) showed the advantages of BiCMOS technology in terms of access time and power dissipation, demonstrated in a 1 Mb DRAM (with transistor-transistor logic) interface. In this research, BiCMOS, in its early days, already showed a 23% lower (better) delay time and 28% lower (better) power dissipation when directly compared to CMOS.

Bipolar transistors, used in BiCMOS, offer speed improvements over CMOS, high gain, and low output impedance (while maintaining relatively low power consumption), and these properties combined are ideal for analog amplifiers, such as PAs and LNAs, which can be used in combination with high input impedance CMOS logic gates, especially practical in mixed-signal design (for example, analog-to-digital converters, SoC, and analog power management).

At the device level, the performance advantage that is offered by BiCMOS is largely due to the exponential transfer characteristics of the bipolar transistor, and can be shown, or derived, by first inspecting the collector current of the BJT in its active region, given by the well-known equation

$$I_C = I_S e^{V_{BE}/V_T} \tag{6.20}$$

where $I_S$ is the saturation current and $V_T$ is the thermal voltage ($kT/q$). If this equation is differentiated, it results in the transconductance, $g_m$, given as

$$g_m = \frac{\partial I_C}{\partial V_{BE}} = \frac{I_C}{V_{BE}} \tag{6.21}$$

and, by comparison, the transconductance of a MOSFET (in strong inversion) is given by

$$g_m = \sqrt{2\mu C_{ox} \frac{W}{L} I_D} \tag{6.22}$$

where $C_{ox}$ is the oxide capacitance, $W$ and $L$ are the width and length of the transistor, respectively, and $I_D$ is the drain current. The transconductance of the MOSFET *only* increases as the square root of the current, but the $g_m$ of the bipolar device exhibits a linear increase with its collector current, consequently leading to higher gain and lower NF (in comparison). Again, this is ideal for LNAs, VCOs, and mm-wave drivers that require high linearity.

Additionally, if a unified process is used to integrate both device types, it leads to higher levels of heterogeneous functional partitioning (Weimann *et al.* 2016). In such cases, the bipolar transistor is used to handle high-frequency and high-linearity signal amplification, whereas the CMOS devices are used to control the digital

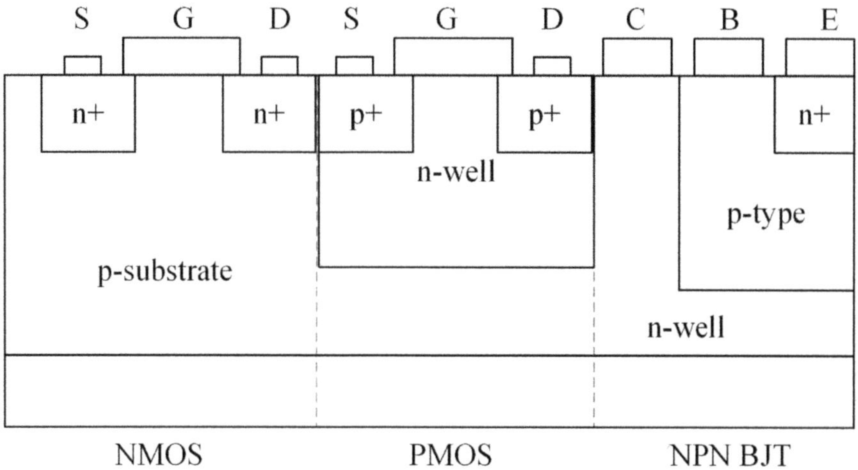


**FIGURE 6.8** The fundamental architectural difference between CMOS MOSFETs (NMOS and PMOS) and BiCMOS (NPN) BJTs.

logic, bias generation, and other high-speed logic interfacing. This combination of devices is especially beneficial in mixed signal SoCs and RF front ends, and allows for both performance gains with high integration density. In modern BiCMOS processes, for example, SiGe BiCMOS, these advantages are further extended by using bandgap-engineered HBTs that can achieve $f_T$ and $f_{max}$ that exceed 300 GHz while retaining full CMOS logic compatibility. Figure 6.8 shows the fundamental difference in fabrication architecture, on wafer-level, between standard CMOS active devices (NMOS and PMOS) and BiCMOS (bandgap-engineered) devices, such as NPN BJTs.

As illustrated in Figure 6.8, the integration of bipolar and CMOS devices within a unified BiCMOS process is fundamentally enabled by an extended and more intricate wafer fabrication sequence. After the initial substrate preparation and epitaxial growth, used to create a high-quality, low-defect silicon foundation, a series of isolation, implantation, and metallization steps is introduced to accommodate both MOSFET and bipolar device regions on the same wafer. The CMOS devices are typically formed in shallow wells near the surface, relying on thin gate oxides and fine channel doping for digital logic operation. In contrast, the bipolar transistors require deeper diffusion profiles and carefully engineered base and emitter regions, often realized through selective epitaxy and precision ion implantation. The process complexity is therefore higher than that of pure CMOS, but this additional effort enables the coexistence of high-gain, high-frequency bipolar devices alongside dense, low-power CMOS logic. The inclusion of SiGe in the bipolar base region further enhances carrier mobility and reduces base resistance, allowing the HBTs in modern BiCMOS technologies to achieve exceptional cutoff frequencies while maintaining full CMOS compatibility.

These continued advancements have elevated BiCMOS from a niche solution into a central enabling technology across multiple industries. It supports a broad spectrum of systems, such as

- Automotive radar
- Optical transceivers
- High-speed instrumentation
- Emerging terahertz front ends

where the seamless co-integration of analog precision and digital control is essential. The steady evolution of SiGe HBTs, combined with deep submicron CMOS, has redefined the practical limits of SoC integration, enabling compact and highly efficient architectures that historically required discrete III–V components.

## DEVICE MODELING FOR RF DESIGN

Modern RF integrated circuits are increasingly limited not by device speed alone, but by the precision with which that device can be modeled, predicted, and verified across bias, frequency, and temperature. As LNAs move deep into mm-wave and sub-THz operation, traditional assumptions, such as lumped capacitances, bias-independent mobility, and simple thermal noise representations, become insufficient. Instead, designers rely on compact models that encapsulate short-channel behavior, substrate coupling, noise correlation, non-quasi-static effects, and even statistical variation across process corners. Accurate models therefore form the bridge between transistor physics and circuit-level performance, enabling bias selection, impedance matching, noise optimization, and linearity enhancement with confidence rather than approximation. In this context, modeling is not a back-end verification step, but a foundational part of LNA design methodology.

### WHY ACCURATE MODELS MATTER

In the previous sections, process-level device enhancements were reviewed, including techniques such as substrate isolation, BOX layers, metal thickness, and more complex process improvements such as SiGe BiCMOS integration. However, regardless of the effectiveness and complexity of these process enhancements, it remains crucial that the devices are accurately modeled in simulation. Only then can these enhancements effectively translate into circuit-level performance improvements that capture the underlying physical effects across all simulation regimes, including DC, AC, and RF. Performance parameters such as NF, gain, matching, and linearity are all heavily dependent on the accuracy and efficacy of the small-signal equivalent circuits and parameters that are captured by accurate device models. Internal parameters such as $g_m$, $r_o$, $C_{gs}$, $C_{gd}$, $R_s$, etc., all impact circuit performance, especially at RF and mm-wave (Ninan *et al.* 2016), and the accuracy of these internal parameters, within the small-signal models, is crucial for simulation-to-silicon correlation (like, for example, layout versus schematic (LVS) verification). As seen earlier in this chapter, substrate coupling and series resistance can significantly degrade $Q$-factor, and if these effects are not properly represented in the device models, the simulations for metrics such as $S_{11}$ and NF (Shi *et al.* 2022) would not match real-world performance.

The accuracy of the small-signal models for LNAs can be mathematically shown if we consider a simplified circuit that is fundamentally formed from the device-level small-signal intrinsic parameters. To indicate this, accurate models must provide frequency-dependent values for each intrinsic term (transconductance, output gain, gate capacitances, series resistance, load, etc.) and should also include effects such as the skin effect in metals, capacitances that depend on the layout considerations, and substrate coupling, as discussed earlier in this chapter. As an example, consider a single-stage common-source (CS) transistor amplifier with source degeneration resistance $R_s$ and an external load ($R_L$) connected to its drain node. A large signal representation of this circuit is given in Figure 6.9.

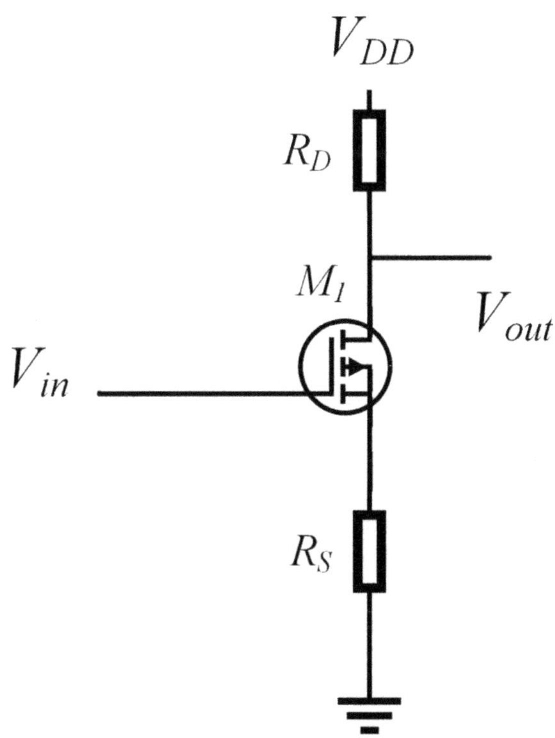


**FIGURE 6.9** A large signal circuit-level representation of a single-stage common-source transistor amplifier with source degeneration.

With reference to Figure 6.9, the low-frequency small-signal voltage gain (from the gate to the drain) is given by

$$A_v = \frac{v_{out}}{v_{in}} \approx -\frac{g_m\left(r_o \parallel R_L\right)}{1+g_m R_s + g_{mb} R_s} \tag{6.23}$$

where $g_{mb}$ is the body-effect transconductance (if present). From a practical perspective, therefore, if the output impedance is much larger than the external load, thus $r_o >> R_L$, then the factor $r_o \parallel R_L$ reduces to only $R_L$. This simplifies the numerator, and the denominator shows explicitly how the source degeneration reduces the voltage gain, and the effective transconductance of the circuit shows that any inaccuracies in $g_m$, $R_s$, or $g_{mb}$ will directly lead to errors in the predicted gain. Notably, at higher frequencies, the effective load includes the frequency-dependent impedance of the matching network as well as the intrinsic capacitances, and to accurately predict gain roll-off (and matching), extremely accurate modeling information on $C_{gs}$ and $C_{gd}$ is needed.

Moving from the device level to the circuit level, model accuracy becomes equally important in the context of PDKs. For example, in LNA design, PDK model accuracy is a determining factor for achieving first-pass silicon success. The accuracy of an LNA prediction depends not only on the intrinsic transistor models but also on the accuracy of the passive component models, layout-extracted parasitics, and EM interactions between structures. High-quality PDKs are typically built from parameter extraction procedures based on measured data, followed by validation against additional test structures to ensure that simulated and measured responses align across frequency, bias, and temperature.

Several factors influence this overall accuracy. The first is the fidelity of the transistor models themselves (whether CMOS, SiGe HBT, or III–V HEMT), as any deviation in their extracted parameters will directly affect gain, noise, and linearity predictions. Another critical aspect is the inclusion of EM simulations to account for layout-dependent parasitics and coupling effects, especially at mm-wave frequencies where interconnect discontinuities and ground return paths can significantly alter circuit behavior. Furthermore, model extraction and validation must be performed over a sufficiently broad operating range to ensure accuracy under different bias, temperature, and power conditions.

The LNA topology and design objectives further shape the required modeling precision. For instance, cascaded configurations tend to be more sensitive to parasitic capacitance accuracy, whereas common-source stages are often dominated by transconductance and source resistance accuracy. Similarly, designs optimized for minimum NF will be particularly sensitive to the accuracy of gate resistance and flicker noise models. In contrast, high-linearity designs may be more affected by the accuracy of output conductance and substrate resistance modeling.

It is also important to note that not all PDKs are developed with the same emphasis. A generic mixed-signal PDK, for instance, may not fully capture high-frequency or low-noise effects, requiring designers to refine or extend the models for their specific RF applications. These refinements may involve incorporating custom noise models, applying empirical correction factors, or performing partial model re-extraction using measured data.

Ultimately, there is always a balance between model complexity and simulation efficiency. More detailed physics-based models can improve accuracy but may increase simulation time substantially, which becomes critical for large RFICs or SoC designs. Nevertheless, ensuring accurate model representation, through validated PDKs, EM-based parasitic extraction, and correlation with measured data, remains fundamental to achieving predictive simulations and first-pass silicon success. Without this, even the most advanced process technologies cannot be fully exploited, as the simulated circuit behavior would fail to reflect the physical reality of the fabricated device.

## BSIM Models for RF and Their Limitations

The Berkeley short-channel IGFET model (BSIM) is a family of transistor models that are used to accurately simulate the behavior of MOSFETs in ICs. BSIM is built to represent a physics-based yet empirically tuned formulation where device current and charge are described by analytical equations that are derived from the semiconductor transport theory and adjusted using parameters that are extracted from experimental data. It is one of the most widely used and approved industry-standard model families in circuit design, which includes complex physical effects such as mobility degradation, current saturation, gate tunneling, quantum effects, and RF effects (Sheu *et al.* 1987). To extract parameters, a typical process involves starting with DC I-V and C-V measurements, so that the basic parameters such as threshold voltage, mobility, oxide thickness, etc., can be determined. This is followed by small-signal S-parameter measurements for high-frequency refinements. The BSIM model then provides the large-signal current and charge equations from which the simulator can derive small-signal parameters such as transconductance, output resistance, and intrinsic capacitances, which are critical for (accurate) RF and analog circuit prediction.

Newer BSIM models are constantly developed to incorporate refinements and improvements (such as BSIM-BULK, formerly BSIM6) (Gupta *et al.* 2019) for newer technologies; however, BSIM has its limitations. These limitations are also more apparent in high-frequency applications (Niknejad *et al.* 2005), where effects like polysilicon gate depletion are not accurately modeled. These issues are also specific to symmetry and scalability for certain geometries, like FinFETs. The BSIM model might not include the series resistance increase at frequencies up to approximately 10 GHz (Suet Fong Tin *et al.* 1998) that are specifically caused by this effect. To improve the accuracy, parasitic effects from layout, such as substrate resistance, must be added as an extended RLC network to the standard model. To address some of these issues and limitations, several researchers and industrial partners have focused on extending BSIM models for RF CMOS.

While the BSIM family remains the de facto standard for CMOS compact modeling, particularly in foundry PDKs, its extensions are continually refined to maintain accuracy at increasingly higher frequencies. For more advanced technologies such as FinFETs and gate-all-around devices, BSIM-CMG and other surface-potential-based formulations are often used, sometimes augmented with parasitic RLC networks to improve RF and mm-wave performance. In BiCMOS technologies, the HICUM model family is generally preferred for high-frequency bipolar transistors and SiGe HBTs. Overall, the exact choice of compact model is largely dictated by the device type and process, and by the degree to which parasitic and substrate effects must be explicitly represented. The focus in the following section, therefore, shifts from these core model formulations to their small-signal equivalents, which form the practical basis for circuit-level RF analysis and design.

## Small-Signal Equivalent Circuits in Nanoscale CMOS and BiCMOS

Compact models and PDKs remain important for designers and researchers to accurately design, model, and simulate their integrated circuits to ensure correct and stable real-world operation. However, even though models such as BSIM and HICUM aim to describe the physical behavior of transistors, designers still have the duty to use their corresponding small-signal equivalent circuits to predict their performance in the frequency domain. These equivalent circuits are ideal for capturing the linearized response of a circuit or component around a particular DC bias point and enable the analytical estimation of key RF and mm-wave metrics such as gain, input and output impedance, and noise parameters such as NF. The equivalent circuit acts as a bridge between the process-level device physics and circuit-level design and analysis. For example, the voltage gain of a common-source MOSFET, if considering the parasitic values, is given by (like the previous equation, but rewritten slightly to intuitively show how capacitance and transconductance interplay at RF)

$$A_v = \frac{v_{out}}{v_{in}} \approx -g_m \frac{\left(r_o \parallel R_L\right)}{1 + j\omega C_{gd}\left(r_o \parallel R_L\right)} \tag{6.24}$$

where $C_{gd}$ is the gate-drain (parasitic) capacitance – showing that if the intrinsic capacitance increases, the voltage gain decreases. In the BiCMOS process, the bipolar transistor model is used to complement the CMOS by offering higher transconductance as well as lower noise at the same bias current, which is particularly advantageous in LNAs, VCOs, and frequency multipliers. For bipolar transistors, a similar expression for the voltage gain exists to compare the differences for the same bias current. The voltage gain for a SiGe HBT integrated in a BiCMOS process is given by

$$A_v = \frac{v_{out}}{v_{in}} \approx -g_m\left(r_o \parallel R_L\right) \tag{6.25}$$

which looks similar, but it is important to note that the output resistance and small signal transconductance are given by

$$r_o \approx \frac{1}{\lambda I_C} \tag{6.26}$$

$$g_m = \frac{I_C}{V_T} \tag{6.27}$$

which (the transconductance) is orders of magnitude higher than that of the CMOS transistor. The exponential I–V characteristic of the HBT leads to this transconductance equation (with $V_T$ around 26 mV at room temperature). In contrast, for a MOSFET, the transconductance is proportional to

$$g_m \propto \sqrt{2\mu C_{ox}\left(\frac{W}{L}\right)I_D} \tag{6.28}$$

and for the same bias current, it is typically an order of magnitude smaller. This directly translates to higher small-signal gain in the HBT, since $A_v$ is directly proportional to $g_m$. For example, considering a bias current of 1 mA, it means that $I_C = 1$ mA yields a $g_m$ of around 38.5 mS at room temperature, while a CMOS device biased at the same current provides only between 2 and 5 mS (depending on the channel length and overdrive voltage). The gain advantages of SiGe BiCMOS come at a cost, however, specifically in terms of higher base resistance, larger circuit-level parasitic capacitances, and reduced integration density when compared to nanoscale CMOS. This means that, specifically at lower frequencies, the gain advantage of the bipolar device is often offset by these limitations (or penalties), and CMOS remains a more practical solution in these cases. In nanoscale CMOS, the continuous scaling of the gate length and the oxide thickness improves the $f_T$ and $f_{max}$, and it does in many ways narrow the gap between the most modern CMOS nodes and SiGe HBTs. As an example, modern advanced CMOS nodes below the 28 nm gate length typically achieve $f_T$s that exceed 200 GHz, while advanced optimized SiGe HBTs in BiCMOS processes can reach $f_T$s of over 500 GHz. Even though there is a significant difference here, it still means that for applications in the tens of GHz range, CMOS-only implementations are also able to meet the design targets without the added cost and complexity of BiCMOS integration. This again shows that there is a fundamental trade-off in performance (gain, noise, and integration cost) that must be considered for each application, target frequency, and circuit type. In low- to mid-frequency RF applications (for example, sub-6 GHz LNAs), a meticulously designed CMOS topology can provide sufficient gain and noise performance when combined with proper impedance matching. However, in higher frequency applications, such as mm-wave VCOs, PAs, and mixers, the better transconductance and lower NF of SiGe HBT become a more appropriate solution, specifically to achieve better linearity, lower phase noise, and improved power efficiency.

As mentioned earlier in this chapter, another important consideration for most integrated circuits is the output impedance and voltage swing. In general, the smaller output resistance of CMOS devices tends to limit voltage gain in high-output-swing applications, whereas HBTs (having a higher Early voltage) maintain higher output impedance and better gain linearity over a wider bias range. Additionally, bipolar transistors also have an exponential input–output relationship, which results in a smoother transconductance modulation profile and lower analog or RF amplification distortion when compared to the abrupt saturation region of a MOSFET.

From a scaling perspective, there are differences between the scaling process of BiCMOS (Raje 1992) and that of traditional CMOS. Understandably, since BiCMOS is inherently a more complex process, its scaling cannot be done at the same pace as that of CMOS nodes (Harame 1997). CMOS scaling is, in a way, straightforward; it is the shrinking of feature sizes, primarily limited by photolithography, and in general, it leads to higher speed transistors with lower power consumption. BiCMOS scaling, however, involves scaling both MOSFETs and bipolar devices simultaneously, which is a more complex and costly exercise since it requires more fabrication steps. Therefore, there exists a *scaling gap* between CMOS and BiCMOS, and this is noticeable when looking at the technology $f_T$. The SiGe HBT in BiCMOS technology has a historical lead in intrinsic transit frequency, which reflects the vertical heterojunction and engineered base that essentially reduces the carrier transit time. CMOS steadily (more frequently and aggressively) increases the $f_T$ through improvements in photolithography, mobility engineering, and stacked device architectures. But even as CMOS approaches the 200–400 GHz $f_T$ regime, SiGe devices typically retain their margin at the highest measured frequencies that are relevant to mm-wave and sub-THz applications.

## Parameter Extraction and Validation

The compact models discussed in this chapter, therefore, BSIM and HICUM, for example (and VBIC), are only as accurate as their extracted parameters. The process of extracting parameters as well as validating them becomes the link, or bridge, between measured device behavior and the mathematical model that is used in circuit simulators. The aim here is to ensure that a single set of parameters, typically *S*-parameters, accurately reproduces the measured DC, AC, and RF responses across various bias conditions and temperatures. The traditional steps to go from low-frequency, or DC, to high-frequency (*S*-parameter) characterization, also sometimes referred to as progressive (polynomial) calibration (Pereira *et al.* 2009), can be divided into five broad steps:

1. DC characterization, sometimes called static region I–V extraction, is the foundation for transconductance and output conductance. During this step, variables such as $I_D$–$V_{GS}$ and $I_D$–$V_{DS}$ are determined for MOSFETs, and for HBTs, measurements for $I_C$–$V_{BE}$ and $I_C$–$V_{CE}$ are determined. Basic parameters such as threshold voltage, mobility, channel-length modulation, series

resistance, saturation current, and forward beta can be calculated using these DC measurements.

2. Capacitive extraction, or sometimes called quasi-static or C-V-based extraction, which affects small-signal and transient behavior. For this step, traditional C-V measurements (or charge-based extraction in BSIM or HICUM) are used to determine parameters such as $C_{gs}$, $C_{gd}$, and junction capacitances $C_{db}$ and $C_{sb}$.
3. S-parameter measurements and de-embedding to achieve RF characterization, where *S*-parameter measurements are done over frequency using on-wafer probes and a carefully calibrated vector network analyzer (VNA). The de-embedding process is also crucial to remove any pad capacitance, interconnect conductance, and contact resistance, typically done by using dummy *open* and *short* structures. After de-embedding is completed, the *S*-parameters can be converted to *Y*- or *Z*-parameters to directly relate them to intrinsic small-signal equivalents ($Y_{de\text{-}embedded} = Y_{measured} - Y_{open}$).
4. Fitting of small-signal parameters. In this step, and from the measured (or converted) *Y*-parameters, the intrinsic model elements like $g_m$, $g_{mb}$, $r_o$, $C_{gs}$, $C_{gd}$, $C_{db}$, $C_{sb}$, etc., are extracted. Different fitting techniques can be used, such as least-squares optimization or gradient-based parameter tuning, and the techniques aim to minimize the difference between measured and simulated *Y*-parameters across the range of frequencies as well as bias points.
5. High-frequency validation to check that the extracted model reproduces the measured gain, phase, NF, $f_T$, and $f_{max}$, and if not, the parasitic values or non-quasi-static (NQS) parameters can be refined. After these steps, importantly, corner validation across the temperature range and process variations must be performed to ensure robust environmental simulations.

Even though these five steps indicate a logical flow of steps toward accurate *S*-parameters, it must also be noted that the steps are not always sequential, but are typically iterative. Also, in modern design environments, the extraction workflow is automated to various degrees, within foundry-specific parameter extraction tools or EDA platforms. The final accuracy depends heavily on the initial DC and capacitance extraction, so steps 1 and 2. Any small deviation in the extracted series resistance or threshold voltage can lead to significant mismatches in the simulated S-parameters, especially at higher frequencies. As a result, the process is often referred to as hierarchical refinement, where each stage (step) constrains and improves the reliability of the next.

Also important to consider during these steps is that each compact model (BSIM, HICUM, etc.) has its own specific parameter hierarchy and interdependencies. As an example, in the compact model BSIM (Sheu *et al.* 1987), values for the mobility degradation, short-channel effects, and DIBL are fitted primarily from the DC measurements, but, for HICUM, the transit time and charge partitioning terms are tuned using RF data. As a result, although the logical flow of measurements and de-embedding looks similar in many ways, there are inherent theoretical differences between CMOS and bipolar devices. Primarily, and relatively easy to remember, is that CMOS parameter extraction is, in general, geometry and bias dependent, whereas HBT parameter extraction is more frequency- and charge-transport (the vertical movement of electrons and holes through the layers of the semiconductor) dependent. Consistency with physical value is critical and is done through validation against independent measurements, for example, against specifically designed test circuits such as common-source or common-emitter amplifier topologies. These validation measurements are used to confirm that the simulated values for the gain and NF correctly correspond to actual measured performance. Additionally, to have confidence that these extracted models not only fit the small-signal data, but also have predictive accuracy for nonlinear (and large-signal) simulations, simulated $f_T$ and $f_{max}$ can be plotted against bias, and compared to measured *S*-parameter roll-off curves.

In the nanoscale regime of device geometries, parameter extraction becomes more complex and faces various additional challenges. With decreased component sizes, parasitic values of gate resistance, fringe capacitance, and substrate coupling become more pronounced. Therefore, the actual boundary between intrinsic and extrinsic elements becomes more difficult to clearly define. Furthermore, NQS effects cause further frequency-dependent deviations that are impossible to capture using traditional and somewhat simplified lumped element models. Modern workflows, therefore, increasingly rely on distributed equivalent circuits, or sometimes physics-informed parameter extraction. In these cases, EM simulations are combined with measured *S*-parameters to completely isolate and parasitic coupling paths. These methods have been well established (and are almost mandatory) to accurately represent component and circuit behavior in mm-wave and sub-THz operations.

Accurate de-embedding of parasitic effects remains a persistent challenge during parameter extraction, especially for LNA and mm-wave designs. Ideally, the goal of this process is to completely isolate the intrinsic device behavior; however, real-world pad capacitances, interconnect inductances, and substrate coupling paths can dominate the measured response in high-GHz measurements; also, these values will be completely layout-dependent and will differ between designs and processes. For example, in a typical 65 nm CMOS or BiCMOS node, gate resistance and pad capacitance can potentially introduce a significant phase delay as well as apparent losses in S-parameters, and if these are not properly de-embedded, they could be falsely attributed as actual intrinsic device values. In LNA designs, in particular, minor inaccuracies in extracted parameters can lead to large (amplified) errors in the simulated input matching and NF. Substrate parasitic modeling, especially in foundry processes that implement low-resistivity silicon, is also difficult since, at high frequencies, the substrate can support lateral wave propagation. This leads to coupling between neighboring devices, followed but incorrect interpretations of gain and phase. Some advanced extraction methods may account for lossy substrate networks, typically, RLC ladders, which are fitted to capture the frequency-dependent impedance that is seen *through* the substrate. Such strategies and complexities become increasingly important when modeling the noise behavior of LNAs, as substrate resistance and coupling can introduce correlated noise sources that eventually increase (degrade) the minimum achievable NF.

Overall, while modern PDKs provide increasingly accurate parameter sets, the precision of compact modeling at mm-wave and sub-THz frequencies still depends heavily on careful measurement technique, robust de-embedding, and a good physical understanding of the layout parasitics. The extraction process for such devices thus moves beyond a purely numerical fitting exercise, and it becomes a multidomain calibration problem, spanning device physics, electromagnetics, and circuit-level validation – more recently, also artificial intelligence (AI).

Recent developments in parameter extraction increasingly integrate AI-assisted optimization and machine-learning regression to reduce manual fitting effort and improve cross-node transferability. Neural-network-assisted extraction, for example, can predict initial parameter sets based on measurement data trends, thereby accelerating convergence in iterative optimization loops. For LNA and mm-wave front-end design, this becomes particularly valuable when compact models need to remain valid over wide temperature and process corners without overfitting to a single bias regime. Another emerging direction is the cross-correlation of EM and device-level simulations, where data from layout-extracted parasitic networks, 3D field solvers, and compact model tuning are fused into a unified optimization framework. Such methods enable better prediction of device behavior in the presence of packaging and interconnect parasitics, and they represent the next evolution of parameter extraction workflows, bridging measurement, modeling, and automated calibration.

## LNA DESIGN TRADE-OFFS IN SCALED CMOS

LNA design in deeply scaled CMOS (and BiCMOS) is extremely dependent on the intricate balance between performance metrics that are also trade-offs of each other; these metrics include:

- NF
- Gain
- Linearity
- Power consumption

All these metrics are inherently coupled and not independent, linked through the device transconductance, bias current, and the effective input and output matching conditions. These relationships become increasingly nonlinear and bias-dependent in nanoscale CMOS, since short-channel effects, parasitic capacitances, and mobility degradation dominate intrinsically. It is therefore crucial to have a clear and functional understanding of these interdependencies to design for optimal trade-offs, typically as a function of the application. As an example, the effective LNA noise factor, approximated as a function of active transistor-level parameters, is written as

$$F = 1 + \frac{\gamma}{\alpha}\frac{1}{g_m R_S} + \frac{\delta R_G}{R_S} + \frac{R_D}{R_S} \tag{6.29}$$

where $R_S$ is the source resistance, typically 50 Ω, $R_G$ and $R_D$ are the gate and drain resistances, respectively, $\gamma$ is the thermal noise in the channel, and $\alpha$ is used to capture the gate-induced noise correlation, therefore, it is a constant factor that varies based on the process. This equation for noise factor is directly tied to the achievable NF to the transconductance efficiency, therefore $g_m/I_D$, and the design of the matching network. Furthermore, the power gain of a common-source LNA stage (if both the device transconductance and load impedance are considered) is expressed as (related to (6.24))

$$A_v = \frac{v_{out}}{v_{in}} = g_m\left(r_o \parallel R_L\right)\left|\frac{Z_{in}}{Z_{in} + R_S}\right| \tag{6.30}$$

where $Z_{in}$ is the combined input impedance. This shows the (dual) dependence on $g_m$, and therefore also the bias current, and the impedance matching network. If the bias is increased, the gain will also increase; however, at the same time, this raises power consumption and device temperature, both of which are important when considering biasing for linearity versus noise optimization, as well as power consumption versus thermal effects. Another aspect to consider is linearity, where a key figure of merit is the input-referred third-order intercept point (IIP3), a parameter that can also be approximated as a function of the transconductance and its nonlinearity coefficients, with

$$IIP3 \approx \frac{4}{3}\frac{V_{ov}}{\left|a_3 / a_1\right|} \tag{6.31}$$

where $V_{ov}$ is the overdrive voltage, and $a_1$ and $a_3$ are coefficients from the Taylor expansion (Tehrani and Ludeman 1993) of the transistor transconductance. The intrinsic input impedance of a transistor, especially at deeply scaled nodes (for example, 28 nm to 7 nm CMOS nodes), is further complicated due to parasitic capacitances (specifically the gate–source and gate–drain capacitances), leading to complexity in the coupling between its transconductance, matching, and linearity. The parasitics at these nodes heavily affect the transistor input impedance, where

$$Z_{in}\left(j\omega\right) \approx \frac{1}{j\omega C_{gs}} + \frac{g_m}{\omega^2 C_{gd}^2} \tag{6.32}$$

which shows that the input impedance is not resistive (or constant) and it varies significantly with both frequency and bias point. This means that the choice of the bias point to obtain a specific $g_m$ simultaneously shifts the optimum noise impedance $Z_{opt}$ and, as a result, influences the matching network. This limits the designer to freely choose an optimum gain as well as an optimum noise, independently. It also means that the noise-optimum impedance can be expressed as a proportionality in

$$Z_{opt} \propto \frac{1}{g_m} \tag{6.33}$$

where the maximum power gain effectively requires

$$Z_{in} = R_S \tag{6.34}$$

and these two equations are inherently and fundamentally incompatible (except if $Z_{opt}$ = 50 Ω). Therefore, designing for minimum NF or maximum gain requires different strategies in these heavily scaled nodes.

## System-Level Co-Optimization

In nanoscale CMOS, LNA performance is not determined, or dictated, by single mechanisms, and is an amalgamation of various first, second, and even third-order effects, particularly parasitic effects that emanate from device-level process parameters, and from increases in operation frequency. Some of the key mechanisms discussed in this section include the following:

- Coupling between impedance matching networks
- Device biasing considerations on transconductance
- Thermal behavior, especially at mm-wave frequencies
- Long-term stability and reliability for "hard-working" and small active devices

But the fundamental quantities that determine NF, gain, and linearity are already heavily impacted by transconductance, parasitic capacitances, and biasing currents; it becomes increasingly difficult to independently design for one metric without significant impact on the others, more so than in larger nodes and at lower frequencies.

First, it is important to note that the noise-optimum impedance as described earlier ($Z_{opt}$) is inversely proportional to $g_m$. At the same time, gain maximization requires matching of the input to the source resistance (usually 50 Ω). Since the intrinsic device impedance of nanoscale CMOS active devices is not resistive and varies strongly with frequency and bias point, it is unlikely that both parameters can be optimized simultaneously. Although this is generally true for all amplifiers, the effects are bigger in nanoscale CMOS. Practically, minimum NF requires a mismatch toward $Z_{opt}$, whereas maximum gain requires the traditional conjugate matching between source resistance and device input impedance.

Second, if ID increases, so does gm, which lowers NF, but increases gain, as expected, but, additionally, these same actions will reduce overdrive headroom, which negatively affects the IIP3, and push the device into strong inversion, where mobility degradation and parasitic capacitances increase (therefore worsen). Biasing, therefore, has three (simultaneous) outcomes, such that

$$Better\,NF \Rightarrow Better\,gain \Rightarrow Worse\ linearity$$

The third consideration is power consumption versus the inherent thermal effects, since any improvement in NF (for example, with a higher $g_m$) or linearity (for example, with a higher $V_{ov}$) will need a higher bias current, which will have second-order effects, including:

- Higher self-heating of the device
- Higher junction temperatures
- A degradation in $g_m$, and therefore in NF and circuit gain

Finally, device reliability becomes a longer-term consideration, since higher bias current and electric field stress will accelerate degradation effects such as

- Hot carrier injection (HCI), which is a result of large drain–source electric fields, and over time, the IIP3 will shift
- Bias temperature instability (BTI), which affects the threshold voltage and shifts $Z_{opt}$ as well as the requirements of the matching networks during operation, as the temperature varies
- Variability in the process, especially in sub-28 nm nodes, specifically because $g_m$ becomes much more sensitive to changes in $V_{th}$

# OUTLOOK

## Beyond FinFET: Nanosheet, Gate-All-Around, 2D Materials in CMOS Flows

FinFETs, as described in this chapter, were specifically developed to mitigate the performance degradation as much as possible, caused by short-channel effects in traditional (planar) MOSFETs as transistors continue to scale down and shrink. These transistor structures essentially replaced the two-dimensional channel with a three-dimensional "fin" that is wrapped by the gate, on multiple sides, providing (among other advantages) significantly better electrostatic control. Below the 28 nm node, the distance between the source and drain terminals became so small that the electric fields of the source and the drain compromised the control on the gate. This led to undesirable effects such as DIBL, increased off-state leakage current, reduced threshold voltage, and degraded subthreshold slope. With its advanced gate control, full channel depletion, reduced DIBL, lower leakage current, higher on/off ratio, and the elimination of channel doping, FinFETs address these challenges and extend Moore's law for several technology nodes. However, the effectiveness of FinFETs starts to diminish around the 5 nm node, and below, primarily due to the fin height and width limitations. To further progress with scaling, the industry has started to focus its efforts on GAAFETs.

In GAAFETs, the successor to the FinFET, the transistor gate surrounds the channel on all four sides, and the channel is often implemented as stacked horizontal nanosheets or nanowires. These devices allow $g_m$ scaling without the direct penalty in parasitic $C_{gs}$, which improves $Z_{opt}$ without resorting to aggressive matching networks, ideal in LNA applications. These modern, next-generation semiconductors give the gate superior electrostatic control over the channel and enable both higher performance and lower power consumption, compared to traditional MOSFETs and even FinFETs. As a result, GAAFETs offer reduced leakage current, quicker on/off switching, improved scalability by mitigating short-channel effects, and flexible design alternatives, particularly when using nanosheets, which allow researchers and engineers to adjust the width of nanosheets. As a result, by varying the width of the nanosheets, the transistor can be optimized for either high performance (wider sheets) or lower power consumption (narrower sheets). It is important to note that GAA is not a single structure, but is more a classification of a class of transistors. There are several variants, but the most common are

- Nanosheets
- Nanowires

Nanosheets specifically use horizontally stacked sheets as the channel, and in this design (sometimes referred to as a multi-bridge channel MOSFET, or MBCFET), allow for greater control over the width of the channel as well as the drive current. Nanowires are a variation with rounder horizontal nanowires as the channel, but still provide good electrostatic control with high levels of flexibility. As with most new technologies, GAASFETs have several challenges as they find their way into mass commercial adoptions. These transistors present significant manufacturing challenges, such as the precise fabrication and uniform etching of delicate nanostructures. Major foundries are still actively using and integrating this technology, especially in nodes of 3 nm and smaller, as overcoming their limitations and challenges is likely to provide major advantages for performance computing in 5G, 6G, AI, and other advanced applications.

Another technology that deserves credit for advancing, or at least maintaining Moore's law, and is driving LNA front-end development is monolayer 2D semiconductors (Kis 2021; Nandan *et al.* 2025), such as $MoS_2$, $WS_2$, and black phosphorus. To overcome the issues of extremely small channels and challenging electrostatic control of traditional CMOS transistors, researchers are exploring the potential of atomically thick channel materials, such as monolayers of the transition metal dichalcogenide molybdenum disulfide ($MoS_2$).

These 2D semiconductors hold great potential for ultra-scaled transistors, and the gate length of $MoS_2$ transistors has been scaled to 1 nm and 0.3 nm (300 pm) using single-walled carbon nanotubes and graphene, respectively. However, there are still considerable challenges in scaling the channel length of these short-gate transistors, and the processing difficulties, particularly in alignment, are extremely challenging. From the perspective of the LNA, there are various device-level advantages of 2D transistors, and these include:

- Atomic-scale thickness and tight gate control will decrease the parasitic capacitances such as $C_{gs}$ and $C_{gd}$, which can limit high-frequency gain.
- Carrier transport becomes more efficient in ultra-thin channels and therefore leads to improved scaling of gm per unit current.

- Thin layers couple more effectively to the substrate and have the potential to lower the thermal budget and reduce self-heating.
- In LNAs, the overdrive voltage and biasing limits are relaxed and therefore the linear region is higher, resulting in better IIP3.

However, even these advantages come with significant practical challenges, challenges that increase complexity, cost, and design interpretations, and specifically make it more difficult to achieve:

- Wafer-scale uniformity
- Low contact resistance
- High yield
- CMOS-compatible workflows in fabrication
- Integrating 2D structures in RF front-end modules

The 2011 *Nature* paper (Radisavljevic *et al.* 2011) represents the first credible demonstration that $MoS_2$ can behave like a high-performance transistor channel rather than a novelty material. Using a channel thickness of just 6.5 Å, the authors demonstrated that aggressive electrostatic control can be achieved while maintaining a mobility of ~200 cm²/V·s after deposition of a high-κ $HfO_2$ dielectric. The device exhibited a bandgap of ~1.8 eV, an $I_{on}/I_{off}$ ratio greater than $10^8$, and ohmic contacts using gold, achievements that surpass what has been possible with graphene (gapless) or ultra-thin silicon (thicker than $MoS_2$ and requiring far more processing complexity). For RF front ends and LNAs, this implies that 2D semiconductors could offer extremely low parasitic capacitances, improved $g_m$ per current density, and biasing stability due to the inherent bandgap. However, the remaining challenges, contact resistance, wafer-level uniformity, and integration into RF module flows, currently prevent these devices from competing with nanosheet CMOS (Table 6.3).

Monolayer 2D semiconductors ($MoS_2$, $WS_2$, black phosphorus) remain one of the more interesting attempts to keep transistor scaling relevant for RF front ends. By pushing device thickness down to a single atomic layer, researchers have demonstrated channel lengths from 1 nm to sub-nanometers using carbon nanotube and graphene

**TABLE 6.3**
**Device comparison of modern technologies that further Moore's law**

| Device | Short-channel control | RF transconductance efficiency | Parasitics |
|---|---|---|---|
| FinFET | Good | Improving, but still limited at mm-wave | Moderate |
| GAA nanosheet | Excellent | Better $g_m/I_D$ than FinFET | Lower |
| 2D materials | Ideally perfect | Potentially highest | Unknown (contact resistance challenges) |

gates. For LNAs, the appeal is straightforward: extreme electrostatic control, inherently low parasitic capacitances, and the possibility of maintaining strong transconductance at very low bias currents. The thin channel structure also improves thermal coupling to the substrate, reducing local temperature rise during high-gain operation. In principle, this combination could support higher linearity and more relaxed biasing margins, something traditional CMOS fights for at these dimensions.

As research advances toward monolayer channel materials and unconventional device architectures, the focus inevitably shifts from scaling limits to long-term technological relevance. Despite the promise of these emerging devices, CMOS remains the only platform with proven manufacturability, supply-chain maturity, and repeatable RF performance at scale. The following section examines how CMOS is positioned within future wireless generations, specifically 6G and beyond, and evaluates whether it remains the foundation for next-generation RF systems.

## Future Role of CMOS in 6G/7G and Beyond

There have been several fundamental shifts as technology has evolved from 5G. 5G pushes the communications systems into the sub- and mm-wave bands. It uses multiple frequency ranges, including the sub-6 GHz bands up to the 24–40 GHz (FR2) bands. It offers significantly higher speeds compared to previous generations, as well as ultra-low latency for a mobile broadband connection, and the ability to connect an extremely large number of devices simultaneously. Furthermore, 5G offers high capacity and improved network slicing for customized virtual networks, all while maintaining energy efficiency. The move to 5G, with its massive multiple-input-multiple-output (MIMO) and carrier aggregation, increases the complexity of the RF front-end. CMOS technology remains crucial for keeping the resulting chips small and cost-effective. 6G proposals are targeting sub-THz and THz ranges (100–300 GHz, potentially up to 1 THz) to achieve even higher data rates (up to 1 Tbps) and much wider bandwidth than previous generations. Although this will, in theory, enable a greater variety of services and applications, it also presents new challenges that require forward-thinking in the enabling technologies. These extremely short wavelengths have higher path loss and increased complexity in antenna design and RF integrated circuits. While III–V semiconductors have better high-frequency performance when compared to CMOS, CMOS remains the favored technology for its cost-effectiveness, high integration density, and low power consumption, and is ideal for massive and complex antenna arrays and SoC solutions that are required for 6G's ultra-high frequencies and new and novel capabilities. Researchers are using techniques like advanced neutralization and transformer-based matching networks in CMOS designs to achieve the required wide bandwidth and stability for 6G applications. Hybrid approaches may also be used, where CMOS handles most of the system integration, and the more specialized and costly solutions are only used in specific areas where the high-frequency signals are processed. Conceptually, 7G will move to co-integrated sensing and communications, possibly using the 300 GHz to 1 THz bands and the mainstream and commercial applications. CMOS will again be pushed to its limits to achieve this co-integration methodology, and with IoT, edge sensing, and AI-at-the-edge at the center of the sensing integration, RF front ends will need to scale to an extremely large number of devices. Therefore, in most likelihood, CMOS will remain the default platform of future applications, and structured around the four pillars:

1. Manufacturability and yield
2. System integration capability
3. Cost and supply chain
4. Power efficiency

Beyond these advantages, the continued relevance of CMOS in 6G and beyond is not simply a consequence of technological inertia; it is driven by system-level economics and architectural scalability. As operating frequency moves toward sub-THz bands, the radio front end becomes increasingly defined by array size rather than individual device performance. Massive MIMO and fully integrated beamforming architectures demand the replication of entire receive chains, LNAs, phase shifters, and ADCs across hundreds or thousands of elements. In this context, transistor choice becomes secondary to the question of how many front ends can be integrated per unit area, per watt, and per dollar. CMOS, with its unparalleled density and mature ecosystem, uniquely enables this scaling. Competing device technologies may offer better raw noise or gain, but cannot currently deliver the required integration density, packaging maturity, or manufacturing yield at comparable cost. Thus, the competitive metric shifts from "best transistor" to "best platform for system duplication at scale."

If 5G was the era where CMOS proved that mm-wave radios could be mass-produced, 6G will be the era where CMOS decides whether RF front ends scale to arrays of thousands of receivers. 7G (if it materializes) may be defined not by transistors, but by co-integration of sensing, communications, and computation, with CMOS acting as the backbone tying disparate technologies together.

## CONCLUSION

RF CMOS has evolved from a marginal analog technology to a foundation for modern front-end design, and its trajectory is shaped not by a single breakthrough, but by the cumulative negotiation of trade-offs between physics, architecture, and system-level demands. Throughout this chapter, it became clear that the ability to extract meaningful performance from deeply scaled devices depends as much on modeling accuracy, parasitic awareness, and measurement discipline as it does on raw transistor speed. Parameter extraction, compact models, and de-embedding are no longer supported processes; they dictate what the designer can trust, simulate, and ultimately deliver. As device dimensions approach physical limits, NF, linearity, gain, and power efficiency cannot be optimized independently, and the designer must instead think in terms of co-optimized operating points rather than isolated performance targets.

The discussion across LNA design methodology and RF-focused process variants reinforces that scaling unlocks opportunity, but introduces complexity: matching becomes more sensitive to bias, layout parasitics become first-order, and reliability mechanisms such as BTI and HCI shift from long-term risks to active design constraints. Meanwhile, defense, sensing, and communication systems increasingly prioritize wideband operability and dynamic range over minimum noise alone, driving architectures that derive performance from system-aware design rather than device-ideal assumptions.

Ultimately, this chapter establishes a central message: CMOS remains relevant not through perfection, but through balance. Its strength lies in its manufacturability, integration density, and adaptability, attributes that matter as much as $f_T$ or NF when architectures scale to hundreds or thousands of channels. As the following outlook explores the frontier beyond FinFETs, CMOS and its successors must continue to evolve not only as transistors, but as platforms that enable RF systems to be built, replicated, and trusted at scale.

## REFERENCES


Abidi, A.A., 2014. CMOS Microwave and Millimeter-Wave ICs: The Historical Background. In: *2014 IEEE International Symposium on Radio-Frequency Integration Technology.* IEEE, 1–5.

Abou-Allam, E., Manku, T., Ting, M., and Obrecht, M.S., 2000. Impact of Technology Scaling on CMOS RF Devices and Circuits. In: *Proceedings of the IEEE 2000 Custom Integrated Circuits Conference (Cat. No.00CH37044).* IEEE, 361–364.

Allam, M., Anis, M., and Elmasry, M., n.d. Effect of Technology Scaling on Digital CMOS Logic Styles. In: *Proceedings of the IEEE 2000 Custom Integrated Circuits Conference (Cat. No.00CH37044).* IEEE, 401–408.

Banna, S., 2015. Scaling Challenges of FinFET Technology at Advanced Nodes and Its Impact on SoC Design (Invited). In: *2015 IEEE Custom Integrated Circuits Conference (CICC).* IEEE, 1–8.

Barraud, S., Lapras, V., Previtali, B., Samson, M.P., Lacord, J., Martinie, S., Jaud, M.-A., Athanasiou, S., Triozon, F., Rozeau, O., Hartmann, J.M., Vizioz, C., Comboroure, C., Andrieu, F., Barbe, J.C., Vinet, M., and Ernst, T., 2017. Performance and Design Considerations for Gate-All-around Stacked-NanoWires FETs. In: *2017 IEEE International Electron Devices Meeting (IEDM).* IEEE, 29.2.1–29.2.4.

Borkar, S., 2006. Electronics beyond Nano-Scale CMOS. In: *Proceedings of the 43rd Annual Conference on Design Automation – DAC '06.* ACM Press, 807.

Brodic, D., 1991. Comparison of the CMOS and BiCMOS Simple Architecture Operational Amplifier. In: *[1991 Proceedings] 6th Mediterranean Electrotechnical Conference.* IEEE, 299–302.

Choi, H.-W., Choi, S., and Kim, C.-Y., 2020. A CMOS Band-Pass Low Noise Amplifier with Excellent Gain Flatness for mm-Wave 5G Communications. In: *2020 IEEE/MTT-S International Microwave Symposium (IMS).* IEEE, 329–332.

da Silva, M.B., Wirth, G.I., Tuinhout, H.P., Duijnhoven, A.Z., and Scholten, A.J., 2023. Random Telegraph Noise in Analog CMOS Circuits. *IEEE Transactions on Circuits and Systems I: Regular Papers*, 70 (6), 2229–2242.

Davari, B., Dennard, R.H., and Shahidi, G.G., 1995. CMOS Scaling for High Performance and Low Power-the Next Ten Years. *Proceedings of the IEEE*, 83 (4), 595–606.

del Alamo, J.A., Antoniadis, D.A., Lin, J., Lu, W., Vardi, A., and Zhao, X., 2015. III-V MOSFETs for Future CMOS. In: *2015 IEEE Compound Semiconductor Integrated Circuit Symposium (CSICS).* IEEE, 1–4.

Dewey, G., Hudait, M.K., Kangho Lee, Pillarisetty, R., Rachmady, W., Radosavljevic, M., Rakshit, T., and Chau, R., 2008. Carrier Transport in High-Mobility III–V Quantum-Well Transistors and Performance Impact for High-Speed Low-Power Logic Applications. *IEEE Electron Device Letters*, 29 (10), 1094–1097.

Dunwell, D., and Frank, B., 2007. Accumulation-Mode MOS Varactors for RF CMOS Low-Noise Amplifiers. In: *2007 Topical Meeting on Silicon Monolithic Integrated Circuits in RF Systems*. IEEE, 145–148.

Floyd, B.A., Reynolds, S.K., Pfeiffer, U.R., Zwick, T., Beukema, T., and Gaucher, B., 2005. SiGe Bipolar Transceiver Circuits Operating at 60 GHz. *IEEE Journal of Solid-State Circuits*, 40 (1), 156–167.

Forsberg, T., 2018. *Efficient mm-Wave Transmitter Design in CMOS Technology*. Faculty of Engineering, LTH.

Gupta, C., Goel, R., Agarwal, H., Hu, C., and Chauhan, Y.S., 2019. BSIM-BULK: Accurate Compact Model for Analog and RF Circuit Design. In: *2019 IEEE Custom Integrated Circuits Conference (CICC)*. IEEE, 1–8.

Harame, D., 1997. High Performance BiCMOS Process Integration: Trends, Issues, and Future Directions. In: *Proceedings of the 1997 Bipolar/BiCMOS Circuits and Technology Meeting*. IEEE, 36–43.

Hemantha, G.R., Priya, A.S., Suman, J.V., Rao, T.V.J., Priyadarshini, G.M.A., and Mallam, M., 2024. Characterization and Modeling of Gate-All-around FET (GAA FET) for Low-Power and High- Performance Applications. In: *2024 International Conference on Advances in Modern Age Technologies for Health and Engineering Science (AMATHE)*. IEEE, 1–5.

Hu, Y., and Chi, T., 2023. A Systematic Approach to Designing Broadband Millimeter-Wave Cascode Common-Source with Inductive Degeneration Low Noise Amplifiers. *IEEE Transactions on Circuits and Systems I: Regular Papers*, 70 (4), 1489–1502.

Hyltin, T.M., 2000. The Beginnings of Solid State Radar. *IEEE Transactions on Aerospace and Electronic Systems*, 36 (3), 1016–1019.

Jenei, S., Decoutere, S., Van Huylenbroeck, S., Vanhorebeek, G., and Nauwelaers, B., 2001. High Q Inductors and Capacitors on Si Substrate. In: *2001 Topical Meeting on Silicon Monolithic Integrated Circuits in RF Systems. Digest of Papers (IEEE Cat. No.01EX496)*. IEEE, 64–70.

Kim, S., Lee, H., and Jeong, Y., 2024. Cost-Effective Co-Optimization of RF Process Technology Targeting Performances/Power/Area Enhancements for RF and mmWave Applications. *Electronics*, 13 (13), 2513.

Kis, A., 2021. How We Made the 2D Transistor. *Nature Electronics*, 4 (11), 853–853.

Ko, C.-L., Kuo, M.-C., Hsu, C.-M., and Kuo, C.-N., n.d. A CMOS Dual-Mode RF Front-End Receiver for GSM and WCDMA. In: *Proceedings of 2004 IEEE Asia-Pacific Conference on Advanced System Integrated Circuits*. IEEE, 374–377.

Kobal, E., Siriburanon, T., Chen, X., Nguyen, H.M., Staszewski, R.B., and Zhu, A., 2022. A Gm-Boosting Technique for Millimeter-Wave Low-Noise Amplifiers in 28-nm Triple-Well Bulk CMOS Using Floating Resistor in Body Biasing. *IEEE Transactions on Circuits and Systems I: Regular Papers*, 69 (12), 5007–5017.

Koga, Y., and Kurita, K., 2019. SOI Wafer Fabricated with Extra Thick Deposited BOX Layer Using Surface Activated Bonding at Room Temperature for Customized Power Devices. In: *2019 6th International Workshop on Low Temperature Bonding for 3D Integration (LTB-3D)*. IEEE, 5.

Kumar, M.J., Gupta, S.K., and Venkataraman, V., 2006. Compact Modeling of the Effects of Parasitic Internal Fringe Capacitance on the Threshold Voltage of High-k Gate-Dielectric Nanoscale SOI MOSFETs. *IEEE Transactions on Electron Devices*, 53 (4), 706–711.

Lai, R., Mei, X.B., Deal, W.R., Yoshida, W., Kim, Y.M., Liu, P.H., Lee, J., Uyeda, J., Radisic, V., Lange, M., Gaier, T., Samoska, L., and Fung, A., 2007. Sub 50 nm InP HEMT Device with Fmax Greater than 1 THz. In: *2007 IEEE International Electron Devices Meeting*. IEEE, 609–611.

Lai, Y.-S., Yei, J.-W., Fu, Z.-H., and Lin, K.-Y., 2023. A CMOS Low Noise Amplifier with Gm-Boosting Technique for Multi-band 5G Systems. In: *2023 Asia-Pacific Microwave Conference (APMC)*. IEEE, 351–353.

Lambrechts, W., and Sinha, S., 2017. *SiGe-Based Re-engineering of Electronic Warfare Subsystems*. Springer International Publishing Imprint: Springer.

Lee, T.H., 2004. *The Design of CMOS Radio-Frequency Integrated Circuits*. Cambridge University Press.

Lee, T.H., 2007. From Oxymoron to Mainstream: The Evolution and Future of RF CMOS. In: *2007 IEEE International Workshop on Radio-Frequency Integration Technology*. IEEE, 1–6.

Lihui, G., and Jie, J.S.Y., 2004. Integration of Top Metal Layer with Thick Cu/Black Diamond/ spl Trade/ (BD). In: *Proceedings of the 11th International Symposium on the Physical and Failure Analysis of Integrated Circuits. IPFA 2004 (IEEE Cat. No.04TH8743)*. IEEE, 319–322.

Maity, S.K., and Pandit, S., 2021. Performance Assessment of CMOS Circuits Using III-V on Insulator MOS Transistors. *Silicon*, 13 (6), 1939–1949.

Masselink, E.W., Stark, A., Yang, B.B., and Harris, T.R., 2023. Review of Hybrid Integration Techniques for Integrating III-V onto Silicon. In: *2023 IEEE International 3D Systems Integration Conference (3DIC)*. IEEE, 1–7.

Masuda, H., Nakai, M., and Kubo, M., 1979. Characteristics and Limitation of Scaled-Down MOSFET's Due to Two-Dimensional Field Effect. *IEEE Transactions on Electron Devices*, 26 (6), 980–986.

McQuiddy, D.N., Gassner, R.L., Hull, P., Mason, J.S., and Bedinger, J.M., 1991. Transmit/ Receive Module Technology for X-Band Active Array Radar. *Proceedings of the IEEE*, 79 (3), 308–341.

McQuiddy, D.N., Wassel, J.W., LaGrange, J.B., and Wisseman, W.R., 1984. Monolithic Microwave Integrated Circuits: An Historical Perspective. *IEEE Transactions on Microwave Theory and Techniques*, 32 (9), 997–1008.

Mei, X., Yoshida, W., Lange, M., Lee, J., Zhou, J., Liu, P.-H., Leong, K., Zamora, A., Padilla, J., Sarkozy, S., Lai, R., and Deal, W.R., 2015. First Demonstration of Amplification at 1 THz Using 25-nm InP High Electron Mobility Transistor Process. *IEEE Electron Device Letters*, 36 (4), 327–329.

Miura-Mattausch, M., Feldmann, U., Fukunaga, Y., Miyake, M., Kikuchihara, H., Ueno, F., Mattausch, H.J., Nakagawa, T., and Sugii, N., 2014. Compact Modeling of SOI MOSFETs with Ultrathin Silicon and BOX Layers. *IEEE Transactions on Electron Devices*, 61 (2), 255–265.

Moore, G.E., 1998. Cramming More Components onto Integrated Circuits. *Proceedings of the IEEE*, 86 (1), 82–85.

Moran Guizan, C., Baumgartner, P., Heinen, S., and Lauritano, M., 2024. Millimeter-Wave Tunable Impedance Matching Network in an Advanced CMOS Process. *IEEE Transactions on Microwave Theory and Techniques*, 72 (8), 4590–4596.

Mu, J., Chou, X., Ma, Z., He, J., and Xiong, J., 2018. High-Performance MIM Capacitors for a Secondary Power Supply Application. *Micromachines*, 9 (2), 69.

Nandan, K., Naseer, A., Agarwal, A., Bhowmick, S., and Chauhan, Y.S., 2025. Transistors Based on Novel 2-D Monolayer Semiconductors $Bi_2O_2Se$, InSe, and $MoSi_2N_4$ for Enhanced Logic Density Scaling. *IEEE Transactions on Electron Devices*, 72 (1), 516–521.

Niknejad, A.M., Doan, C., Emami, S., Dunga, M., Xi, X., He, J., Brodersen, R., and Hu, C., 2005. Next Generation CMOS Compact Models for RF and Microwave Applications. In: *RFIC Symposium, 2005. Digest of Papers. 2005 IEEE Radio Frequency integrated Circuits*. IEEE, 141–144.

Ninan, B.M., Balamurugan, K., and Devi, M.N., 2016. Design and Analysis of Low Noise Amplifier at 60 GHz Using Active Feedback and Current Re-use Topologies. In: *2016 3rd International Conference on Devices, Circuits and Systems (ICDCS)*. IEEE, 161–167.

Oktyabrsky, S., Yakimov, M., Tokranov, V., Kambhampati, R., Bakhru, H., Koveshnikov, S., Tsai, W., Zhu, F., and Lee, J., 2008. Challenges and Progress in III-V MOSFETs for CMOS Circuits. *International Journal of High Speed Electronics and Systems*, 18 (4), 761–772.

Pahwa, G., Salahuddin, S., and Hu, C., 2022. Critical Importance of Nonuniform Polarization and Fringe Field Effects for Scaled Ferroelectric FinFET Memory. *IEEE Transactions on Electron Devices*, 69 (9), 4900–4908.

Park, B., Jin, S., Jeong, D., Kim, J., Cho, Y., Moon, K., and Kim, B., 2016. Highly Linear mm-Wave CMOS Power Amplifier. *IEEE Transactions on Microwave Theory and Techniques*, 64 (12), 4535–4544.

Pascht, A., Fischer, J., and Berroth, M., 2001. A CMOS Low Noise Amplifier at 2.4 GHz with Active Inductor Load. In: *2001 Topical Meeting on Silicon Monolithic Integrated Circuits in RF Systems. Digest of Papers (IEEE Cat. No.01EX496)*. IEEE, 1–5.

Pedersen, E., 2001. RF CMOS Varactors for 2 GHz Applications. *Analog Integrated Circuits and Signal Processing*, 26 (1), 27–36.

Pengelly, R.S., and Turner, J.A., 1976. Monolithic Broadband GaAs f.e.t. Amplifiers. *Electronics Letters*, 12 (10), 251–252.

Pereira, J.M.D., Postolache, O., and Girao, P.M.B.S., 2009. PDF-Based Progressive Polynomial Calibration Method for Smart Sensors Linearization. *IEEE Transactions on Instrumentation and Measurement*, 58 (9), 3245–3252.

Plouchart, J.-O., Zamdmer, N., J. Kim, Trzcinski, R., Narasimha, S., Khare, M., Wagner, L.F., Sweeney, S.L., and Chaloux, S., 2005. A 243-GHz Ft and 208-GHz Fmax, 90-nm SOI CMOS SoC Technology with Low-Power mm-Wave Digital and RF Circuit Capability. *IEEE Transactions on Electron Devices*, 52 (7), 1370–1375.

Poulain, L., Waldhoff, N., Gloria, D., Danneville, F., and Dambrine, G., 2011. Small Signal and HF Noise Performance of 45 nm CMOS Technology in mmW Range. In: *2011 IEEE Radio Frequency Integrated Circuits Symposium*. IEEE, 1–4.

Qian, Y., Shen, Y., and Hu, S., 2023. Millimeter-Wave CMOS Low-Noise Amplifier with High Gain and Compact Footprint. *IEEE Microwave and Wireless Technology Letters*, 33 (6), 699–702.

Radisavljevic, B., Radenovic, A., Brivio, J., Giacometti, V., and Kis, A., 2011. Single-Layer MoS2 Transistors. *Nature Nanotechnology*, 6 (3), 147–150.

Raje, P., 1992. Design and Scaling of BiCMOS Circuits. In: *Proceedings 1992 IEEE International Conference on Computer Design: VLSI in Computers & Processors*. IEEE Comput. Soc. Press, 234–238.

Rappaport, T.S., Sun, S., Mayzus, R., Zhao, H., Azar, Y., Wang, K., Wong, G.N., Schulz, J.K., Samimi, M., and Gutierrez, F., 2013. Millimeter Wave Mobile Communications for 5G Cellular: It Will Work! *IEEE Access*, 1, 335–349.

Razavi, B., 1997. *RF Microelectronics*. Prentice Hall PTR.

Razavi, B., 2006. A 60-GHz CMOS Receiver Front-End. *IEEE Journal of Solid-State Circuits*, 41 (1), 17–22.

Rimmelspacher, J., Werthof, A., Weigel, R., Geiselbrechtinger, A., and Issakov, V., 2019. Experimental Considerations on Accurate fT and fmax Extraction for MOS Transistors Measured up to 110 GHz. In: *2019 92nd ARFTG Microwave Measurement Conference (ARFTG)*. IEEE, 1–4.

Rotella, F., Howard, D., Racanelli, M., and Zampardi, P., 2003. Characterizing and Optimizing High Q Inductors for RFIC Design in Silicon Processes. In: *IEEE Radio Frequency Integrated Circuits (RFIC) Symposium, 2003*. IEEE, 339–342.

Saini, G., and Rana, A.K., 2011. Physical Scaling Limits of FinFET Structure: A Simulation Study. *International Journal of VLSI Design & Communication Systems*, 2 (1), 26–35.

Schwierz, F., and Liou, J.J., 2020. Status and Future Prospects of CMOS Scaling and Moore's Law – A Personal Perspective. In: *2020 IEEE Latin America Electron Devices Conference (LAEDC)*. IEEE, 1–4.

Seifried, M., Villares, G., Baumgartner, Y., Hahn, H., Halter, M., Horst, F., Caimi, D., Caer, C., Sousa, M., Dangel, R.F., Czornomaz, L., and Offrein, B.J., 2018. Monolithically Integrated CMOS-Compatible III–V on Silicon Lasers. *IEEE Journal of Selected Topics in Quantum Electronics*, 24 (6), 1–9.

Sheu, B.J., Scharfetter, D.L., Ko, P.-K., and Jeng, M.-C., 1987. BSIM: Berkeley Short-Channel IGFET Model for MOS Transistors. *IEEE Journal of Solid-State Circuits*, 22 (4), 558–566.

Shi, D., Deng, D., Zhang, S., Wang, H., Chang, S., Huang, Q., and He, J., 2022. An E-Band High-Gain LNA in 65-nm RF CMOS. In: *2022 IEEE MTT-S International Microwave Workshop Series on Advanced Materials and Processes for RF and THz Applications (IMWS-AMP)*. IEEE, 1–3.

Spessot, A., Parvais, B., Rawat, A., Miyaguchi, K., Weckx, P., Jang, D., and Ryckaert, J., 2020. Device Scaling Roadmap and Its Implications for Logic and Analog Platform. In: *2020 IEEE BiCMOS and Compound Semiconductor Integrated Circuits and Technology Symposium (BCICTS)*. IEEE, 1–8.

Srinivasan, P., 2011. Noise and Reliability in Advanced CMOS Devices for Low Power Applications. In: *Proceedings of the 24th Symposium on Integrated Circuits and Systems Design*. ACM, 228–228.

Tarasov, K., 2025. Exclusive Look at the Creation of High NA, ASML's New $400 Million Chipmaking Colossus [online]. *CNBC*.

Taylor, G.W., 1984. Velocity-Saturated Characteristics of Short-Channel MOSFETs. *AT&T Bell Laboratories Technical Journal*, 63 (7), 1325–1404.

Tehrani, R., and Ludeman, L.C., 1993. Signal Processing Using the Generalized Taylor Series Expansion. In: *Proceedings of 27th Asilomar Conference on Signals, Systems and Computers*. IEEE Comput. Soc. Press, 1446–1449.

Tin, S. F., Osman, A., Mayaram, K., and Hu, C., 1998. BSIM3 MOSFET Model Accuracy for RF Circuit Simulation. In: *Proceedings RAWCON 98. 1998 IEEE Radio and Wireless Conference (Cat. No.98EX194)*. IEEE, 351–354.

Tsuchiya, T., Sato, Y., and Tomizawa, M., 1998. Three Mechanisms Determining Short-Channel Effects in Fully-Depleted SOI MOSFETs. *IEEE Transactions on Electron Devices*, 45 (5), 1116–1121.

Watanabe, T., Kitsukawa, G., Kawajiri, Y., Itoh, K., Hori, R., Ouchi, Y., Kawahara, T., and Matsumoto, T., 1989. Comparison of CMOS and BiCMOS 1-Mbit DRAM Performance. *IEEE Journal of Solid-State Circuits*, 24 (3), 771–778.

Weimann, N.G., Stoppel, D., Schukfeh, M.I., Hossain, M., Al-Sawaf, T., Janke, B., Doerner, R., Sinha, S., Schmückle, F.-J., Krüger, O., Krozer, V., Heinrich, W., Lisker, M., Krüger, A., Datsuk, A., Meliani, C., and Tillack, B., 2016. SciFab: A Wafer-Level Heterointegrated InP DHBT/SiGe BiCMOS Foundry Process for mm-Wave Applications. *Physica Status Solidi (a)*, 213 (4), 909–916.

Wenger, C., Oberbeck, L., Schroeder, U., Mussigmussig, H.-J., Lippert, G., Sorge, R., Schroeder, T., Mane, A.U., Lupina, G., Dabrowski, J., Zaumseil, P., and Fan, X., 2006. High-Quality MIM Capacitors for RF Applications. *IEEE Transactions on Electron Devices*, 53 (8), 1937–1939.

Wong, H., 2018. Compact Modeling and Short-Channel Effects of Nanowire MOS Transistors (Invited). In: *2018 IEEE International Conference on Integrated Circuits, Technologies and Applications (ICTA)*. IEEE, 7–10.

Yamasaki, K., and Hirayama, M., 1986. Determination of Effective Saturation Velocity in n+ Self-Aligned GaAs MESFET's with Submicrometer Gate Lengths. *IEEE Transactions on Electron Devices*, 33 (11), 1652–1658.

Yasuda, Y., Liu, T.-J.K., and Hu, C., 2008. Flicker-Noise Impact on Scaling of Mixed-Signal CMOS with HfSiON. *IEEE Transactions on Electron Devices*, 55 (1), 417–422.

Yudistira, D., Van Campenhout, J., and Kunert, B., 2025. Electrically Injected InGaAs/GaAs Nano-Ridge Laser Fully Processed in a 300 mm CMOS Pilot Line. In: A.A. Belyanin and P.M. Smowton, eds. *Novel In-Plane Semiconductor Lasers XXIV*. SPIE, 19.

Zhang, K., Shi, C., Chen, G., Chen, J., and Zhang, R., 2020. A 64.5-88 GHz Coupling-Concerned CMOS LNA with >10 dB Gain and 5 dB Minimum NF. In: *2020 IEEE/MTT-S International Microwave Symposium (IMS)*. IEEE, 337–340.

Zota, C.B., Morf, T., Müller, P., Convertino, C., Filipp, S., Riess, W., and Czornomaz, L., 2019. III-V-on-CMOS Devices and Circuits: Opportunities in Quantum Infrastructure. In: *2019 IEEE International Electron Devices Meeting (IEDM)*. IEEE, 31.8.1–31.8.4.

# 7 Low-Noise Amplifiers in Scaled CMOS

## *Passive Gain-Boosting, Linearity, and System Trade-Offs*

## INTRODUCTION

The design of low-noise amplifiers (LNAs) in modern CMOS technologies has evolved from a largely device-level exercise into a deeply multidisciplinary challenge that spans device physics, electromagnetic modeling, millimeter-wave circuit techniques, and complex system-level optimization. As wireless standards push into multi-gigahertz and millimeter-wave (mm-wave) bands, the requirements placed on the receiver front-end have tightened to the point where traditional single-inductor, source-degenerated topologies can no longer satisfy the competing demands of noise figure (NF), gain, linearity, bandwidth, impedance matching, power consumption, and layout feasibility. At the same time, the relentless progression of CMOS scaling, while enabling unprecedented digital capability, has eroded many of the analogue advantages historically enjoyed by larger-geometry processes. Transistors in advanced nodes exhibit lower intrinsic gain, reduced breakdown voltage, pronounced parasitic coupling, and limited effective mobility at mm-wave frequencies. As a result, high-performance LNAs must increasingly rely on architectural enhancements, magnetic coupling mechanisms, and passive gain-boosting techniques to overcome inherent device limitations.

Transformers and mutual inductive structures have therefore become indispensable elements of contemporary radio-frequency integrated circuit (RFIC) design. Although occupying more silicon area than their single-inductor counterparts, transformer-based topologies provide several unique and powerful benefits that directly address the shortcomings of nanoscale CMOS. By introducing controlled magnetic coupling, designers gain access to impedance transformation that does not dissipate power, current splitting, and combining mechanisms that naturally extend bandwidth, and passive noise shaping that improves the achievable NF without the penalty of increased bias current. These properties allow transformer-based LNAs to achieve noise and gain performance that would be otherwise unattainable within realistic power and area budgets.

The increasing reliance on transformers is also driven by broader system trends. Future wireless networks, spanning 5G FR2, emerging 6G paradigms, Wi-Fi 6/7, ultra-wideband (UWB), and advanced IoT systems, demand LNAs that operate over wide bandwidths while maintaining tight amplitude and phase consistency. Phased arrays

DOI: 10.1201/9781003459385-7

for mm-wave communications and radar applications require dozens to hundreds of LNA elements, all of which must demonstrate robust matching, uniform gain, and stable noise performance across process corners and operating temperatures. In these environments, transformer coupling offers a natural pathway to symmetry, compact routing, and multi-resonant responses that help balance the interconnected demands of bandwidth, NF, and power. A transformer serves not merely as a passive component but as a shaping element that influences the overall architecture of the receiver front-end.

From a historical perspective, transformer-coupled LNAs represent a significant departure from earlier generations of RFIC design. In traditional sub-6 GHz circuits, inductive source degeneration provided a straightforward means of stabilizing the input impedance, reducing the minimum achievable NF, and setting the LNA's gain-bandwidth characteristics. However, these circuits relied heavily on the assumption that device transconductance was plentiful and that inductors provided reasonably high quality factor ($Q$) over the intended operating range. Neither of these assumptions holds at mm-wave frequencies. As operating frequencies climb into tens or hundreds of gigahertz, the parasitic resistances and capacitances associated with on-chip inductors degrade sharply; the $Q$ of spiral inductors collapses; and the transistor's unity-gain frequency ($f_T$) and maximum oscillation frequency ($f_{MAX}$) are approached or exceeded. Under such conditions, every picohenry and femtofarad contributes to gain roll-off, mismatch, noise degradation, or even outright instability.

Transformer structures mitigate many of these issues by leveraging passive gain boosting and impedance scaling. In practical terms, magnetic coupling allows a transistor to see a transformed impedance that improves its noise match without requiring the large transconductance and therefore large bias current that a single-inductor topology would demand. This enables power-efficient LNAs, which are crucial for battery-operated IoT devices, dense mm-wave phased arrays, and power-constrained mobile systems. Furthermore, the presence of two magnetically coupled windings introduces dual resonances, which automatically flatten the frequency response between them. This characteristic is particularly valuable in wideband receivers, where maintaining a consistent NF and input match over a broad spectrum is central to system-level performance.

The practical implementation of transformers in CMOS, however, introduces its own set of challenges that must be addressed through careful design and layout. On-chip transformers are inherently lossy due to finite metal conductivity, substrate coupling, and limited spacing between windings. Their parasitic capacitances lower the self-resonant frequency of each winding, and the finite coupling coefficient introduces leakage inductance that shapes the high-frequency response. Achieving strong coupling is further hampered by the planar nature of integrated circuits, where vertical magnetic flux is constrained by the substrate and limited metal stack height. Nonetheless, advancements in back-end-of-line technologies, including thicker top metals, low-resistance Cu interconnects, patterned ground shields, and multi-layer interleaving, have substantially improved the feasibility and performance of integrated transformers, even in nodes at or below 22 nm.

Because transformer-based LNAs rely on the interaction between two magnetically coupled inductors, their behavior cannot be captured by the simple intuition developed for single inductors or source-degenerated topologies. Designers must consider how the windings interact across frequency, how noise from each winding reflects through the coupling network, and how the secondary-side device influences the primary-side impedance. The addition of a transformer transforms the LNA from a primarily device-level problem into a multi-domain design exercise that spans electromagnetic simulation, circuit-level optimization, layout codesign, and system-level analysis of bandwidth and noise. Despite these complexities, transformer-coupled LNAs consistently deliver exceptional performance across a wide range of wireless applications.

This chapter provides a comprehensive examination of transformer-based and mutual-inductive techniques for noise reduction, gain enhancement, and bandwidth extension in CMOS LNAs. It begins by revisiting the fundamental small-signal interpretation of mutual inductance and shows how coupled inductors alter the input impedance and noise behavior of the amplifier. The chapter then expands on how transformers introduce passive transconductance boosting, enabling higher gain and lower NF simultaneously, even in highly scaled technologies where transistor intrinsic gain is fundamentally limited. Following this, the discussion shifts to practical considerations, including layout, parasitic interactions, and the role of coupling strength in shaping the LNA's operating regime.

A key theme of this chapter is the recognition that transformer-based LNAs achieve a form of performance scaling that is orthogonal to CMOS device scaling. While transistor improvements have slowed, transformer-based techniques provide a powerful path to continued performance growth without relying on unrealistic device metrics. By manipulating magnetic coupling and frequency-dependent impedance shaping, designers can extract significantly more performance from a given device technology than would otherwise be possible. This is particularly important given the future needs of 5G/6G communications, large-scale beamforming arrays, precision sensing systems, and ultralow-power IoT receivers, all of which impose more stringent requirements on LNA performance than previous generations of wireless systems.

Ultimately, the transformer, an element often considered bulky or antiquated in integrated design, has re-emerged as a cornerstone of advanced RFIC design. It provides an elegant and passive means of overcoming limitations that transistor scaling alone cannot solve, while simultaneously enabling circuits that are more power-efficient, wider-band, and lower-noise than their traditional equivalents. The sections that follow formalize these concepts and establish a unified framework for understanding how magnetic coupling can be strategically exploited to meet the increasingly demanding specifications of modern and future wireless systems.

## ADVANCED TECHNIQUES FOR ULTRA-LOW NF

Transformer-coupled LNA architectures have become one of the most powerful techniques for extending bandwidth, reducing noise, and improving gain in modern

CMOS receivers, particularly as technology scaling forces devices to operate closer to their fundamental noise-linearity limits. By introducing mutual inductance, a transformer allows impedance transformation without power dissipation, enabling noise matching at more favorable $g_m$ values, while simultaneously providing current reuse, passive gain boosting, and intrinsic noise cancellation that conventional inductively degenerated CS stages cannot achieve. The presence of two magnetically linked inductors also reshapes the small-signal input impedance and noise transfer characteristics, allowing the designer to trade coupling coefficient, inductance, and device transconductance against NF, gain, and bandwidth in a controllable manner. As the analysis shows, the transformer modifies the input impedance through a reflected inductance term, introduces additional thermal noise sources that must be referred through $M$, and boosts effective transconductance by a factor proportional to $\omega M$, a key result that explains why transformer-based LNAs frequently outperform single-inductor designs despite on-chip losses. Resonant splitting further broadens the passband, enabling multi-GHz front-ends for mm-wave 5G/6G, UWB, imaging, phased arrays, and cryogenic radio astronomy where low NF, wide impedance windows, and device-noise suppression are critical. The remainder of this subsection formalizes these behaviors mathematically, deriving impedance, noise, and $g_m$-boost expressions from first principles, and demonstrating why mutual inductance has become a cornerstone of advanced RF CMOS LNA design.

## Noise-Cancelling Topologies

Noice-cancelling LNA topologies are most often implemented using common-gate/common-source (CG/CS) circuits, or in some cases, dual-stage topologies. Some of the most common techniques to achieve a form of noise-cancelling include current reuse, $g_m$-boosting, and transformer feedback. These techniques cancel out noise from different stages of the amplification and aim to attain wideband performance, high gain, and improved impedance matching while suppressing noise sources like those from input transistors.

The CG/CS is a fundamental approach where a CG amplifier stage is placed in parallel with a CS stage, both of which receive the input signal to be amplified. Through careful design of the gain and the phase of each stage, the noise from the CS stage can destructively interfere with the noise from the CG stage, effectively cancelling it out (Shin *et al.* 2017, Soundarya *et al.* 2024). The CG stage is often used for its wideband input matching capabilities, whereas the CS stage is added to boost the gain further and cancel out the noise of the first stage. The basic CS and CG LNA circuits are commonly used in LNAs, especially in CMOS RF IC designs (Allstot *et al.* 2004). In the CS topology, its superior noise performance is achieved due to the inductive degeneration, which is ideally noiseless, and since the RF input signal is pre-amplified by the input-matching series resonant network. The CG stage can provide wideband input matching primarily because of its inherent lower sensitivity to parasitic capacitances. Figure 7.1 depicts the (a) CG and (b) CS LNA topologies.

As shown in Figure 7.1, the (a) CG and (b) CS LNA stages can be analyzed to determine how each of these stages differs in input matching capabilities, gain, NF,

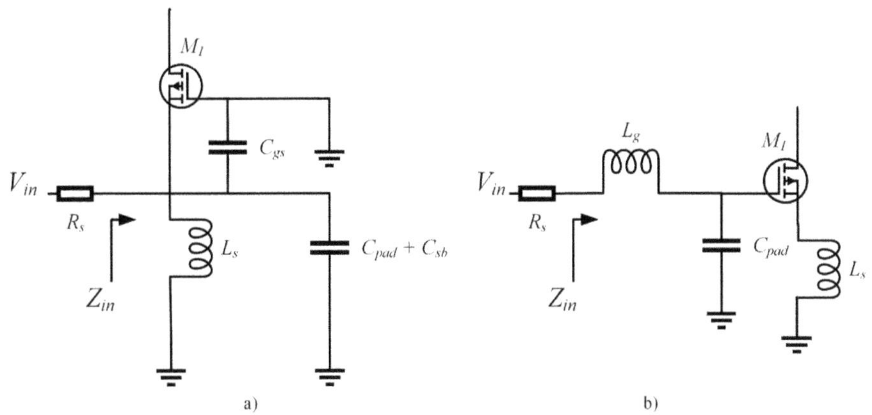


**FIGURE 7.1** The fundamental CMOS (a) CG and (b) CS LNA topologies.

and reverse isolation and stability. These analyses are presented in detail in Allstot *et al.* (2004), and only the fundamental conclusions that relate to the work in this chapter are briefly presented here. Firstly, differences in input matching between the two stages provide insight into how wideband matching is achieved when the stages are used in combination. As is well-known, input impedance matching should be done at specific values, such as 50 Ω, for example, to match the preceding band-select filter stage. The LNA should be designed to add the minimum amount of noise in the RF signal path while maintaining and synthesizing the specified input impedance. For the CG stage in Figure 7.1a, the input impedance can be simplified to approximately

$$Z_{in} = \frac{1}{g_{m1}} \tag{7.1}$$

where $g_{ml}$ is the transconductance of the MOSFET transistor, *M1*. However, the input impedance of the CS stage in Figure 7-1b cannot be simplified to the same extent, and is written as

$$Z_{in} = s\left(L_g + L_s\right) + \frac{1}{sC_{gs}} + \left(\frac{g_{m1}}{C_{gs}}\right)L_s \tag{7.2}$$

where $L_g$ is the inductance on the gate of $M_1$, $L_s$ is the degeneration inductance on the source of $M_1$, and $C_{gs}$ is the intrinsic gate-source capacitance of the same transistor. This representation of the input impedance of the CS stage can be slightly simplified to

$$Z_{in} \approx s\left(L_g + L_s\right) + \frac{1}{sC_{gs}} + \omega_T L_s \tag{7.3}$$

and this representation helps to see that the values of the gate and source inductors should be chosen to resonate with the gate-source capacitance at the operating frequency, set to 50 Ω, at $\omega_T$. According to Allstot *et al.* (2004) the fundamental difference between the input matching networks is that the CS stage is series resonant, whereas the CG stage is parallel resonant. As a result, the quality factors of these two stages exhibit some fundamental differences, as can be indicated through their equations, being

$$Q_{CS} = \frac{1}{2\omega_0 C_{gs} R_s} > 1 \tag{7.4}$$

where $\omega_0$ is the resonant frequency, and

$$Q_{CG} = \frac{\omega_0 C_{gs} R_s}{2} < 1 \tag{7.5}$$

Furthermore, the sensitivity of the input impedance to parasitic components is proportional to the quality factor of the matching network, according to Huang *et al.* (1998), which leads to the conclusion that the CG LNA stage, which inherently has a lower $Q$ parallel resonant network, is more robust against variations that arise from the production process, temperature gradients, and voltage instabilities. The parasitic capacitance at the input of Figure 7.1a can also be naturally absorbed into the LC tank as seen from its input.

The analysis of these two stages, as presented in Allstot *et al.* (2004) follows a similarly effective strategy to highlight the differences in the gain performance of the two stages. It does so by first presenting the (effective, so $G_m$) transconductance of the CS stage, in Figure 7.1b, as being

$$G_{m,CS} = Qg_{m1} \tag{7.6}$$

which, from its small-signal equivalent circuit, can be derived as

$$G_{m,CS} = \frac{g_{m1}}{\omega_0 C_{gs}\left(R_S + \omega_T L_s\right)} \tag{7.7}$$

such that

$$G_{m,CS} = \frac{\omega_T}{\omega_0 R_S\left(1 + \frac{\omega_T L_S}{R_S}\right)} \tag{7.8}$$

and, importantly, if the input is matched to $R_S$, it follows that

$$G_{m,CS} = \frac{1}{2R_S}\left(\frac{\omega_T}{\omega_0}\right) \tag{7.9}$$

it shows that the effective transconductance (if matched to $R_S$) is only a function, or ratio, of $\omega_T$ and $\omega_0$. This means that the value is effectively independent of the small-signal transconductance of the MOSFET, $M_1$. Under ideal and perfect matching conditions, the CG LNA shows a different result, and its effective transconductance is derived to be

$$G_{m,CG} = \frac{1}{2} g_{m1} \tag{7.10}$$

or

$$G_{m,CG} = \frac{1}{2R_S} \tag{7.11}$$

Since the value of $\omega_T/\omega_0$ is typically in the range of 5–10 (Allstot *et al.* 2004), it shows that the gain of the CS stage is effectively higher by this factor, compared to that of the CG stage.

The NF of the CG/CS also has advantages due to the topology itself, taking the best features of each stage and combining them towards a well-performing LNA. The CS amplifier with its inductive degeneration has a major advantage in that its resistive input impedance is theoretically noiseless. This differs from other topologies, where there is often a noisy resistor placed in the signal path to create a 50 Ω impedance and the terminating node. The small-signal equivalent circuit analysis of a CS LNA topology can be used to show that, with inductive degeneration, the noise factor, $F$, of the topology is derived from

$$F = \frac{\overline{i^2_{n,out}}}{\overline{i^2_{R_s}}} \tag{7.12}$$

which, from the small-signal equivalent, is shown to be

$$F = 1 + \frac{\gamma}{\alpha}\frac{1}{Q}\left(\frac{\omega_0}{\omega_T}\right)\left[1 + \frac{\delta\alpha^2}{5\gamma}\left(1+Q^2\right) + 2|c|\sqrt{\frac{\delta\alpha^2}{5\gamma}}\right] \tag{7.13}$$

and is a relatively complex derivation, and it must also be considered that

$$\overline{i_{R_s}^2} = \frac{4kT\Delta f}{R_S} \tag{7.14}$$

and

$$\overline{i_d^2} = 4kT\gamma g_{d0}\Delta f \tag{7.15}$$

and

$$\overline{i_g^2} = 4kT\gamma g_g\Delta f \tag{7.16}$$

with the quality factor defined by

$$Q = \frac{1}{\omega_0 C_{gs} R_S} \tag{7.17}$$

and

$$\overline{i_g i_d^*} = c\sqrt{\overline{i_g^2 i_d^2}} \tag{7.18}$$

and $c$ is equal to $j0.395$, where $\alpha$, $\gamma$, and $\delta$ are bias-dependent parameters that are defined in Yuhua Cheng *et al.* (2002), and $\omega_0$ and $\omega_t$ are the operating and unity current gain frequencies, respectively, as also used and defined in earlier equations. Although relatively complex derivations, these equations can effectively be used to show the impact of the quality factor, $Q$, on the NF in the LNA if the CS topology is implemented. From (7.13), the NF comprises three factors being channel noise, gate noise, and correlated noise. The contribution of the channel noise can be reduced simply by increasing the $Q$, but this is a passive component-specific limitation. In direct contrast, though, the gate noise is enhanced by the same quality factor. As a result, an optimum $Q$ for a given overdrive voltage and $f_T$ must be found to minimize the NF, and these values have been determined, provided in Allstot *et al.* (2004), and represented here as

$$F_{min,CS} = 1 + \frac{\gamma}{\alpha}\left(\frac{\omega_0}{\omega_T}\right)\frac{2\delta\alpha^2}{5\gamma}Q_{opt} \tag{7.19}$$

and this optimum $Q$ can be determined by

$$Q_{opt} = \sqrt{1 + 2|c|\sqrt{\frac{5\gamma}{\delta\alpha^2}} + \frac{5\gamma}{\delta\alpha^2}} \tag{7.20}$$

To achieve a high transit frequency, $f_T$, the most common method is to use the minimum gate length of the technology node. As a result, the optimum transistor width can be calculated if the optimum quality factor for the minimum NF is known. The CG topology noise factor is

$$F_{CG} = 1 + \frac{\gamma}{\alpha} \tag{7.21}$$

which shows that the value is constant with respect to the transit and unity gain frequencies, in contrast to the linear proportionality to these values of the CS stage. According to Allstot *et al.* (2004), the reverse isolation and stability of the CS topology are degraded due to the feed-forward path between its input and output, created by the presence of $C_{gd}$. Conversely, the CG topology does not exhibit this Miller effect due to $C_{gd}$ and as a result offers better reverse isolation and stability (and input matching).

Another topology that offers noise cancellation is the current conveyor (CC) (Smith and Sedra 1968), introduced in 1968, where current conveyors are implemented to cancel out thermal noise at the input of the MOSFET while essentially acting as a voltage-to-current converter. The CC, like a traditional operational amplifier (op-amp), uses current as the primary active parameter, as opposed to voltage, as most op-amps do. The CC allows for high bandwidth as well as accuracy by minimizing voltage variations at sensitive nodes. Importantly, it is a three-terminal device having two input terminals (*X* and *Y*) and one output terminal (*Z*). It performs a unity-gain voltage transfer from *Y* to X, which mathematically relates to $V_X = V_Y$. It then transfers the current from X to Z, with the output current being equal to the input current at *X*; therefore, $I_Z = I_X$. It can be described by the following hybrid equation

$$\begin{bmatrix} i_Y \\ v_x \\ i_z \end{bmatrix} = \begin{bmatrix} 0 & a & 0 \\ 1 & 0 & 0 \\ 0 & \pm 1 & 0 \end{bmatrix} \begin{bmatrix} v_Y \\ i_x \\ v_z \end{bmatrix} \tag{7.22}$$

where the symbols indicate instantaneous quantities. If $a = 1$, it represents the first-generation conveyor, denoted as CCI, whereas if $a = 0$, it represents the second-generation conveyor, denoted CCII. The plus sign in the equation for $i_Z$ represents a positive current conveyor, denoted CCI+ or CCII+ (the output current is in phase with the input current), whereas a minus sign represents a negative current conveyor, so CCI- or CCII- (the output current and the input current are 180° out of phase). The second-generation conveyor CCII is the most popular. The CC exhibits a high input impedance at terminal *X* and a low impedance at terminal *Z*. It is usually made up of current mirrors and voltage followers (Vanaja and Tantry 2021), and it is a fundamental design building block of low-noise op-amps, as described earlier. There is already a significant body of work that focuses on implementing these circuits in scaled CMOS technologies, specifically for low-voltage and low-power applications. In modern ICs, the CC still provides several advantages over the traditional op-amp topologies, including

- higher bandwidth with improved gain-bandwidth products and slew rates,
- higher dynamic range and better linearity, especially in ICs with varying line lengths,
- reduced complexity in circuit design with fewer external passive components, and
- lower voltage operation as supply voltage typically decreases as CMOS nodes scale down.

Another beneficial factor of the CC is that it can be implemented in both BJT and CMOS technologies, with the original presentation in Smith and Sedra (1968) given with bipolar transistors. This makes the circuit ideal for use in hybrid technologies such as SiGe BiCMOS. As mentioned earlier in this section, these circuits can cancel out the thermal noise at the MOSFET, the primary contributor of noise to the circuit. In Diaz-Chinea *et al.* (2015) this noise-cancelling technique is described based on the circuit-level implementation. The noise-cancelling technique is based on the circuit where the transistor is in a CG configuration, where the noise generated in this MOSFET is represented as a current source, which produces a voltage at its source and a fully correlated voltage at its drain, in antiphase. The signals pass through CS inverting stages, which cancel out the noise. The input signal is amplified via two separate paths: first, an inverting CG–CS path, and secondly, an inverting CS path, so the signals are effectively summed at the output. To achieve perfect noise cancelling, both these paths should present the same, matched gain. The transistors in these paths, therefore, need to be biased independently. These circuit elements can be integrated into an LNA design, which combines the CC-based LNA and the noise-cancelling technique. The CC-based LNA, as an input stage, provides a degree of freedom in satisfying the input matching condition compared to a simple CG stage (Diaz-Chinea *et al.* 2015).

Another noise-cancelling LNA topology, or strategy, is current-reuse $g_m$-boosting, a form of CG amplifier (Walling *et al.* 2007), which, compared to the CS amplifier, uses less current and area, primarily due to the reduced number of inductors. There are also other techniques to perform transconductance boosting or making the transistors act as if they have a higher transconductance, even at a constant bias current. These include capacitor cross-coupling and transformer-based $g_m$-boosting. The gain of the current-reuse $g_m$ boosting circuit gain is somewhat limited due to the finite quality factor of the on-chip CMOS inductor, which may be a limitation in some applications (Rafati *et al.* 2019), but its effective noise is lower when compared to traditional implementations. The $g_m$-boosting topology is presented in Figure 7.2.

In the topology presented in Figure 7.2, the transistor gate terminal is not connected to a DC bias voltage, but an inverting amplification is introduced between the source and the gate nodes of the MOSFET. As a result, the effective transconductance, when looking into the source terminal, is boosted from $G_m \approx g_{m1}$ to $G_m = (1 + A)g_{m1}$. In this boosted amplification, $A$ is the gain that exists between the source and the gate.

In Allstot *et al.* (2004), it is then shown that the noise factor of this circuit strategy becomes

$$F = 1 + \frac{\gamma}{\alpha} \frac{1}{(1+A)^2 g_{m1} R_S} \tag{7.23}$$

which is slightly different from the traditional CG LNA, where

$$F = 1 + \frac{\gamma}{\alpha} \frac{1}{g_{m1} R_S} \tag{7.24}$$

by the factor $(1 + A)_2$ in the denominator. Thus, input matching would require that

$$(1+A) g_{m1} = \frac{1}{R_S} \tag{7.25}$$

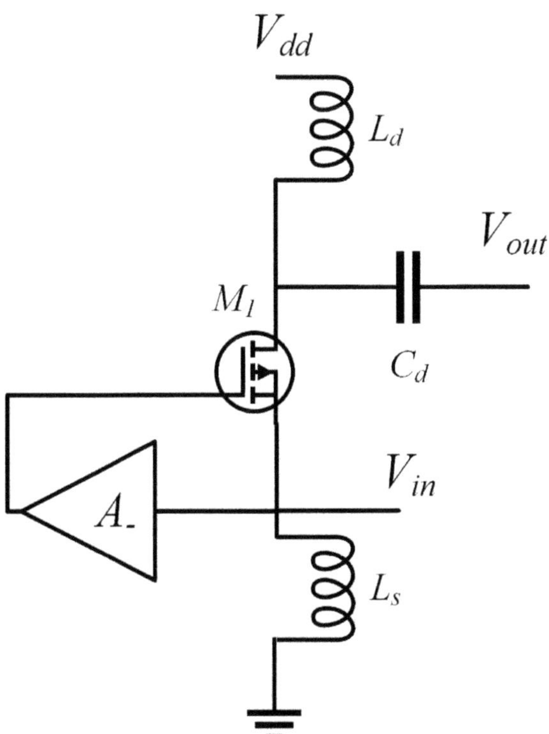


**FIGURE 7.2** A common-gate LNA that implements $g_m$-boosting.

which would result in a noise factor of

$$F = 1 + \frac{\gamma}{\alpha}\frac{1}{1+A} \tag{7.26}$$

which clearly shows a reduction in noise factor by a factor of $1 + A$. The gm-boosting LNA is particularly well-suited for mm-wave CMOS implementations due to its low noise contribution, as is presented in Kobal *et al.* (2022) where the physical layout of the transistors takes advantage of the triple-well geometry that can be found in a variety of CMOS processes. This process adds a third well (typically a deep *n*-well) to the standard dual-well process, and this isolates the transistor from the *p*-type substrate. This isolation offers further advantages and enhancements for the configuration, such as improved noise immunity for analogue RF circuits, better latch-up immunity, and an ability to create and operate isolated high-voltage devices.

While the techniques discussed in this section focus on minimizing the noise contribution of the LNA and improving sensitivity, an LNA must also maintain adequate signal integrity under large-signal excitation. In modern CMOS and BiCMOS implementations, the aggressive scaling of device geometries increases the intrinsic transconductance efficiency but simultaneously introduces stronger nonlinearities in the drain current and worsens voltage headroom limitations. As a result, although noise-optimized topologies such as CS, CG, or noise-cancelling structures can achieve state-of-the-art NFs, their linearity performance often degrades due to the higher sensitivity to device overdrive and the inherent cubic nonlinearity of the MOSFET's transconductance characteristic. Therefore, the next design challenge is not only to minimize noise but to ensure that the amplifier maintains a sufficiently high third-order intercept point (IIP3) and compression characteristics, such as the 1 dB compression point ($P_{1dB}$), without compromising gain or noise. The following section introduces several linearity-enhancement techniques applicable to scaled CMOS LNA design, focusing on derivative-superposition, $g_m$ linearisation, and active feedback methods.

## Linearity Enhancement Techniques in Scaled CMOS LNAs

The linearity of an LNA determines its ability to handle large or closely spaced input signals without generating intermodulation distortion or compression. In scaled CMOS and BiCMOS processes, where supply voltages are reduced, and device transconductance characteristics become increasingly nonlinear, maintaining linearity has become a critical design objective alongside noise and gain optimization. Nonlinear distortion in the LNA arises primarily from the higher-order terms in the device transconductance function, which produce unwanted harmonics and intermodulation products when multiple tones are present at the input. The third-order distortion, originating from the cubic term of the drain current expansion, is typically the dominant contributor and is quantified by the IIP3. Since the mechanisms that improve noise performance, such as higher transconductance or strong inversion biasing, often exacerbate nonlinearity, the designer must employ dedicated linearization techniques to achieve an acceptable tradeoff. Several circuit-level approaches have been developed to mitigate these nonlinear effects, including derivative superposition (DS), feedback linearization, capacitive cross-coupling, and adaptive biasing strategies, which are discussed in the following subsections.

To quantify these enhancements, the IIP3 and $P_{1dB}$ are briefly redefined. Firstly, the IIP3 is a figure of merit that is critically important to define in scaled, high-frequency, CMOS circuits (Aya *et al.* 2009), especially amplifiers and mixers. To design an ultra-low noise LNA, the input power of the third-order intermodulation distortion product should theoretically and hypothetically be equal to the fundamental output signal. If the IIP3 is higher, it indicates that the device, or amplifier, is more linear and produces less distortion when amplifying multiple signals at once. The nonlinearities of a MOSFET are due to its $i_d - v_{gs}$ relationship being nonlinear. This nonlinearity, which is essentially directly related to its transconductance $g_m$, results in higher-order distortion components as the amplitude of the input signal increases. This is because $g_m$ is a nonlinear function of the bias point and the instantaneous gate voltage. The nonlinearities are mathematically described when considering the Taylor series around the bias point, which is given by

$$i_d = g_m v_{gs} + \frac{1}{2} g_{m2} v_{gs}^2 + \frac{1}{6} g_{m3} v_{gs}^3 + \ldots \tag{7.27}$$

where

$$g_m = \frac{\partial i_d}{\partial v_{gs}} \mid V_{GS} = V_{GS0} \tag{7.28}$$

and

$$g_{m2} = \frac{\partial^2 i_d}{\partial v_{gs}^2} \mid V_{GS} = V_{GS0} \tag{7.29}$$

and

$$g_{m3} = \frac{\partial^3 i_d}{\partial v_{gs}^3} \mid V_{GS} = V_{GS0} \tag{7.30}$$

where the first term in (27) corresponds to the fundamental component of the drain current, which effectively defines the small-signal gain of the transistor. The second term in this equation produces even-order distortion components, and these are typically rejected if a differential architecture is used. The third term in this equation generates odd-order distortion components that are relatively close to the fundamental frequency and manifest as in-band distortion in RF applications. For this reason, the third-order intercept point, or IIP3 as previously defined, is primarily defined by this dominant factor, $g_{m3}$. The third-order distortion product amplitude is proportional to the transconductance and gate-source voltage by

$$amplitude \propto g_{m3} v_{gs}^3 \tag{7.31}$$

while the fundamental tone directly scales with $g_m v_{gs}$. As the input amplitude to the amplifier is increased, the third-order term also increases, but much faster than the linear term, resulting in a gradual degradation of the signal. To determine the IIP3, the linear fundamental response and the cubic distortion response are extrapolated on a log-log scale, whereas the intersection defines the IIP3. Importantly, this intersection typically does not occur in the normal operating region of the LNA; it does, however, provide valuable insight into how rapidly distortion increases with input power. Essentially, a higher IIP3 indicates that the nonlinearities are smaller than the fundamental term, and the active device can accommodate a larger input dynamic range before the distortion becomes too prominent. In contrast, the 1dB compression point characterizes large-signal compression under single-tone excitation. This means that as the input amplitude increases, $g_m$ will gradually shift away from its small-signal value. This causes the output voltage (or current) to deviate from the expected linear gain. The $P_{1db}$ point is easily defined as being the input power level where the small-signal gain decreases by 1 dB. This figure of merit, therefore, defines the gain compression due to saturation of the transconductance. Together, these two figures of merit provide a comprehensive indication of the amplifier's linearity and large-signal robustness. In advanced CMOS nodes, both parameters tend to degrade due to reduced voltage headroom and stronger short-channel effects, making linearization strategies such as derivative-superposition essential for maintaining performance.

In deep-submicron CMOS technologies, there are several effects that are caused by the scaling itself, which effectively increase (or degrade) the linearity of the active devices. Furthermore, the reduced supply voltage limits the swing of the drain source and causes the device to enter the triode region earlier. This is especially noticeable under large signal conditions. The higher electric fields, especially in short-channel devices, increase mobility degradation and velocity saturation, both of which distort the transconductance of the MOSFET. Although technology scaling is effective in improving or increasing the $f_T$ and $f_{max}$, this same enhancement is accompanied by higher-order derivatives, and this worsens the large-signal linearity. CMOS scaling, therefore, requires additional circuit-level linearization techniques that compensate for the nonlinear transconductance terms. A common method to achieve this is implementing the derivative-superposition method.

DS (Gao *et al.* 2015, Mecwan and Devashrayee 2017) in LNAs involves adding an auxiliary amplifier path in parallel with the main amplifier path. The derivative word comes from the mathematical principle that, by controlling the bias current (and other parameters of the auxiliary path), its contribution can be made to effectively cancel the derivative of the distortion in the main path to improve overall linearity. DS remains one of the most used and widely adopted linearization techniques in CMOS LNAs; this is primarily because it addresses linearity at the source, which is the nonlinear dependence of the transistor drain current on the instantaneous gate voltage. Through its deliberate combination of transistor devices that are biased at different operating points, it is possible to exploit the higher-order derivatives of the drain current and gate-source voltage characteristics and the fact that these have different polarization and amplitudes across the entire bias range. If the device sizes are carefully chosen and the bias voltages allow for partial cancellation of the undesired second- and third-order derivatives, both the IIP3 and $P_{1db}$ can be improved without large penalties in NF or power consumption.

Although originally introduced in early RF CMOS literature as a relatively simple linearisation approach, DS has since evolved into multiple forms and is now routinely incorporated into both narrowband and broadband LNA architectures. Modern variations exploit changes in $g_{m3}$ curvature across bias, multi-branch compensation, body-bias manipulation, and adaptive-bias operation. Regardless of the implementation, the underlying physical concept remains consistent: identify where the nonlinear transconductance derivatives change sign or magnitude, and design a composite device such that the composite third-order term is reduced.

At its core, the first-order transconductance ($g_m$) adds constructively, and the higher-order derivatives, especially $g_{m3}$ (which dominates the third-order distortion), can be forced to cancel if multiple devices are summed in parallel. From the Taylor series expansion in (27), the expression becomes even more useful if it is recognized that the same expansion applies to every device that is placed in the parallel network. Therefore, if we consider, for example, two active devices, $M_1$ and $M_2$, each having its own bias point and carefully selected geometry, the total drain current for this network can be expressed as

$$i_{d,tot} = \left(g_{m1} + g_{m2}\right)v_{gs} + \frac{1}{2}\left(g_{m2,1} + g_{m2,2}\right)v_{gs}^2 + \frac{1}{6}\left(g_{m3,1} + g_{m3,2}\right)v_{gs}^3 + \dots \quad (7.32)$$

and the DS condition is conceptually quite simple, whereas

$$g_{m3,1} + g_{m3,2} \approx 0 \quad (7.33)$$

and this will reduce the cubic composite term and, as a result, increase the IIP3. It also yields a significant reduction in the odd-ordered distortion terms, and because $g_{m3}$ changes its sign across the transition from weak inversion to moderate inversion, the appropriate choice of $V_{GS}$ and device size allows cancellation while maintaining a reasonably high combined $g_m$.

If the fundamental output amplitude is plotted as a function of the input power, on a logarithmic scale, the cubic term will grow at three times the slope of the linear term, as can be expected. If these lines are extrapolated, and the intercept point is determined, this point corresponds to a hypothetical input level where the cubic distortion is equal to the extrapolated linear output power. Although this interception is hypothetical and not realistic, its location still provides a comparative measure of how quickly distortion will increase as the input amplitude grows. At the same time, the onset of gain compression is closely linked to the deviation of $g_m$ from its small-signal value. If the swing of the input becomes large, the device will enter a regime where the incremental gain diminishes due to mobility reduction, channel-length modulation, and velocity saturation. Therefore, $P_{1dB}$ reflects how early these large-signal nonlinearities start showing up and increasing. Techniques that flatten the effective $g_m(v_{gs})$ curve usually improve both IIP3 and $P_{1dB}$; however, the improvements in IIP3 are typically more pronounced. Figure 7.3 is a circuit-level representation of the two-branch DS LNA topology.

From Figure 7.3, the primary transistor, $M_1$, is biased for optimal noise and gain, whereas the auxiliary transistor, $M_2$, is biased with a lower gate voltage, so a lower current density. Both transistors are driven by the same RF input signal, and their drain currents are summed to form the effective transconductance. $M_1$ usually operates near the peak $g_m/I_D$ point to achieve its low NF, while $M_2$ is biased near the region where its cubic nonlinearities exhibit the opposite curvature. Since the auxiliary branch can be significantly smaller than the main device, DS improves linearity with minimal impact on noise contribution, input capacitance, and power. Advantages of this topology are that it is relatively easy to implement, it exhibits robust operation, and it is compatible with inductive-source-degenerated LNAs, common-gate LNAs, as well as distributed broadband designs.

Even though this DS topology is conceptually easy to implement and it does not increase component count drastically, the practical implementation of these circuits does require a form of empirical tuning. In most CMOS PDKs, the $g_{m3}$ term is not directly provided as a model parameter, but the curvature of $i_d(v_{gs})$ can be extracted using periodic steady-state (PSS) as well as periodic AC (PAC) simulation, or through

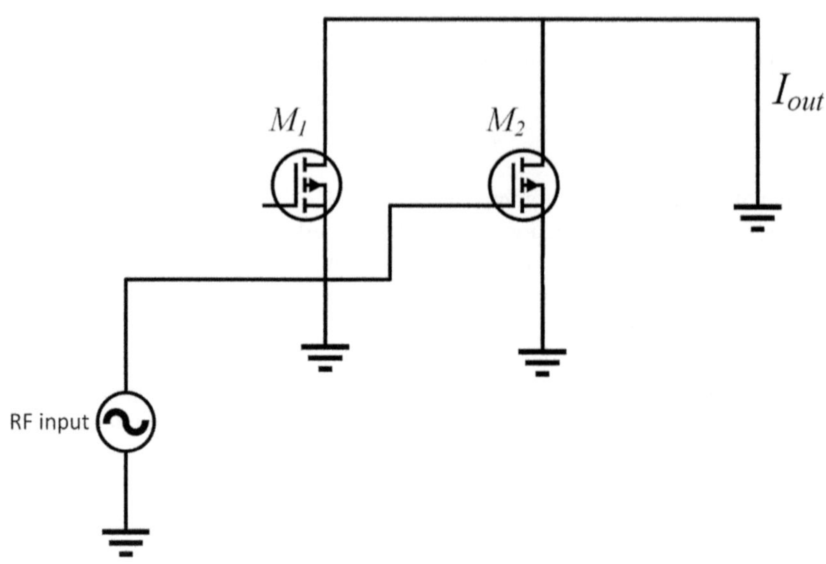


**FIGURE 7.3** A two-branch DS LNA topology.

DC sweeps of the IV characteristics. From (33), a convenient sizing rule emerges, such that

$$\frac{W_2}{W_1} \approx \left| \frac{g_{m3,1}}{g_{m3,2}} \right| \tag{7.34}$$

and since $g_{m3,2}$ at the lower $V_{GS}$ tends to be smaller in magnitude, it follows that the auxiliary device width, $W_2$, is typically around 5% to 30% smaller than the transistor in the main branch. Table 7.1 illustrates example operating points to achieve DS cancellation in a two-branch topology.

Table 7.1 summarizes the key device parameters used in the classical two-branch derivative-superposition (DS) structure. The main transistor, $M_1$, is biased close to the optimum point for minimum NF and maximum transconductance, typically with a gate–source voltage in the range of 0.7–1.0 V. This biasing also results in a relatively high current density (≈ 0.40 mA/μm), producing a large small-signal transconductance $g_m$ and a correspondingly positive third-order transconductance coefficient $g_{m_3}$.

In contrast, the auxiliary transistor, $M_2$, operates at a significantly reduced $V_{GS}$, generally between 0.5 and 0.75 V, placing it in weak-to-moderate inversion. As reflected in the table, this biasing leads to a substantially lower current density (≈ 0.10 mA/μm) and therefore a smaller fundamental transconductance. Crucially, under these conditions, $M_2$ exhibits a negative $g_{m_3}$, which is the defining requirement for third-order distortion cancellation when its drain current is summed with that of $M_1$. The combination of a high-gain, noise-optimised main branch and a low-gain, nonlinearity-shaping auxiliary branch enables the DS technique to maintain the noise and gain performance of $M_1$, while selectively reducing the third-order distortion that limits IIP3.

The final row reiterates this functional partitioning: $M_1$ is responsible for the LNA's noise and gain performance, while $M_2$ contributes primarily to linearity correction, with minimal impact on the overall transconductance or NF when properly scaled and biased.

**TABLE 7.1**
**Example operating points to achieve DS cancellation in a two-branch topology**

| Parameter | $M_1$ (main device) | $M_2$ (auxiliary device) | Comments |
|---|---|---|---|
| $V_{GS}$ | ~ 0.7 – 1.0 V | ~ 0.5 – 0.75 V | Auxiliary in weak to moderate inversion |
| $I_D/W$ | ~ 0.40 mA/μm | ~0.10 mA/μm | Lower $g_m$ density for $M_2$ |
| $g_m$ | High | Low | Summed for gain |
| $g_{m3}$ | Positive | Negative | Enables cancellation |
| Function | NF + gain | Linearity correction | - |

To improve the intermodulation performance, the effective cubic coefficient of the DS is reduced, and the relationship between the composite cubic term and the IIP3 can be approximated as

$$IIP3 \propto \frac{g_m}{\left|g_{m3,eff}\right|} \tag{7.35}$$

and when $g_{m3,eff}$ is reduced by a factor of between 2 and 5, a typical improvement of around 6–12 dB can be expected in practice. It is, however, also possible to implement more elaborate DS structures using three active devices or body-bias modulation, and these implementations can potentially lead to improvements of over 15 dB. However, this improvement does come at the cost of increased complexity, additional power usage, increased area on chip, and complexity in matching. A similar simplification can be performed on the 1 dB compression point since the DS also tends to flatten the transconductance characteristics. A qualitative relationship between the 1 dB compression point and the transconductance is

$$P_{1dB} \propto \frac{1}{\left|g_{m2}\right| + \left|g_{m3}\right|} \tag{7.36}$$

and a DS added to a conventional inductively degenerated common-source LNA typically yields an improved $P_{1dB}$. As with many circuits, the DS also has several variations, or extensions, each with different performance goals based on the required application. These variations include the following:

- the multi-branch DS (such as a three-device topology), which can suppress both the third- and fifth-order nonlinearities,
- DS implemented using body-bias modulation, which can be performed in CMOS processes with isolated wells (such as SOI or bulk processes) where a varied body bias is applied to the main and auxiliary devices to shift their effective threshold voltages. This gives more control and a finer resolution to tuning of $g_{m3}$, which can also simplify the bias network,
- distributed DS where the DS is distributed across multiple input fingers and each finger can be biased at a slightly different operating point. This method effectively spreads the input capacitance to reduce peaking in the frequency response, and
- adaptive (sometimes referred to as dynamic) DS, which only enables the auxiliary device to be activated if the operating conditions require high linearity.

To indicate the advantages and limitations of the DS topology, Table 7.2 lists the applicable parameters that are influenced by this design.

DS is a powerful technique for improving LNA linearity by directly shaping the nonlinear transconductance profile of the transistor. By combining devices biased at different operating points, higher-order nonlinearities, especially the cubic term

**TABLE 7.2**
**Summary of the DS topology advantages and limitations**

| Figure of merit | Advantage | Limitation |
|---|---|---|
| IIP3 improvements | Can be in the 10s of dB | Requires accurate bias control |
| P1dB improvement | Approximately 1 to 5 dB | Less pronounced than IIP3 |
| Noise figure | Minimal degradation | Auxiliary devices contribute some noise |
| Power consumption | Slightly increased | An additional bias network is needed |
| Implementation | Not complex and compact | Sensitive to parasitics and temperature variations |
| Scalability | Easy to scale | Complexity of input capacitance |

responsible for IIP3 degradation, can be significantly reduced. DS offers a compelling balance among linearity performance, power consumption, and design simplicity, making it particularly well suited for scaled CMOS technologies where supply voltage constraints and short-channel effects exacerbate intrinsic nonlinearity.

With careful bias optimization, device sizing, and layout planning, DS provides measurable improvements in both IIP3 and $P_{1dB}$ without compromising NF. Its versatility extends across narrowband, broadband, and even mm-wave LNAs, and multiple architectural variants provide designers with flexibility among complexity, linearity, and bandwidth. In modern RF front-ends where coexistence with strong blockers is essential, DS remains a foundational linearization tool and a natural complement to noise-cancelling and gain-boosting topologies explored earlier in this chapter.

## Transformer-Based and Mutual-Inductive Techniques for Noise and Bandwidth Enhancement

Transformer-based LNA topologies have emerged as one of the most effective ways to simultaneously improve noise performance, increase bandwidth, stabilize input matching, and enable scalable design across a wide range of CMOS technology nodes. Unlike single inductors, which provide a one-dimensional reactive impedance, transformers introduce mutual inductance, current splitting, and voltage step-up/step-down capabilities. These properties can be exploited to lower the effective noise contribution of the input device, shape the frequency response of the amplifier, and relax several biasing and matching constraints that would otherwise limit performance at mm-wave and sub-6 GHz frequencies.

At a fundamental level, an on-chip transformer is simply a pair of inductors with a controlled degree of magnetic coupling. However, the introduction of the coupling term significantly alters the small-signal model of the LNA input stage. The impedance seen at the primary is no longer just $j\omega L_1$, but a transformed impedance proportional to the square of the turns ratio and the coupling coefficient $k$. As a result, transformers provide impedance scaling without dissipating power, enabling noise matching at a more favorable device transconductance. This is particularly useful in advanced CMOS nodes where channel-length scaling forces devices to operate at high current densities to reach the required $g_m$, thereby degrading both noise and linearity.

Another compelling advantage of transformer-coupled structures is that they naturally implement current reuse and noise cancellation. When the RF input signal is applied to the primary, the induced current in the secondary can either be added to or subtracted from the main device current, depending on orientation and coupling polarity. This allows transformer-based LNAs to realize a form of intrinsic noise shaping: the noise of the input transistor can be partially cancelled through magnetic coupling, while the signal is constructively reinforced. The result is a lower minimum NF than what the same transistor could achieve in a traditional inductively degenerated CS stage.

In addition to noise benefits, on-chip transformers are powerful tools for bandwidth enhancement. In multi-resonant LNA structures, the primary and secondary inductances interact to form dual-pole or complex-conjugate zero responses that broaden the usable frequency range. This has made transformer-based LNAs a dominant choice for wideband 5G/6G front ends, mm-wave phased arrays, and broadband low-noise receivers in imaging and sensing applications. At high frequencies, these techniques compensate for the decreasing $Q$ of on-chip inductors in scaled technologies by enabling broader and more controlled impedance environments.

On the layout side, improvements in back-end-of-line (BEOL) stack thickness, patterned ground shields, and low-loss top metals have made it possible to integrate transformers even in highly scaled nodes such as 22 nm and below. Although the achievable coupling coefficient $k$ remains modest on silicon (typically 0.5–0.8), careful geometry selection, stacked windings, interleaved structures, and symmetric layouts can significantly improve performance. Designers also exploit center-tapped primaries, floating secondaries, and autotransformer configurations to enable differential operation, additional current reuse, or low-power biasing schemes.

In summary, transformer-based techniques form an essential class of noise- and bandwidth-enhancement strategies in modern CMOS LNAs. Their ability to transform impedances, enable noise cancellation, increase gain, widen bandwidth, and reduce NF without incurring significant power overhead makes them uniquely suited to the demands of modern wireless, sensing, and cryogenic applications. The following subsections formalize the small-signal models, derive the noise and gain transformations, and present example topologies illustrating how mutual inductance is used in practice.

It is useful to first derive and interpret the small-signal model of a transformer in CMOS LNAs (Zhang *et al.* 2022), both the primary and secondary equivalent impedances, to better understand how impedance scaling reduces the NF. LNAs that are transformer-coupled rely on the interaction between two linked inductors (magnetic) to

- manipulate the input impedance,
- generate signal gain, and
- shape the noise contribution of the (input) transistor.

It is therefore a key requirement to review the small-signal model of the transformer as it is typically presented as a black box, but it often outperforms inductively degenerated CS stages, both in terms of noise and bandwidth. To derive this model, consider two individual inductors that are placed in proximity of each other and therefore exhibit a mutual inductance of

$$v_1 = j\omega L_1 i_1 + j\omega M i_2 \tag{7.37}$$

and

$$v_2 = j\omega L_2 i_2 + j\omega M i_1 \tag{7.38}$$

where $L_1$ and $L_2$ are the self-inductances of each inductor, $M$ is the mutual inductance between these, and $k$ is the coupling coefficient, given by

$$k = \frac{M}{\sqrt{L_1 L_2}} \tag{7.39}$$

and $i_n$ and $v_n$ are the current and voltages at each inductor. A circuit-level representation of this model is shown in Figure 7.4.

Theoretically, in an ideal transformer model, the following would hold:

$$v_1 = n v_2 \tag{7.40}$$

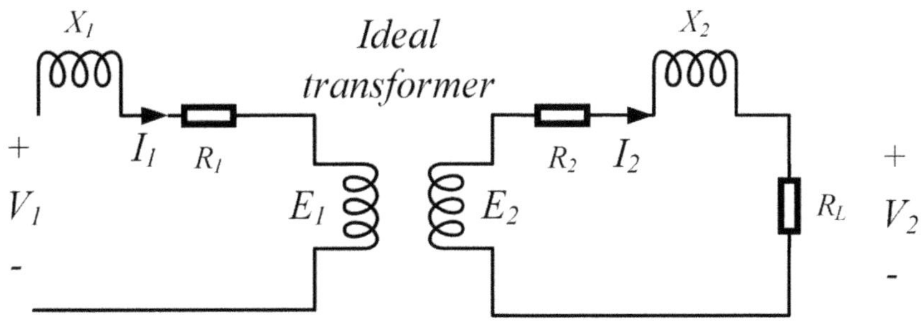


**FIGURE 7.4** The ideal equivalent circuit representation of a CMOS LNA transformer.

and

$$i_2 = ni_1 \tag{7.41}$$

however, on-chip CMOS transformers (Zhang *et al.* 2022) are not ideal (as is the case with virtually every real-world component), and the nonunity coupling coefficient introduces leakage currents, which translate to leakage inductances. Thermal noise is added by the series windings, and additional resonances are formed due to the parasitic capacitances.

In CMOS RF LNAs, the transformer is mostly loosely coupled and consists of either

- cross-coupled spiral inductors,
- stacked (or interleaved) metal layers, or
- patterned ground shields.

where typical values of the coupling coefficient for each of these technologies (a value of less than 1), in CMOS, can be found in literature or through practical experimentation. To model these nonideal transformers as small-signal equivalents, an adapted representation and two-port (De Conti *et al.* 2015) model is shown in Figure 7.5.

In Figure 7.5, $R_{s1}$ represents the series resistance on the primary side, and on the same side, $L_1$ is the series inductance, and $C_{p1}$ is the primary-to-substrate capacitance. On the secondary side, $R_{s2}$, $L_2$, and $C_{p2}$ indicate similar components. $M$ represents the mutual inductance of either a dependent source or of an ideal transformer. In Figure 7.5, $P_{GS}$ is the patterned ground shield. Using these component definitions, the primary-side input impedance is described by

$$Z_{in} = R_{s1} + j\omega L_1 + \frac{\omega^2 M^2}{R_L + j\omega L_2} \tag{7.42}$$

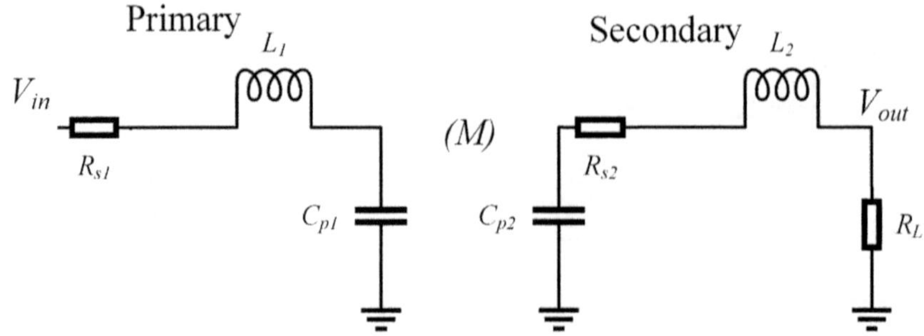


**FIGURE 7.5** The two-port nonideal equivalent circuit representation of a CMOS LNA transformer.

and if we consider a purely resistive load $R_L$, which is probable if this node is the input of a CS transistor, so an approximated $1/g_m$, it follows that we can simplify to

$$Z_{in} \approx j\omega L_1 + \frac{\omega^2 M^2}{j\omega L_2} \tag{7.43}$$

which is the key expression that shows the impedance transformation. Because

$$M = k\sqrt{L_1 L_2} \tag{7.44}$$

it follows that

$$Z_{in} \approx j\omega L_1 + \frac{\omega^2 k^2 L_1 L_2}{R_L} \tag{7.45}$$

and the real part is given by

$$\Re\{Z_{in}\} = \frac{\omega^2 k^2 L_1 L_2}{R_L} \tag{7.46}$$

This demonstrates that it is possible to achieve good noise matching without aggressively engineering the transistor transconductance by adjusting the coupling coefficient $k$, the inductances, and the effective load resistor $R_L$, which is the inverse of $g_m$.

A similar strategy can be implemented to derive and define the noise transformation and noise contributions towards a sensible transformer noise efficiency as a function of gain. The first step is to determine the series resistance noise of the primary winding, which has a finite resistance ($R_{s1}$) that generates thermal noise, as

$$\overline{v_{n1}^2} = 4kTR_{s1} \tag{7.47}$$

where $v_{n1}$ is the RMS noise voltage generated by the resistance of the primary winding, $k$ is Boltzmann's constant, and $T$ is the absolute operating temperature. Because this noise contribution appears directly across the primary winding nodes, it forms a noise floor baseline that applies to all transformer-coupled LNAs. Since there is no coupling-dependent scaling factor in this equation, the term is particularly important if the transformer is fabricated on-chip, since metal losses here can be substantial. Similarly, the secondary windings will contribute their own thermal series resistance noise, described as

$$\overline{v_{n2}^2} = 4kTR_{s2} \tag{7.48}$$

but this noise contribution is not located on the primary side of the transformer, and to effectively express its contribution at the input, requires to be defined at the circuit level; the value must be reflected, or referred, through the transformer. The transformer reflects impedances by the square of the turns ratio, or equivalently by the ratio of the mutual inductance ($M$), a noise voltage source on the secondary, transforms according to

$$\overline{v_{n2,ref}^2} = 4kTR_{s2}\left(\frac{\omega M}{R_L}\right)^2 \tag{7.49}$$

where ω is the angular frequency ($2\pi f$) and $R_L$ is the load resistance, which is connected to the secondary node, or it could also be seen as the effective resistance that is seen into the gate or LC network of the active device. As a result, the voltage that appears across the secondary resistance is referred to the primary side with the mutual inductance value, and the transfer ratio of this voltage, from the secondary to the primary side, is defined by

$$\frac{V_1}{V_2} = \frac{\omega M}{R_L} \tag{7.50}$$

and the noise power will therefore scale with its square value. This expression shows that the noise grows with frequency, due to the $\omega^2$ factor, and that $M$ strongly affects the noise reflection near resonance, especially for on-chip inductors. Optimization of $M$ can directly result in better-performing LNAs. Importantly, the winding loss on the secondary side can dominate the total input-referred noise, even though it is not physically close to the RF input terminal. The MOSFET itself at the secondary side contributes additional thermal channel noise, and the expression for the drain/gate-referred channel noise of a MOS transistor is given by

$$\overline{v_{n,gm}^2} = 4kT\gamma\frac{1}{g_m} \tag{7.51}$$

where γ is the excess noise factor of the MOSFET, a value that is different for short- or long-channel transistors. This noise value is commonly expressed as a voltage noise that appears at the gate or drain of the transistor, depending on the model that is used. Since the transistor is modeled as a device on the secondary side, the same scaling factor should be used to refer it to the input (primary side), such that

$$\overline{v_{n,gm,ref}^2} = 4kT\gamma g_m\left(\frac{\omega M}{g_m}\right)^2 \tag{7.52}$$

which simplifies to

$$\overline{v_{n,gm,ref}^2} = 4kT\gamma\frac{\omega^2 M^2}{g_m} \tag{7.53}$$

and shows that the noise contribution increases by the square of the angular frequency and mutual inductance/coupling, but decreases with the transconductance, $g_m$. Therefore, practically in CMOS LNAs, the NF can be reduced (improved) by improving the coupling coefficient, and poor coupling leads to the secondary-side noise dominating the circuit. Finally, the effective input-referred transistor noise scales as a function of

$$\overline{v_{n,gm,ref}^2} \propto \frac{M^2}{g_m} \tag{7.54}$$

which indicates that if $M$ is increased, the requirement for a very large gm is decreased, and the circuit can achieve a lower NF at lower power. Additionally, if $g_m$ is fixed, the input-referred noise can be lowered by increasing $M$, which effectively provides some noise leverage with additional benefits such as inherent filtering and electrostatic isolation. This is one of the primary reasons why transformer-coupled LNAs often perform better than inductively source-degenerated topologies; the magnetic coupling gives a noise scaling factor without requiring additional DC power consumption.

It is also important to note that a transformer not only provides impedance matching and magnetic coupling, but it also fundamentally changes how the signal current at the primary side interacts with the active device on the secondary side. This phenomenon is often referred to as passively realized gain boost, a concept that is unique to transformers. To clearly indicate this benefit, the voltage at the secondary side must be studied as a function of the current at the primary side. Also, how this generated voltage results in an effective transconductance boost that improves both gain and noise performance, simultaneously. Consider the real part of (38), this induced voltage is crucial as it is the signal stimulus to the MOSFET input, on the secondary side of the transformer. In most CMOS LNAs that use a transformer front-end, the secondary winding directly drives the gate of a MOSFET that is in the common-source configuration, and this means that the transistor senses the induced voltage and converts it to a drain current through its own transconductance, therefore

$$i_d = g_m v_2 \tag{7.55}$$

and the induced voltage expression can be substituted, leading to

$$i_d = g_m\left(j\omega M i_1\right) \tag{7.56}$$

and if rewritten;

$$i_d = j\omega M g_m i_1 \tag{7.57}$$

which indicates that the drain current at the secondary side is proportional to the primary current, and scaled by passive values $j\omega M$ and an active value, $g_m$. It is also possible to quantify how effectively the primary current ($i_1$) generates the output current ($i_d$) by defining the effective transconductance as

$$g_{m,eff} = \frac{i_d}{i_1} \tag{7.58}$$

and through substitution:

$$g_{m,eff} = j\omega M g_m \tag{7.59}$$

which shows that the transformer multiplies the transistor transconductance by a factor $j\omega M$ to provide passive gain boosting that inherently increases with operating frequency, magnetic coupling, and internal transconductance. The expression for the mutual inductance can also be substituted for

$$g_{m,eff} = g_m\left(j\omega k\sqrt{L_1L_2}\right) \tag{7.60}$$

which is a convenient expression to show how the effective transconductance is directly dependent on the

- frequency (higher RF frequencies increase effective gain),
- coupling (better coupling increases gain),
- both windings' inductance (transformer geometry impacts gain), and
- active device gain (transistor sizing remains an important design characteristic).

In summary, the transformer provides a unique and passive mechanism that multiplies the transistor gm by a frequency- and geometry-dependent factor. The circuit therefore experiences higher gain, lower noise, and reduced power consumption; the tradeoffs being larger die area and therefore an increase in cost, higher reliance on on-chip inductors that typically have lower $Q$ and limited bandwidth due to self-resonance. On-chip transformers also have relatively complex structures considering their self-capacitance and parasitic coupling that influence the operating frequency of the LNA, especially since a transformer inherently exhibits two natural resonant frequencies that strongly impact the LNA bandwidth, gain shape, impedance transformation ratio, and the achievable NF. Both the primary and secondary coil behaves as a parallel resonant LC tank, formed by their inductance and parasitic (typically, winding-to-substrate) capacitance, such that

$$\omega_{sr,1} = \frac{1}{\sqrt{L_1C_{p1}}} \tag{7.61}$$

and

$$\omega_{sr,2} = \frac{1}{\sqrt{L_1C_{p2}}} \tag{7.62}$$

These frequencies effectively represent the points where each winding stops behaving as an inductor only and starts becoming capacitive. In RF CMOS, this point onwards also leads to rapid $Q$ degradation. To mitigate or accommodate these frequencies, the LNA operating frequency should be significantly above (higher than) these frequencies. In mm-wave, this constraint becomes one of the dominant layout challenges. It becomes useful to define the upper and lower resonant frequencies if we first consider that the magnetic coupling effectively splits the ideal single LC resonant circuit into a pair of resonant frequencies, where

$$\omega_{coupled} \approx \omega_0 \sqrt{1 \pm k} \tag{7.63}$$

and from here, the upper resonant frequency is

$$\omega_+ = \omega_0 \sqrt{1+k} \tag{7.64}$$

and the lower resonant frequency is

$$\omega_- = \omega_0 \sqrt{1-k} \tag{7.65}$$

showing how a stronger coupling ($k$) leads to greater resonant splitting, which increases design flexibility. The practical impact on LNA design is that with the two resonant peaks, bandwidth (and input matching) is expanded since the gain envelope between these is flatter. These advantages are especially useful in 802.11 ax/ay LNAs, 5G FR2 LNAs, UWB receivers, and mm-wave phased arrays.

## WIRELESS COMMUNICATIONS

Modern wireless communication systems, ranging from emerging 5G/6G links to dense mm-wave phased arrays and ultra-low-power IoT devices, place unprecedented pressure on the RF front-end and particularly on the LNA. As frequencies move into the tens or hundreds of gigahertz, traditional LNA limitations become amplified: free-space path loss rises, atmospheric absorption becomes significant, antenna gains narrow, and the LNA's NF increasingly dictates overall system sensitivity. At the same time, deeply scaled CMOS nodes, while essential for integrating digital and RF subsystems, introduce reduced supply voltages, lower intrinsic gain, higher parasitics, and substantial gate and metal resistance, all of which degrade the achievable NF in conventional inductor-based LNAs. These constraints shift the design emphasis toward passive noise-cancelling and transformer-based architectures, which can recover lost transconductance, suppress gate-resistance and substrate-induced noise, and provide impedance transformation without consuming additional power. Such benefits are valuable in very different application domains: in 5G/6G receivers, they help overcome physical propagation limits; in phased arrays, they improve per-element NF, gain uniformity, and calibration effort; and in IoT systems, they enable meaningful sensitivity at extremely low currents and with low-cost, low-$Q$ passive structures. Across all these systems, the LNA is no longer an isolated block but a repeated, system-critical element whose noise behavior, gain distribution, and layout symmetry influence the performance of the entire receiver. This section, therefore, examines how the mechanisms described earlier, noise cancellation, magnetic coupling, gm-boosting, and passive impedance translation, translate directly into practical advantages for the major classes of modern wireless applications.

## 5G/6G Receiver Front-End NF Constraints

In wireless communications, specifically at RF frequencies and sub-6 GHz transmissions, the receiver sensitivity is often dominated by thermal noise and moderate antenna gains. However, in the 28–300 GHz regime, where 5G FR2, 6G (Hong *et al.* 2021), and sub-THz candidate bands exist, additional physics-related limitations further attenuate signals that make communication at these frequencies challenging. These constraints, as outlined in Lambrechts and Sinha (2016, 2017), include

- atmospheric (oxygen) absorption peaks at around 60 GHz and water vapor absorption between 120 and 180 GHz,
- free-space path loss, which is proportional to $f^2$,
- beamforming, which becomes mandatory, and
- dominant receiver NF as a system-level limitation.

Therefore, RFIC LNAs for 5G and 6G must be designed with extremely sensitive NF, and, in a sense, an advanced scaled CMOS node, especially 28 nm and smaller, makes achieving this even more challenging. This is because, as CMOS nodes shrink, the transconductance also reduces, power supply voltages are lower, and parasitics further increase, due to the shrinking node as well as the increased operating frequency. Advanced CMOS nodes such as FinFETs and nanosheet FETs operate with very limited effective mobility at mm-wave, and the transconductance per unit current is significantly lower when compared to older bulk processes. Furthermore, the gate resistance of these transistors dominates the noise at mm-wave. $R_g$ injects direct noise into the input of the LNA, where

$$NF_{R_g} \propto \frac{4kTR_g}{\left|Z_{in}\right|^2} \tag{7.66}$$

so as the operating frequency drastically increases, the input impedance also follows suit, and the gate resistance noise rapidly increases. On-chip spiral inductors are also very lossy with Q's of between 5 and 10, which means that with single inductor matching (to limit die-area), the NF becomes difficult to manage. The transistor gate-source and gate-drain capacitances can also exceed 100 fF if the transistors are geometrically wide, a requirement for high-gain, and matching becomes extremely challenging without additional tuning networks, which inherently increase overall loss. A single-inductor source-degenerated CS CMOS LNA has built-in tradeoffs in NF, gain, bandwidth, power supply voltage, and current use. As a result, conventional

LNA topologies are rarely used in 5G and 6G front-ends. Magnetic coupling is therefore an ideal strategy to provide passive noise cancellation as it injects antiphase noise currents from the gate resistance, source degeneration, matching networks, and lossy substrate paths. These LNAs can achieve a lower NF than any of their building blocks and transformer LNAs, although physically larger, often show single-digit noise improvements when compared to equivalent single-inductor designs. To meet the demands of 6G, future wireless systems also need to be reconfigurable and intelligent (Lambrechts *et al.* 2024). Next-generation networks are characterized by their distributed and heterogeneous nature and necessitate ubiquitous public and private artificial intelligence (AI) services. As proposed in Lambrechts *et al.* (2024), the roll-out of these services should include both technology and nontechnology facets, a purposeful step towards augmented artificial intelligence ($A^2I$).

## MM-WAVE LNAs FOR PHASED ARRAYS

A phased array, as per its definition, requires numerous LNAs to work in unison; often, these systems require an LNA for each antenna element. The performance of each LNA is therefore crucial as it affects the sensitivity of the entire array, the beamforming signal-to-noise ratio (SNR), calibration requirements, the levels of the side-lobes, and the direction-of-arrival (DoA) estimation accuracy. In these systems, the LNA is not merely a subsystem requirement but forms an integral part of the system-level specifications. For example, in an $N$-element phased array, the effective NF is approximated by

$$NF_{array} \approx NF_{LNA} + \frac{1}{G_{LNA}} \tag{7.67}$$

and the total gain must be high enough to dominate additional elements such as the gain shifters, signal combiners, and even distribution networks. However, the strategy of beamforming imposes even more limitations and constraints on the system, since the gain must be matched across all LNAs with little variation, the phase as well, consistency in bandwidth across all elements is required, and power consumption must be reasonable across numerous active gain elements and circuits. Again, transformer LNAs become a suitable proposition in phased arrays. Their noise-cancelling capabilities can improve the NF for each element, and even the slightest improvement per element adds up and translates to better beamforming gain, reduced complexity of calibration, and better angular resolution. The gain per bias current advantages discussed in this chapter mean that, because power consumption is better managed, the overall thermal budget can also be lowered. Furthermore, since beamforming is sensitive to mismatches, and transformers offer an inherent form of layout symmetry, additional advantages are that inductances are equalized across the system, so are parasitic capacitances, and phase uniformity improves. In dense mm-wave arrays, considering physical layout, the routing of RF lines and the placement of inductors become very challenging and time-consuming. The transformer topology mitigates some of these challenges since it already combines two inductors in one footprint, which not only reduces area but also simplifies routing with stacked metals, inherently provides isolation between RF and local oscillator lines, and performs impedance matching and gain boosting simultaneously.

## IoT Devices: Ultra-Low Power vs NF Tradeoffs

Generally, Internet-of-things (IoT) devices rely heavily on robust, low-power modulation schemes such as FSK, LoRa, or DSSS, as these schemes present low SNR thresholds. However, the front-ends of these IoT devices still require low NF, relatively high gain, and very low power consumption, three tradeoffs in traditional LNA design. Again, transformer-based and noise-cancelling LNA designs can decrease some of the design limitations towards specifications that better suit IoT LNA requirements. IoT receivers also operate in a fundamentally different way when compared to 5G or 6G receivers, or even phased arrays, as discussed in the previous section. Although some constraints overlap, the atmospheric absorption, beamforming, or link distance constraints are less prominent in IoT. For these devices and applications, strict energy budgets, intermittent operation, lifetimes, and low cost are the primary considerations, and often, limitations. Considering that many IoT devices operate in the sub-GHz or 2.4 GHz bands, they still require moderate gain with minimal power consumption. CMOS scaling leads to lower power supply voltages, hence headroom for gain, and this causes some limitations. Like traditional receivers, IoT also suffers from reduced transconductance at low (drain) currents, especially at scaled CMOS, where short-channel effects dominate. Furthermore, in attempts to keep power consumption low, matching networks are often single-staged and simple, which can increase the NF due to lossy elements at the input of the LNA or other active devices. IoT SoCs also often have their RF front-ends situated directly next to digital microcontrollers, clock generators, or switching antennas. The contribution of EM noise to these nodes significantly adds to the system noise, which lowers sensitivity, and this is where passive noise cancellation mechanisms can be integrated into the LNA stage. Here, $g_m$-boosting, impedance translation, and noise-cancelling work in slightly different mechanisms than in mm-wave or sub-THz applications. Lower frequency IoT devices can use transformer designs to artificially increase the effective transconductance while still maintaining a relatively small absolute transconductance, leading to lower current through the active MOSFETs. Thus, although the $\omega M$ is lower than in mm-wave, it still yields a measurable noise benefit without increasing power consumption. IoT devices, often integrated into low-cost packaging technologies, rely on low inductor Q values, increasing the complexity of matching networks. Transformer designs can decrease some of the matching network complexity through their inherent turn ratio and coupling coefficient dependence, making designs more robust to process variations, temperature drift, package-level variations, and absorbing some of the shortcomings of low-cost printed circuit board implementations. Even when IoT SoCs support multiple communication standards in one package, like Bluetooth, Zigbee, sub-GHz ISM, or even proprietary protocols, coupled inductors naturally provide either dual-resonant or wideband frequency responses for matching networks. The built-in impedance transformation of these topologies also mitigates antenna impedance variations arising from process differences or temperature drift, lowering the need for adaptive matching networks that require more power from these systems.

## SPACE AND RADIO ASTRONOMY

### Sensitivity Requirements and Noise Temperature

Radio astronomy is a discipline that operates at the limits of electromagnetic sensing since its power levels at the receiver input typically fall below the −200 dBW mark. This is significantly lower than the typical values experienced in terrestrial wireless systems. In this domain, the conventional RF metrics of NF become less intuitive, and sensitivity is qualified in terms of noise temperature (Kerr 1999). Noise temperature is a physically interpretable figure of merit, a temperature in Kelvin, that describes the equivalent thermal agitation at the point of the receiver. It is therefore the physical temperature of a passive system with an available noise power per bandwidth that is equal to that of the actual system. To quantify the noise in an electronic system, especially in radio astronomy with extremely low signal levels, the relationship between NF and noise temperature, $T_e$, follows

$$NF = 1 + \frac{T_e}{T_0} \tag{7.68}$$

where $T_0$ is a reference temperature. The equivalent noise temperature is a theoretical value that describes the noise generated by the system and not the physical temperature of the system itself. It is the temperature of a passive system that would generate the same amount of available noise power per unit of bandwidth as the actual system. Another mathematical expression that is often used is

$$T_{sys} = T_{rx} + T_{sky} + T_{geo} + T_{cmb} \tag{7.69}$$

where $T_{rx}$ is the noise contribution from the receiver, usually dominated by the LNA, $T_{sky}$ is the background temperature of the sky, $T_{geo}$ is the ground spillover, and $T_{cmb}$ is the cosmic microwave background at 2.7 K. The receiver noise could be less dominant in astronomical observations below 5 GHz, where the background temperature can exceed hundreds of Kelvins. However, in high-frequency deep-space spectroscopy with frequencies between 20 and 40 GHz, the receiver temperature component becomes larger, and a variable that can be controlled through careful and purposeful design. Even though the NF is rarely used in radio astronomy, a conversion to receiver noise temperature is needed when designing CMOS-based LNAs. This conversion is done by

$$T_{rx} = T_0 (NF - 1) \tag{7.70}$$

and $T_0$ is specified at 290 K. One technique that is often used to lower the NF, and therefore the noise temperature, is to operate devices at cryogenic temperatures. But, since CMOS implementations, especially when scaled, are inherently noisy due to their lower gm and higher gate resistance, they often implement noise cancellation, transformer coupling, or impedance scaling even more aggressively when compared to terrestrial applications. In radio telescopes, the radiometer equation governs the RMS sensitivity to a flux density source, a fundamental metric of detectability, through

$$\Delta S = \frac{2k_B T_{sys}}{A_{eff}\sqrt{B\tau}} \tag{7.71}$$

where $A_{eff}$ is the effective area of reception, or it could also be the total area of an arrayed system, $B$ is the bandwidth, $k_B$ is Boltzmann's constant in J/K, and $\tau$ is the radiometric integration time. This equation shows the relatively weak dependence on bandwidth and integration time, and that $T_{sys}$ can also be carefully optimized to improve the detectability. In front-ends that rely heavily on CMOS LNAs, it is crucial to maintain gain and noise stability, sometimes over hours of integration time, and it is useful to describe the derivative of system temperature with respect to the LNA gain variation with

$$\frac{\partial T_{sys}}{\partial G_{LNA}} = -\frac{T_{rx}}{G_{LNA}^2} \tag{7.72}$$

where GLNA is the power gain of the LNA. This illustrates that small values of gain drift can lead to calibration artefacts. The negative sign indicates that an increase in gain reduces the apparent contribution of internal noise, and over long integration times, gain drift can lead to

- calibration bias,
- artefacts in beam-formed images, and
- spatially correlated noise in maps.

Applications such as the square kilometer array (SKA) (Schaubert *et al.* 2003) therefore require CMOS LNAs with advanced engineering of their thermal gain stability, bias point repeatability, flicker-noise mitigation, and temperature tracking calibrations. Consequently, noise temperature is not merely a receiver parameter but the dominant constraint on scientific performance, establishing the front-end as the primary determinant of detection sensitivity in modern radio telescopes.

## SQUARE KILOMETER ARRAY (SKA) AS A BENCHMARK LOW-NOISE SYSTEM

The SKA is a prominent modern radio astronomy system engineering project where a single scientific instrument is used for a specific and specialized mission, pushing each receiver element to its practical limits. The traditional LNA emerged in this project as an element that determines the performance for observing time, survey speed, and discovery potential. From the perspective of the LNA, the SKA-Mid (350 MHz – 15.4 GHz) requirements highlight two crucial metrics: the array sensitivity and the system equivalent flux density (SEFD). Its sensitivity, defined by

$$\frac{A_{eff}}{T_{sys}} \tag{7.73}$$

in $m^2/K$ has a design target in the order of $10^3$ to $10^4$ $m^2K^{-1}$ (Sokolowski *et al.* 2022). Note that a lower SEFD gives greater sensitivity for equivalent observing time. On the system level, improvements in $T_{rx}$ directly translate to changes in the observing time. Using the radiometer equation (assuming source-limited observations), the observation or integration time ($\tau$) to reach a target SNR scales with

$$\tau \propto \left(\frac{T_{sys}}{A_{eff}}\right)^2 \tag{7.74}$$

which shows that even small improvements (reductions) in $T_{sys}$ result in a large change in required observation time. In an array of $N$ elements, per-element improvements translate into a collective sensitivity boost of the order $\sqrt{N}$ times in the detection rate. Noise improvements, such as transformer-based noise cancelling in CMOS, can save months of observing time across large surveys, and this is why front-end engineering is crucial in SKA. SKA receivers contain numerous amplification stages from the receiving antenna towards the internal circuitry. It requires low-loss transitions at the first stage, which is cryogenically cooled to provide the lowest possible $T_{rx}$. CMOS ASICs are also used for additional gain at ambient temperature (before digitization), where the final stage of beamforming in the back end requires stable and well-matched low-distortion analogue inputs. The first amplification stage in the SKA uses cryogenic LNAs (HEMT/HEBT) as opposed to traditional CMOS LNAs, since the noise contribution from earlier stages is amplified in subsequent stages, based on

$$T_{sys} = T_1 + \frac{T_2}{G_1} + \frac{T_3}{G_1 G_2} + \ldots \tag{7.75}$$

The high integration density and low cost of CMOS make it more ideal for the amplification after the first (cryogenic) stage, used for beamforming ASICs, channelization, and moderate-gain warm amplifiers. In most applications, especially true in the SKA, the engineering challenge is a three-way tradeoff between cost, sensitivity, and operational complexity. Both transformers and noise-cancelling topologies are used in various SKA systems, including the room temperature element LNAs, where cryogenic cooling is either impractical or costly.

Finally, the SKA highlights a philosophy that increasingly defines large-scale scientific receivers: no single technology serves all goals. Cryogenic HEMTs deliver the minimum NF, CMOS enables scalable beamforming and large-$N$ integration, and passive transformer networks provide matching while reducing losses at interfaces. The resulting architecture blends these technologies into a global noise and gain budget rather than optimizing each component in isolation. This system-driven approach is now characteristic of modern radio astronomy, remote sensing, and deep-space communications, and it motivates the continued development of noise-cancelling and transformer-based CMOS front-ends, not as replacements for cryogenic LNAs, but as complementary building blocks within increasingly complex receiver ecosystems.

## DEFENCE AND AEROSPACE APPLICATIONS

The demands of defence-grade RF receivers diverge sharply from those in radio astronomy, even though both domains involve weak and uncertain signals. The dominant concern in military and aerospace systems is not only sensitivity, but survivability in the presence of hostile and often intentional interference. In these environments, the LNA is not just a low-noise device: it becomes a mission-critical front-end protection element. This is why the performance landscape expands beyond NF to include linearity, resilience to jamming, and the ability to maintain reliable operation across extreme temperature cycles, platform vibrations, and radiation exposure.

### Radar and EW Receivers: NF and Linearity

Considering a typical RF front end in radar or electronic warfare (EW) applications, the minimum detectable signal (MDS) is determined by the expression

$$MDS = kT_{sys}B + margin \tag{7.76}$$

and even though the noise temperature is still a defining factor, the large-signal handling becomes a more dominant and limiting factor. EW receivers are expected to detect low signals in the presence of various in-band interference, some of which could be potential dummy signals. The most relevant performance metrics in these systems shift from NF to the ratio between sensitivity and linearity, and in practice, a cascaded function can be used to express this limitation, by the spurious-free dynamic range

$$SFDR = \frac{2}{3}\left(IP3 - NF - 10\log B\right) \tag{7.77}$$

which defines the ability of the system to extract (usable) information and intelligence in the presence of powerful transmitters. As opposed to radio astronomy, which is typically operated in the small-signal thermal noise regime, EW relies on SFDR as a critical performance metric. CMOS LNAs in military receivers that use transformer coupling and feed-forward or feedback noise-cancelling are not designed primarily for NF, but rather for their high linearity per unit power, as well as the ability to shape input impedance over wide bandwidths. These LNAs therefore follow a strategy of design-for-linearity over design-for-noise performance. The tradeoff between wideband and narrowband optimization is particularly pronounced in aerospace and defense applications. Wide bandwidth essentially allows for multi-functional applications, multi-mode radar implementations, cognitive sensing, and remote threat sensing and classification. For such applications, narrowband matching for NF is often traded for flat gain and predictable group delay over wide bandwidths, moderate NF with high IP3, reconfigurable matching networks, and, importantly, resistance to interference. Mathematically, these preferences can be described using a simple equation such as

$$Z_{in}(\omega) \approx Z_0 \tag{7.78}$$

over large GHz bands, rather than a tuned, narrow matching strategy. Achieving this will involve tradeoffs, but CMOS, especially on-chip transformer networks, provides a good solution with advantages in wideband matching without excessive loss or the need for discrete filtering. Narrowband systems like missile guidance or single-mode radar do still rely on tuned LNAs that are optimized for minimum NF with maximum gain, typically around a single operating center frequency. A main difference between these (wideband versus narrowband in military applications) is that in the narrowband, the spectral environment around the signal can be controlled, but in EW applications, this environment can be extremely crowded. Military engineering, especially EW, therefore, not only relies on single specifications like noise, but rather on maintaining functionality under various interference scenarios.

## CONCLUSION

Across the diverse domains examined, from deep-space radio astronomy and SKA-scale sensor networks to wideband radar, electronic warfare, phased arrays, and mm-wave multi-standard 6G front ends, one theme repeatedly emerges: the low-noise amplifier determines the reach, fidelity, and ultimate information capacity of the receiver chain. In radio astronomy, where signals arrive barely above the cosmic microwave background, sensitivity is defined more by receiver noise temperature than collecting area, and CMOS LNA architectures now rival prior GaAs and HEMT standards when cryogenically optimized. In military radar and EW systems, by contrast, NF is traded against high linearity and dynamic range, where the ability to survive jammers, self-interference, and large in-band pulses dictates device sizing, bias regime, and whether the architecture leans narrowband for optimum NF or broadband for tactical flexibility. Across both worlds, scaling into deep sub-micron CMOS has reshaped the landscape: $f_T$ and fmax continue to increase, but device noise and breakdown limitations tighten, making matching-network design, input impedance synthesis, and distributed or transformer-based coupling just as important as the transistor itself.

The chapter demonstrated that modern LNA design is no longer a single-objective problem; it is a multidimensional optimization across noise, gain, linearity, power, and bandwidth, where each application sits at a different point in the performance space. The SKA exemplifies noise-temperature minimization at the extreme limit of physics; EW receivers demonstrate survivability, blocking, and intermodulation resilience under hostile spectral conditions; emerging wideband sensing demands a multi-octave input match without NF collapse; and transformer-based mutual inductive coupling provides a unifying platform that scales across these environments. Through rigorous formulation of impedance transformation, noise-referencing through $M$, $g_m$-boost, resonant splitting, and frequency-dependent coupling, we showed that magnetic structures enable performance once considered unattainable for CMOS, low noise without excess power, broadband behavior without lossy matching, and architecture-level noise shaping through passive fields rather than brute biasing.

In closing, the evolution of LNAs from simple inductively degenerated CS stages to multi-resonant, magnetically coupled, system-aware front-ends represents more than an incremental improvement: it is a paradigm shift driven by the demands of next-generation sensing. Whether capturing neutral hydrogen at the edge of the observable universe or detecting a hostile emitter buried beneath intentional interference, the same principles apply: noise must be minimized, linearity preserved, bandwidth strategically shaped, and every femtojoule of signal recovered. CMOS has reached a point where electromagnetic structure, not just semiconductor behavior, dictates ultimate capability. The work presented here positions LNAs not as isolated blocks, but as strategic system enablers, devices that bridge physics and infrastructure, silicon and sky, technology, and discovery.

## REFERENCES

Allstot, D.J., Li, X., and Shekhar, S., 2004. Design Considerations for CMOS Low-Noise Amplifiers. In: *2004 IEE Radio Frequency Integrated Circuits (RFIC) Systems. Digest of Papers*. IEEE, 97–100.

Aya, M., Thierry, T., Yann, D., and Jean-Baptiste, B., 2009. IIP3 Optimization through Body Biasing in Low Noise Amplifiers. In: *2009 Joint IEEE North-East Workshop on Circuits and Systems and TAISA Conference*. IEEE, 1–4.

Cheng, Y., Chen, C.-H., Matloubian, M., and Deen, M.J., 2002. High-Frequency Small Signal AC and Noise Modeling of MOSFETs for RF IC Design. *IEEE Transactions on Electron Devices*, 49 (3), 400–408.

De Conti, A., Oliveira, V.C., Lopes, R.T.S., Brochado, R.O., and Silvino, J.L., 2015. Two-Port Wideband Models of a Single-Phase Distribution Transformer with Center-Tapped Secondary. In: *2015 International Symposium on Lightning Protection (XIII SIPDA)*. IEEE, 221–225.

Diaz-Chinea, D., Garcia-Vazquez, H., San Miguel Montesdeoca, M., Khemchandani, S.L., and del Pino, J., 2015. A Wide-Band Noise-Cancelling CMOS LNA Based on Current Conveyors. In: *2015 Conference on Design of Circuits and Integrated Systems (DCIS)*. IEEE, 1–5.

Gao, W., Chen, Z., Liu, Z., Cui, W., and Gui, X., 2015. A Highly Linear Low Noise Amplifier with Wide Range Derivative Superposition Method. *IEEE Microwave and Wireless Components Letters*, 25 (12), 817–819.

Hong, W., Jiang, Z.H., Yu, C., Hou, D., Wang, H., Guo, C., Hu, Y., Kuai, L., Yu, Y., Jiang, Z., Chen, Z., Chen, J., Yu, Z., Zhai, J., Zhang, N., Tian, L., Wu, F., Yang, G., Hao, Z.-C., and Zhou, J.Y., 2021. The Role of Millimeter-Wave Technologies in 5G/6G Wireless Communications. *IEEE Journal of Microwaves*, 1 (1), 101–122.

Huang, Q., Piazza, F., Orsatti, P., and Ohguro, T., 1998. The Impact of Scaling Down to Deep Submicron on CMOS RF Circuits. *IEEE Journal of Solid-State Circuits*, 33 (7), 1023–1036.

Kerr, A.R., 1999. Suggestions for Revised Definitions of Noise Quantities, Including Quantum Effects. *IEEE Transactions on Microwave Theory and Techniques*, 47 (3), 325–329.

Kobal, E., Siriburanon, T., Chen, X., Nguyen, H.M., Staszewski, R.B., and Zhu, A., 2022. A Gm-Boosting Technique for Millimeter-Wave Low-Noise Amplifiers in 28-nm Triple-Well Bulk CMOS Using Floating Resistor in Body Biasing. *IEEE Transactions on Circuits and Systems I: Regular Papers*, 69 (12), 5007–5017.

Lambrechts, J., and Sinha, S., 2016. *Microsensing Networks for Sustainable Cities*. Springer International Publishing.

Lambrechts, J.W., Sinha, S., Sengupta, K., Bimana, A., Kadam, S., Bhandari, S., Preez, J.D., Shao, Z., Huang, X., Liu, Z., Karahan, E.A., Blundo, T., Allam, M., Ghozzy, S., Zhou, J., Fang, W., and Valliarampath, J., 2024. Intelligent Integrated Circuits and Systems for 5G/6G Telecommunications. *IEEE Access*, 12, 21402–21419.

Lambrechts, W., and Sinha, S., 2017. *SiGe-Based Re-engineering of Electronic Warfare Subsystems*. Springer International Publishing Imprint: Springer.

Mecwan, A.I., and Devashrayee, N.M., 2017. Linearity Improvement of LNA Using Derivative Superposition: Issues and Challenges. In: *2017 7th International Conference on Cloud Computing, Data Science & Engineering – Confluence*. IEEE, 759–763.

Rafati, M., Qasemi, S.R., Nejati, A., and Amiri, P., 2019. A gm-Boosting 3–5 GHz Noise-Cancelling LNA. In: *2019 27th Iranian Conference on Electrical Engineering (ICEE)*. IEEE, 376–379.

Schaubert, D.H., van Ardenne, A., Cracye, C., Boryssonko, A.O., and Bij de Vaate, J.G., 2003. The Square Kilometer Array (SKA) Antenna. In: *IEEE International Symposium on Phased Array Systems and Technology, 2003*. IEEE, 351–358.

Shin, M.-J., Lee, Y., Lee, C., and Lee, D.-H., 2017. A Wideband Noise-Cancelling CG-CS LNA with Transformer Source Coupling. In: *2017 IEEE International Symposium on Radio-Frequency Integration Technology (RFIT)*. IEEE, 177–179.

Smith, K.C., and Sedra, A., 1968. The Current Conveyor—A New Circuit Building Block. *Proceedings of the IEEE*, 56 (8), 1368–1369.

Sokolowski, M., Tingay, S.J., Davidson, D.B., Wayth, R.B., Ung, D., Broderick, J., Juswardy, B., Kovaleva, M., Macario, G., Pupillo, G., and Sutinjo, A., 2022. What Is the SKA-Low Sensitivity for Your Favourite Radio Source? *Publications of the Astronomical Society of Australia*, 39, e015.

Soundarya, V., Kirubaraj, A.A., and Senith, S., 2024. Performance and Evaluation of CG-CS LNA over Gain and Noise Optimization. In: *2024 7th International Conference on Devices, Circuits and Systems (ICDCS)*. IEEE, 262–266.

Vanaja, N., and Tantry, S., 2021. Analysis of Different Generations of Current Conveyors Using CMOS and Its Simulation. In: *2021 International Conference on Intelligent Technologies (CONIT)*. IEEE, 1–6.

Walling, J.S., Shekhar, S., and Allstot, D.J., 2007. A gm-Boosted Current-Reuse LNA in 0.18μm CMOS. In: *2007 IEEE Radio Frequency Integrated Circuits (RFIC) Symposium*. IEEE, 613–616.

Zhang, Y., Wei, Z., Tang, X., Zhang, L., and Huang, F., 2022. A 76.5–92.6 GHz CMOS LNA Using Two-Port kQ-Product Theory for Transformer Design. *IEEE Microwave and Wireless Components Letters*, 32 (10), 1187–1190.

[illegible]

[illegible]

[illegible] International Publishing.

[illegible], Y., [illegible], Sengupta, K., Rizzo, [illegible], [illegible], Zhao, Z., Thomas, [illegible], [illegible], T., Alam, M., [illegible], [illegible], and [illegible] 2023. Intelligent Integrated Circuits and Systems for 5G/6G Telecommunications. IEEE Access, [illegible].

[illegible], W., and [illegible] 2017. RF Circuit Design [illegible] Springer International Publishing, Springer.

McEwan, A. L. and Devashayam, R. M. 2017. Capacity Improvement of [illegible] Using Derivative Superposition: Issues and Challenges. In: 2017 [illegible] Conference [illegible] Computer, Data Science & Engineering [illegible]. IEEE, [illegible].

[illegible], M., [illegible], S. T., [illegible], A., and [illegible], D. 2016. A [illegible] for Blocking-[illegible] Noise Cancelling LNA. In: 2016 [illegible] Conference [illegible]. IEEE, 176–179.

Schmickl, D. M., van Ardenne, A., [illegible], C., [illegible], [illegible], and [illegible] 2023. The Square Kilometre Array [illegible]. In: IEEE International Symposium on [illegible]. IEEE, [illegible].

[illegible], Lee, Y., [illegible] 2011. A Wideband Noise-Cancelling CMOS [illegible] Low Noise [illegible] Coupling. In: [illegible].

[illegible] of the IEEE, 56(8), [illegible].

[illegible], [illegible], Davidson, [illegible], [illegible], R., [illegible], [illegible], [illegible], M., [illegible], G., [illegible], and [illegible], A. 2022. [illegible] Publications of the Astronomical Society of Australia, 39, [illegible].

[illegible], V., [illegible], A., and [illegible], S. 2023. Performance and Evaluation of CG [illegible] CS LNA [illegible] Gain and Noise Optimization. In: 2023 [illegible] International Conference [illegible]. IEEE, [illegible].

[illegible] 2021. Analysis of Different [illegible] Low-Noise Amplifier (LNA) [illegible]. In: [illegible] International Conference [illegible]. IEEE, 1–6.

[illegible]

[illegible]

# Index

## N



## P



## Q



## R